Sustainable Development and Biodiversity

Volume 20

Series editor

Kishan Gopal Ramawat
Botany Department, M.L. Sukhadia University, Udaipur, India

This book series provides complete, comprehensive and broad subject based reviews about existing biodiversity of different habitats and conservation strategies in the framework of different technologies, ecosystem diversity, and genetic diversity. The ways by which these resources are used with sustainable management and replenishment are also dealt with. The topics of interest include but are not restricted only to sustainable development of various ecosystems and conservation of hotspots, traditional methods and role of local people, threatened and endangered species, global climate change and effect on biodiversity, invasive species, impact of various activities on biodiversity, biodiversity conservation in sustaining livelihoods and reducing poverty, and technologies available and required. The books in this series will be useful to botanists, environmentalists, marine biologists, policy makers, conservationists, and NGOs working for environment protection.

More information about this series at http://www.springer.com/series/11920

Vijay Rani Rajpal · Deepmala Sehgal
Avinash Kumar · S. N. Raina
Editors

Genetic Enhancement of Crops for Tolerance to Abiotic Stress: Mechanisms and Approaches, Vol. I

Springer

Editors
Vijay Rani Rajpal
Department of Botany
Hansraj College
University of Delhi
Delhi, India

Deepmala Sehgal
CIMMYT Headquarters
El Batán, Veracruz, Mexico

Avinash Kumar
Vinoba Bhave University
Hazaribag, Jharkhand, India

S. N. Raina
Amity Institute of Biotechnology
Amity University
Noida, Uttar Pradesh, India

ISSN 2352-474X ISSN 2352-4758 (electronic)
Sustainable Development and Biodiversity
ISBN 978-3-319-91955-3 ISBN 978-3-319-91956-0 (eBook)
https://doi.org/10.1007/978-3-319-91956-0

Library of Congress Control Number: 2018951914

This Springer imprint is published by the registered company Springer Nature Switzerland AG
The registered company address is: Gewerbestrasse 11, 6330 Cham, Switzerland

Preface: Volume I

The changing climate change scenarios have gripped humanity for a long time and are expected to worsen in the coming decades. Agriculture is already feeling the effects of climate change by reduced crop productivity, heavy yield losses, scarcity of water for farming, reduced rate of precipitation, and the list goes on. In staple crops, particularly wheat, rice, maize, soybean, barley, and sorghum, research has shown about 30% of the yearly variation in agricultural yields due to changes in rainfall and temperature.

Of all the threats that agriculture is exposed to due to climate change, abiotic stresses such as drought (water deficit), extreme temperatures (cold, frost, and heat), and/or salinity (sodicity) are the most devastating ones, causing more than 50% of crop yield losses. Mineral (metal and metalloid) toxicity is an additional abiotic factor, which is becoming a big threat for both major and minor crops. Thus, improving tolerance to these abiotic stresses is a global plant breeding target. A lot of research has been conducted to investigate plants' responses to these stresses at the structural, physiological, transcriptional, and molecular level and on the resistance mechanisms allowing them to adapt and survive these stressful events. A major research target has also been cross talk among various mechanisms, in case of multiple stresses faced by plants.

Precise analysis of proteome and metabolome is essential for understanding the fundamentals of stress physiology and biochemistry. Scientists have utilized 'omics' platforms to unravel the influence of abiotic stresses on levels of different protein groups and metabolite classes and to pinpoint candidate genes underneath. In addition, chromatin modifications, nucleosome positioning, and DNA methylation have been recognized as important components in plants' adaptations to stresses. The potential of improving stress tolerance in crops by enhancing the stress memory through the activation of priming responses or the targeted modification of the epigenome has been a burning research topic.

This book provides a consolidated and an updated account of the research being conducted in above-mentioned areas by plant scientists all over the world. It is an invaluable resource for researchers and educators in the areas of tools and technologies to unravel plant's responses to abiotic stresses. The outcomes presented on staple crops will be useful to a broad community of scientists working in similar areas and can provide useful leads to build strategies to generate abiotic stress tolerant varieties. Students will find this book handy to clear their concepts and to get an update on the research conducted in various crops at one place.

New Delhi, India — Vijay Rani Rajpal
El Batán, Mexico — Deepmala Sehgal
Hazaribag, India — Avinash Kumar
Noida, India — S. N. Raina

The original version of the book was revised: Book editor's affiliation has been updated in Copyright page. The correction to the book is available at https://doi.org/10.1007/978-3-319-91956-0_12

Acknowledgements

We sincerely thank all authors not only for agreeing to contribute chapters despite their hectic schedules and other work commitments but also for putting in their sincere efforts in providing up-to-date information. We are thankful to Prof. K. G. Ramawat for inspiring us to take up this assignment. All the three editors, Vijay Rani Rajpal, Deepmala Sehgal, and Avinash Kumar are thankful to their Mentor and Guide Prof. S. N. Raina for guidance and inspiration always. Vijay Rani Rajpal is also thankful to the two other co-editors for their active involvement from inception to compilation and editing process through the course of this book. The book would not have been possible without their active involvement and whole-hearted support. The editors Deepmala Sehgal and Avinash Kumar are also thankful to senior editors Profs. S. N. Raina and Vijay Rani Rajpal who had been their teachers and Ph.D. supervisors too for providing them the great opportunity to be an editor.

Last but not least, editors gratefully acknowledge their families for their understanding, patience, and emotional support. Our sincere thanks to the whole Springer team who was tirelessly involved in the production process. We particularly appreciate Dr. Valeria and Dr. Ineke for their continued support.

We are very hopeful that this book will attract readers who are crop scientists and to even undergraduates and postgraduates of agricultural universities and institutes that are interested in the genetic improvement of crop plants using modern tools.

Contents

Contributors

Nasheeman Ashraf Plant Biotechnology Division, CSIR-Indian Institute of Integrative Medicine, Srinagar, Jammu and Kashmir, India

Shoib Ahmad Baba Plant Biotechnology Division, CSIR-Indian Institute of Integrative Medicine, Srinagar, Jammu and Kashmir, India

Renu Deswal Molecular Plant Physiology and Proteomics Laboratory, Department of Botany, University of Delhi, Delhi, India

Om Prakash Dhankher Stockbridge School of Agriculture, University of Massachusetts, Amherst, MA, USA

Manoj K. Dhar Genome Research Laboratory, School of Biotechnology, University of Jammu, Jammu, India

T. R. Ganapathi Nuclear Agriculture and Biotechnology Division, Bhabha Atomic Research Centre, Mumbai, India

Brijesh K. Gupta Plant Stress Biology, International Centre for Genetic Engineering and Biotechnology, New Delhi, India

Deepti Jain Department of Plant Molecular Biology, Interdisciplinary Centre for Plant Genomics, Delhi University, South Campus, New Delhi, India; Department of Botany, Bangalore University, Bangalore, India

Rohit Joshi Plant Stress Biology, International Centre for Genetic Engineering and Biotechnology, New Delhi, India

Prakriti Kashyap Molecular Plant Physiology and Proteomics Laboratory, Department of Botany, University of Delhi, Delhi, India

Sanjana Kaul Genome Research Laboratory, School of Biotechnology, University of Jammu, Jammu, India

J. P. Khurana Department of Plant Molecular Biology, Interdisciplinary Centre for Plant Genomics, Delhi University, New Delhi, India

Rakesh Kumar ICAR-Indian Institute of Wheat and Barley Research, Karnal, India

Smita Kumar CSIR-National Botanical Research Institute, Council of Scientific and Industrial Research (CSIR-NBRI), Lucknow, India

H. M. Mamrutha ICAR-Indian Institute of Wheat and Barley Research, Karnal, India

Manjari Mishra School of Life Sciences, Jawaharlal Nehru University, New Delhi, India

Shikha Mittal Division of Genetics, ICAR-Indian Agricultural Research Institute, New Delhi, India

Girish Chandra Pandey Department of Bioscience and Biotechnology, Banasthali University, Banasthali, India

Ashwani Pareek Stress Physiology and Molecular Biology Laboratory, School of Life Sciences, Jawaharlal Nehru University, New Delhi, India

Prashanti Patel Nuclear Agriculture and Biotechnology Division, Bhabha Atomic Research Centre, Mumbai, India

Suprasanna Penna Nuclear Agriculture and Biotechnology Division, Bhabha Atomic Research Centre, Mumbai, India

Asha Rani Department of Molecular Biology, Biotechnology and Bioinformatics, Chaudhary Charan Singh Haryana Agricultural University, Hisar, India

Suchismita Roy School of Life Sciences, Jawaharlal Nehru University, New Delhi, India

Davinder Sharma ICAR-Indian Institute of Wheat and Barley Research, Karnal, India

Indu Sharma ICAR-Indian Institute of Wheat and Barley Research, Karnal, India

Rahul Sharma Genome Research Laboratory, School of Biotechnology, University of Jammu, Jammu, India

M. N. Shiva Kameshwari Department of Botany, Bangalore University, Bangalore, India

Mohan B. Singh Plant Molecular Biology and Biotechnology Laboratory, University of Melbourne, Parkville, VIC, Australia

Nidhi Singh Division of Genetics, ICAR-Indian Agricultural Research Institute, New Delhi, India

Rajender Singh ICAR-Indian Institute of Wheat and Barley Research, Karnal, India

Richa Singh Department of Molecular Biology, Biotechnology and Bioinformatics, Chaudhary Charan Singh Haryana Agricultural University, Hisar, India

Sneh L. Singla-Pareek Plant Stress Biology, International Centre for Genetic Engineering and Biotechnology, New Delhi, India

Jyoti Taunk Department of Molecular Biology, Biotechnology and Bioinformatics, Chaudhary Charan Singh Haryana Agricultural University, Hisar, India

Nepolean Thirunavukkarasu Division of Genetics, ICAR-Indian Agricultural Research Institute, New Delhi, India

Ratan Tiwari ICAR-Indian Institute of Wheat and Barley Research, Karnal, India

Prabodh Kumar Trivedi CSIR-National Botanical Research Institute, Council of Scientific and Industrial Research (CSIR-NBRI), Lucknow, India

Karnam Venkatesh ICAR-Indian Institute of Wheat and Barley Research, Karnal, India

Parivartan Vishal Genome Research Laboratory, School of Biotechnology, University of Jammu, Jammu, India

Karuna Yadav Nuclear Agriculture and Biotechnology Division, Bhabha Atomic Research Centre, Mumbai, India

Neelam R. Yadav Department of Molecular Biology, Biotechnology and Bioinformatics, Chaudhary Charan Singh Haryana Agricultural University, Hisar, India

Ram C. Yadav Department of Molecular Biology, Biotechnology and Bioinformatics, Chaudhary Charan Singh Haryana Agricultural University, Hisar, India

Chapter 1
Functional Genomics Approach Towards Dissecting Out Abiotic Stress Tolerance Trait in Plants

Rohit Joshi, Brijesh K. Gupta, Ashwani Pareek, Mohan B. Singh and Sneh L. Singla-Pareek

Abstract Plant functional genomics has revolutionized not only the methodologies for identification and elucidation of key genes' function but also in designing strategies for improving tolerance towards abiotic stresses. Leveraging various approaches has demonstrated the robustness and versatility in their application to study gene/genome function and engineering abiotic stress tolerance in plants. With the emergence of novel high throughput technologies in this area, functional genomics can contribute immensely in understanding the gene regulatory networks operating under stress, thereby benefiting crop improvement programs. This chapter provides recent findings in the field of functional genomics, thus offering several efficacious methodologies such as next generation sequencing, genome-wide hybridization, gene-inactivation and genome-editing-based strategies in addition to metabolite analysis for discovery as well as validation of the candidate genes. Further, methodologies such as gene expression microarrays, insertional mutagenesis, map-based cloning and various genomic-assisted methods are evaluated critically and discussed in the light of integration of the information obtained through functional genomics with practical application in crop breeding.

Keywords Functional genomics · Mutants · Crops · Transcriptomics
Gene-inactivation · Genome-wide hybridization · Genome-editing

---

Rohit Joshi and Brijesh K. Gupta have equally contributed to this work.

R. Joshi · B. K. Gupta · S. L. Singla-Pareek (✉)
Plant Stress Biology, International Centre for Genetic Engineering and Biotechnology,
Aruna Asaf Ali Marg, New Delhi 110067, India
e-mail: sneh@icgeb.res.in

A. Pareek
Stress Physiology and Molecular Biology Laboratory, School of Life Sciences,
Jawaharlal Nehru University, New Delhi 110067, India

M. B. Singh
Plant Molecular Biology and Biotechnology Laboratory, University of Melbourne,
Parkville, VIC 3010, Australia

V. R. Rajpal et al. (eds.), *Genetic Enhancement of Crops for Tolerance to Abiotic Stress: Mechanisms and Approaches, Vol. I*, Sustainable Development and Biodiversity 20, https://doi.org/10.1007/978-3-319-91956-0_1

1.1 Introduction

Abiotic stresses such as cold, heat, waterlogging, drought, metal toxicity, salinity and sodicity reduce plants' growth and yield by as much as 50% in both natural and agricultural systems (Nakabayashi and Saito 2015). Improving tolerance to abiotic stresses, therefore, has become a major objective in plant breeding programs globally (Pareek et al. 2010). It has been estimated that a global increase in food production of 44 metric tons will be required each year to fulfill the food demand of rapidly increasing population, which will reach close to 10 billion by 2050 (Bohra et al. 2015; Wang et al. 2016). Plant's responses towards simultaneous occurrence of abiotic stresses, such as drought, heat and salinity, have gathered attention in various genomics studies (Singh et al. 2015a; Joshi et al. 2015b; Kushwaha et al. 2016). However, multigenic nature of abiotic-stress-tolerance trait(s) along with the lack of proficient selection techniques primarily hampers effective breeding strategies for abiotic stress tolerance (Ford et al. 2015). Furthermore, several reports have indicated differences between quantitative trait loci (QTL), some being linked to tolerance at one stage of plant's development while other linked to tolerance during some other stages (Yang et al. 2013). Dissection of the genetic basis of intra-specific variation in traits conferring abiotic stress tolerance will be useful for selecting and creating positive variations within the species. However, limited success has been achieved through traditional approaches such as inter- or intra-generic hybridizations, induced mutations and/or somaclonal variations (Chinnusamy et al. 2004; Bhullar and Gruissem 2013).

Recent advances in genomics and molecular biology have contributed significantly to the breeding programs by rapid identification and characterization of genes and genomic regions conferring abiotic stress tolerance. In this direction, one of the powerful approaches for gene discovery could be the exploration of naturally occurring genetic diversity between landraces and their wild relatives (Dwivedi et al. 2016). Thus, understanding the molecular basis of genetic diversity may help in identifying the key differences, which regulate the differential expression of same set of genes in contrasting genotypes. This may aid in unraveling the novel mechanisms underpinning abiotic stress tolerance in crops (Mickelbart et al. 2015).

Recent genomics-based approaches combined with high throughput tools have led to a revolution in crop improvement approaches. These advanced technologies directly affect the applicability of crop improvement methods by translating the entire genomic regions deciphering molecular responses of plants (Bohra and Singh 2015; Edwards 2016; Gupta et al. 2015). Forward and reverse genetics approaches together elucidate the genes and their products involved in expression, signal transduction and stress tolerance (Urano et al. 2010). Since 1980s, functional genomics leapt from being hypothetical or innovative concept to a widely accepted part of science in the year 2000. In the post genomic era, extensive utilization of functional genomics tools has increased our knowledge of the complex networks operating during stress tolerance and adaptation. These functional genomics strategies combined with phenomics will improve our understanding towards gene complementation, transcript regulation, protein complex formation and their evolutionary pathways regulating abiotic stress tolerance traits. After the initiation of whole genome sequencing programs in 1990s, astonishing advancements in DNA sequencing technologies have brought breakthroughs in this area (Wheeler and Wang 2013). Already completed genome sequences of various model organisms including protists (Armbrust et al. 2004), fungi (Wood et al. 2002; Galagan et al. 2003) and eukaryotic plants (Li et al. 2014a, b; Hirakawa et al. 2014; Varshney et al. 2017) have confirmed the feasibility and efficacy of sequencing large genomes. Further, functional genomics provides the next step towards the biological revolution assigning the function to previously identified genes at organizational level that can control the genetic pathways defining the physiology of an organism (Rahman et al. 2016).

Several interrelated strategies enable the survival of tolerant genotypes under abiotic stresses. However, these strategies are less evolved in agricultural crop species, perhaps due to crop domestication. Abiotic stress tolerance in these plants can be achieved at the molecular level by engineering genes regulating chaperone production, osmoprotectant accumulation, reactive oxygen species (ROS) scavenging mechanisms and/or efficient transporter systems for exclusion or compartmentation of ions (Jan et al. 2013; Gupta and Huang 2014). In addition, several genes and their products act simultaneously at transcriptional and translational levels (Joshi et al. 2015b; Gupta et al. 2015; Guo et al. 2016). Functional validation of these genes can help in untangling the stress tolerance network and also in designing various functional markers for marker-assisted breeding.

Genetic transformation approaches offer a rapid way to improve plant stress tolerance. With the advent of high throughput techniques, functional genomics strategies went through a paradigm shift from single gene discovery to many thousands. Development of expressed sequence tags (ESTs) from cDNA libraries of abiotic stress-treated seedlings of plants as well as their complete genome sequence information provides an additional resource for gene discovery. In addition, strategies including promoter trapping, mutagenesis and gene complementation have led to the identification of key gene pools and hence, have provided valuable inputs towards the functional characterization of stress responsive genes and their underlying mechanisms (Hasanuzzaman et al. 2015). In this chapter we discussed current strategies in the field of functional genomics for improving abiotic stress tolerance in plants. Further, we discuss the role of model species and mutant populations in molecular mapping of abiotic stress tolerance determinants for crop improvement.

1.2 Stress Networks and Signaling Pathways Operative Under Abiotic Stresses in Plants

Stress perception as well as its signaling are the two critical components determining the adaptive response of the plant under unfavorable environmental conditions (Muthurajan and Balasubramanian 2009; Gupta et al. 2015). Osmotic and oxidative stresses induced in plants are a common consequence of abiotic stresses sharing many intermediate components of their signaling cascades (Rejeb et al. 2014). Thus, signaling sensors are now becoming main targets for genetic engineering, as they are the principle transducing elements right from the perception of the signal. One of the important stress sensors in higher plants is the Two Component System (TCS) which consists of histidine kinase (HK) sensor and response regulator (RR) (Pareek et al. 2006; Singh et al. 2015b). The investigations on different plant species such as maize and rice confirmed the role of TCS members in response to abiotic stresses (Liu et al. 2014a; Sharan et al. 2017). During abiotic stress, few of the members of TCS family show up-regulation i.e., *AHK1*, *OsHK3*, *GmHK7*, *GmHP3*, *GmHP6*, *GmRR1*, while others are down regulated i.e., *AHK2*, *AHK3*, *AHK4*, *AHP1*, *AHP3*, *AHP5*, *ARR8*, *ARR9*, *OsHK4*, *GmHK10*, *GmHK12 and GmPHP2* (Le et al. 2011; Nishiyama et al. 2013; Gahlaut et al. 2014).

Through yeast-two-hybrid assay, it was revealed that under cold and salt stress, Mitogen Activated Protein Kinase (MAPK) pathway involves MAPK/ERK kinase kinase-1 (MEKK1) which acts upstream to MAP kinase kinase-1 (MKK1), MAP kinase kinase-2 (MKK2), MAP kinase-4 (MPK4) and Mitogen-activated protein kinase-6 (MPK6) (Sinha et al. 2011). In *Arabidopsis*, the signals received by 80 MAPKKKs are transduced downstream from 10 MAPKKs to 20 MAPKs providing an opportunity for crosstalk at different points (Sinha et al. 2011). Similarly, under drought stress, it was reported that AtMEKK1 and AtMPK3 in *Arabidopsis* and OsMSRMK2 and OsMAPK5 in rice show higher expression (Sinha et al. 2011; Ara and Sinha 2014). Pitzschke et al. (2014) revealed MYB44 transcription factor as the interacting partner of MKK4, which in turn interacts with another MPK3-regulated transcription factor VIP1. These results further confirm that MAPK cascade is playing a central point of crosstalk during stress signaling (Pitzschke 2015; Wen et al. 2015). In addition, another important component during osmotic stress signaling pathway is the calcium-dependent protein kinase (CDPK). Overexpression of rice *OsCDPK7* gene was found to confer tolerance against salt, drought and chilling stress (Boudsocq and Sheen 2013).

Various stresses can occur individually, or in combination with others, at any developmental stage of plant and these vary by location and time, which can negatively affect photosynthetic efficiency and alter the source-sink relationship. Further, it can affect the remobilization of solutes, which is a limiting factor for grain weight and yield. Combinations of various traits contribute towards overall plant tolerance against abiotic stresses (Roy et al. 2011). However, it is still unknown how certain plants maintain yield under abiotic stress conditions

(Tripathi et al. 2012). Identifying key regulatory elements playing roles during multiple stress interactions through gene expression profiling is an important aspect of functional genomics. A number of transcription factors (TFs) differentially regulated during environmental stresses have already been analyzed using genome-wide transcriptome analysis (Hoang et al. 2014; Joshi et al. 2016a). These TFs show a very complex expression pattern, which suggests that stress resistance and tolerance are regulated by an extremely intricate gene regulatory network at transcriptional level. Amongst all, bZIP (Basic Leucine Zipper), MBF1 (Multiprotein bridging factor 1), WRKY, MYB (myeloblastosis) and NAC (NAM, ATAF1,2 and CUC2) transcription factors are the largest transcriptional regulators controlling growth, development, physiological processes, and abiotic stress responses in plants (Sahoo et al. 2013; Baloglu et al. 2014).

Rasmussen et al. (2013) employed microarray analysis to detect plant responses to multiple stress exposures, in combination, or individually and found that 25% of transcripts showed similar responses during individual stresses, but act differentially under stress combinations. Twenty-three transcripts were found to be specifically upregulated in the transcriptome analysis of *Arabidopsis* plants using triple combination of heat stress, drought and virus infection (Prasch and Sonnewald 2013). Of these, DREB2A (Dehydration-responsive element-binding protein 2A) and GBF3 (G-box-binding factor 3) were upregulated, whereas Rap2-9 (Related to APETALA2-9) was strongly down regulated. Transcript profiling of *Arabidopsis* plants revealed 43 drought, cold and salinity stress-inducible transcription factor genes including DREB, ERF (Ethylene Responsive Factor), zinc finger containing factors, MYBs, bHLHs (basic helix-loop-helix), bZIPs, NAC and WRKY (Umezawa et al. 2006). Similarly, transcript expression of whole WRKY family of rice showed 17 WRKY genes to be highly induced in both leaf and root under drought stress (Tripathi et al. 2014). Yang et al. (2011) showed that ABI5-Like1 (ABL1) gene regulates ABA and auxin responses by altering ABRE-containing WRKY genes' response in rice. In addition, it was reported recently that among the 9 members of AREB/ABFs in *Arabidopsis*, AREB1/ABF2, AREB2/ABF4 and ABF3 (Abscisic acid responsive elements-binding factor 3) are highly upregulated by osmotic stress and ABA treatments in vegetative tissues (Yoshida et al. 2014). Similarly, through 24K Affymetrix Genechip array, a total of 514 CBF2 (Centromere-binding factor 2) genes were identified under cold stress in *Arabidopsis*, including co-regulated genes like zinc finger proteins (CZF1 and CZF2), MYB73, RAV1 (related to ABI3/VP1 1), ZAT10 and ZAT12 (Vogel et al. 2005; Park et al. 2015). A genome-wide analysis of paper mulberry in response to cold stress showed that 794 TFs, belonging to 47 families were involved in the cold stress response (Peng et al. 2015). Among the differentially expressed TFs, one bHLH, two ERFs and three CAMTAs were involved in signal transduction at early stages followed by 5 bHLH, 14 ERFs, one HSF, 4 MYBs, 3 NACs, and 11 WRKYs in providing cold resistance. The late responsive group consisted of 3 ARR-B, C3H, 6 CO-like, 2 G2-like, 2 HSFs, 2 NACs and TCP. These results indicated towards a much greater cross-talk among different stresses during signaling processes. The key regulators among this complex network are bHLH, bZIP, MYB and AP2 transcription factor families (Peng et al. 2015).

1.3 Functional Genomics Approaches

In the present scenario, direct introduction of genes through genetic engineering is coming up as a more rapid and reliable technique for improving stress tolerance in plants, in comparison to traditional breeding and marker-assisted selection approaches (Bohra et al. 2015). Current engineering strategies aim to functionally characterize the critical genes participating in either signaling or biochemical pathway to understand their distinctive roles in plant development and physiology (Teotia et al. 2016). The products of these genes are either stress-induced proteins or enzymes for osmoprotectant or scavengers of ROS that directly or indirectly provide tolerance against different environmental stresses (Joshi and Chinnusamy 2014; Khan et al. 2015). In addition, various transcription factor genes controlling the expression of different stress regulatory proteins are also unveiled (Wang et al. 2016). It is now necessary to study the abiotic stress tolerance in a collective manner on a genome-wide scale, which can be further utilized for elucidation of abiotic stress networks. With the availability of various omics tools including genomics, transcriptomics, and proteomics, major progress has been made for understanding the interaction and complexity of the stress adaptive mechanisms and their respective signaling pathways (Liu et al. 2014b). By using transcript profiling and allocation of small responsive elements in promoter regions, the determination of regulatory regions in chromatin structure, and the distribution of *cis*-regulatory elements and transcription factors can be predicted computationally.

One of the major challenges in the post-genomic era is to understand the function of genes. Recent high throughput biotechnological advances have facilitated the discovery of new genes and their functions. Unraveling gene functions and their interactions with other regulatory networks have long been exploited for generation of improved varieties (Akpınar et al. 2013). While the functions of several genes are still unknown, their function can often be correlated in association with other known genes, which provide even better understanding for the whole signaling network. We are now able to obtain a complete overview at the cellular level through transcript, protein and metabolite profiling. These approaches allow a deeper understanding of the complex cellular functioning during different physiological processes (Cramer et al. 2011).

Reconstruction of complex networks at whole genome level is achieved by characterizing and quantifying from genotype to phenotype (Feist and Palsson 2008). Understanding only the basic function of the gene in an organism does not provide an insight to its specific role under stress conditions. Sequence analysis of *Arabidopsis* showed that 13 and 20% of the genes are implicated either in signal transduction or in stress/defense responses, respectively (Mahalingam et al. 2003). Another exhaustive screening of more than 1,500 TFs revealed that almost 40 TFs

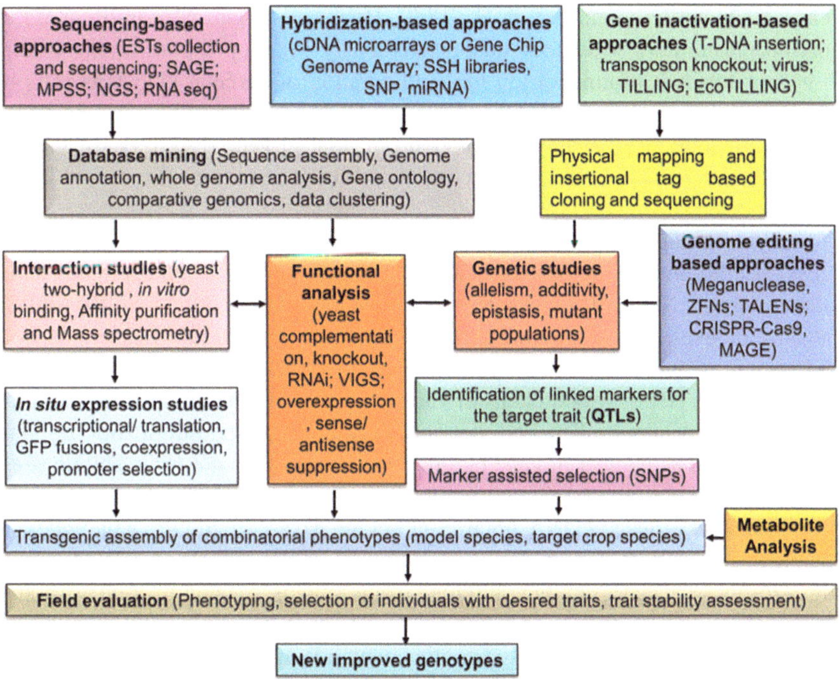

Fig. 1.1 Flow chart of the overview of functional genomics approach for plant improvement

were involved in improving stress tolerance in *Arabidopsis* (Nelson et al. 2007). The major outcome of the current plant genome research is the functional characterization of almost 54% of higher plant genes by comparing them with other known sequences (Sofi and Trag 2006). Parallel studies on the functional genomics in other organisms will also contribute significantly to understand their gene functions in coming years. Functional genomics strategies mainly utilize methodologies, which are sequence-based, hybridization-based, gene inactivation-based or genome-editing based (Fig. 1.1) as discussed below.

1.3.1 Sequencing-Based Approaches

One of the major approaches used to discover abiotic stress expressed gene catalogue is based on ESTs generated from various cDNA libraries expressing transcripts from several stresses in different tissues and developmental stages (Rahman et al. 2016). These libraries have been successfully developed to identify several specific and stress-responsive transcripts, but they under-represent rare transcripts or unexpressed transcripts under certain conditions. EST libraries are major focus of functional studies because they provide easier strategies for gene discovery and genome annotation (Varshney et al. 2006). However, to get more information on polymorphism, EST sequences must be cautiously overlapped onto similar contigs to gather detailed information on the configuration of parental cDNA as in polyploid species like wheat (Rudd 2003). Despite all these factors, EST sequencing is a convincing strategy with already reported potential in gene discovery by aligning with collinear genotypes exposed to control and stress conditions (Ergen and Budak 2009).

Several attempts have been made using model plant species to characterize stress-specific transcripts using EST sequencing in higher glycophytes under salinity stress. National Center for Biotechnology Information (NCBI) dbEST database indexes rapidly growing libraries containing several ESTs generated from various crops and other plant species. However, large-scale cDNA sequencing programs from stress-treated plants of various species at different growth stages are still essential to enrich plant EST datasets. Additionally, the information on gene number as well as number of gene families playing significant roles in abiotic stress responses can be established by clustering the sequences of ESTs obtained through respective stress-treated cDNA libraries (Li et al. 2014a, b). Similar gene-indexing database Swissprot provides important information associated with stress-responsive genes among different plants and is frequently used to assign putative functions to stress-responsive genes (Sreenivasulu et al. 2007). In addition, data clustering produces consensus contigs, which is a more reliable approach than ESTs. Extensive attempts have been made in glycophytes, such as *Arabidopsis*, rice and halophytes to compare the abundance of expressed ESTs in their respective cDNA libraries (Wang et al. 2004; Baisakh et al. 2008; Li et al. 2014a, b). Extensive EST sequencing is still in progress for developmental stage-specific, tissue-specific and stress-specific cDNA libraries obtained from *Arabidopsis* and rice. Analyzing these EST databases will pave our way to specify stress regulated genes that can assist further to unravel the underlying regulatory metabolic pathways (Rahman et al. 2016).

Another approach which enables simultaneous quantitation of thousands of transcripts is SAGE (Serial Analysis of Gene Expression), in which mRNA is oligo (dT)-trapped and reverse transcribed to form cDNA, then small sequence tags are extracted and ligated to form long concatemeric chain and sequenced, leading to complete quantification of gene expression (Vega-Sánchez et al. 2007). Due to the recent advancements in next-generation DNA sequencing technologies, SAGE analysis has emerged as a high throughput, sensitive and cost-effective approach in comparison to Sanger sequencing approaches (Cheng et al. 2013). During the past several years, SAGE has been extensively used in plants with the availability of extensive EST databases of different species (Breyne and Zabeau 2001). Additionally, by combining 5′ RACE (Rapid amplification of cDNA ends) and SAGE (Serial analysis of gene expression) analysis, transcription start sites were also identified (Wei et al. 2004). Later on, several modifications such as SuperSAGE and DeepSAGE became available, in which the tag size is expanded providing greater efficiency to the annotation (Nielsen et al. 2006; Matsumura et al. 2012). Previous studies using SAGE in plants not only revealed new expressed regions in the plant genome but also implied their novel functions including stress response in crops (Cheng et al. 2013).

Massively Parallel Signature Sequencing (MPSS) is also a powerful method enabling the parallel analysis of millions of transcripts on a genome-wide scale (Akpınar et al. 2013). In MPSS, transcription profiling is done using similar tag-based approach, where tagged PCR products obtained from cDNA are amplified so that each mRNA molecule produce ~100,000 of PCR products with a unique tag that are ligated to microbeads and sequenced (Kudapa et al. 2013). After several rounds of ligation-based sequencing, a 16–20 bp sequence signature is identified from each bead resulting into ~1 million sequence signatures. Because of high throughput analysis and longer tags, MPSS can detect novel transcripts particularly in species lacking whole genome sequence, in addition to identifying genes efficiently (Hamilton and Buell 2012). MPSS has also been utilized in small RNA expression studies (Nobuta et al. 2007) along with mRNA transcription studies in plants which are much correlated with abiotic stress responses (Sunkar et al. 2007). Publicly available plant MPSS database (http://mpss.udel.edu/) contains expression data for several genotypes, including economically important crops such as soybean, maize and rice (Nakano et al. 2006). In addition, NGS platforms have expanded genome-wide sequence expression analysis, in which sequencing of RNA populations and quantification of transcripts can be achieved through RNA-seq (Sánchez-León et al. 2012). The efficiency of Illumina-based digital gene expression system for high-throughput transcriptome sequencing has been demonstrated in crops under abiotic stress conditions in different tissues (Tao et al. 2012; Pandey et al. 2014).

1.3.2 Hybridization-Based Approaches

In response to abiotic stress, plants respond and adapt by altering physiological and biochemical processes resulting in altering responses of thousands of genes. Transcriptome analysis using gene chips and microarray technology provides an important experimental opportunity to unravel key biological processes and to provide information about unknown functional genes conferring abiotic stress tolerance (Gul et al. 2016). In principle, DNA sequences of complete genes of an organism are placed on microchips and used as substrates for hybridization for quantifying expression of different genes in a sample (Joshi et al. 2012). This gives the complete quantitative information about the relative expression of genes corresponding to their response towards various abiotic stresses along with the fold change in different developmental processes like germination, vegetative and flowering stages (Wu et al. 2015). In contrast to sequence-based approaches, array-based technique is a targeted approach where sequence is required to design probes (Rahman et al. 2016). Extensive microarray expression data already exists in public domain (www.genevestigator.com/gv/plant.jsp) with complete genome sequences of several model species including *Arabidopsis* and rice (Hruz et al.

2008; Urano et al. 2010). These gene expression databases provide deeper insight of the complex gene regulatory pathways under various stress responses. Furthermore, genes encoding several regulatory and functional proteins are now known, and the complex mechanisms of multi-gene regulation under abiotic stress response are partly deciphered. Several technical limitations including cross-hybridization and background noise etc. affect microarray analysis investigating stress responsive genes. Through oligo microarray, several model plants and economic crops have been analyzed, including *Arabidopsis* (Richards et al. 2012), rice (Jung et al. 2013), wheat (Quijano et al. 2015), corn (Allardyce et al. 2013), soybean (Le et al. 2012) and tomato (Martínez-Andújar et al. 2012).

Another strategy for RNA hybridization and comparative gene expression in tissues/genotypes is the GeneChip Genome Array. Several studies in model crops have employed these GeneChip Genome Arrays to detect expression of several genes at the same time in the whole genome (Verdier et al. 2013; Wu et al. 2015). In contrast to microarrays, gene chips are created by synthesizing several hundred thousand oligonucleotides on a miniature support using photolithography (Joshi et al. 2012). Further, by using this technique, it is feasible to visualize gene chips that represent an entire plant genome. For example, in soybean, gene chip array characterized genome-wide expression pattern, and identified drought-responsive candidate genes (Saxena et al. 2011). During recent years, a large amount of genome data has been obtained in rice using various chips with different specifications, including BGI/Yale 60K chip (Ma et al. 2005), Agilent 44K chip (Ghaffar et al. 2016), NSF45K chip (Jung et al. 2008), Affymetrix 57K chip (Russell et al. 2012) and NimbleGen (Fenart et al. 2013).

1.3.3 Gene Inactivation Based Approaches

Though the reports pertaining to genome-wide expression analysis in diverse plants are increasing on an exponential rate, only a few studies have focused on over-expression or suppression of these differentially expressed genes for their functional characterization. Currently, two main approaches are being utilized to knockout the desired genes, namely T-DNA insertion mutation and TILLING (Targeted Induced Local Lesions In Genomes). TILLING enables high-throughput genome-wide analysis of point mutations in target genomes to generate novel mutant alleles for crop improvement (Lee et al. 2014). It is applicable to the genomes of almost all species of plants including diploids and allohexaploids (Chen et al. 2014). The TILLING populations can be traditionally screened for phenotypic or genotypic variations under abiotic stresses (de Lorenzo et al. 2009).

Another modified method, called EcoTILLING, is also high-throughput, time-saving and cost-effective technique, developed to identify SNPs and small indels (Bajaj et al. 2016). EcoTILLING is applicable in polyploid species for differentiating among alleles of paralogous and homologous genes (Akpınar et al. 2013). It not only provides information on allelic variants for various genes but also helps in unravelling the complexity of abiotic stress tolerance pathways. Recently, it has been used to detect SNPs involved in salt stress response in domestic rice genotypes (Negrão et al. 2011). Naredo et al. (2009) detected several SNPs in both lowland and upland rice cultivars involved in drought stress tolerance. Similarly, 46 INDELs (insertions/deletions) and 185 SNPs (single nucleotide polymorphisms) were identified using EcoTILLING while conducting allele mining for drought related genes in 96 barley genotypes (Cseri et al. 2011). Similarly, using EcoTILLING approach 1133 novel SNP allelic variants were discovered from diverse coding and regulatory sequence components of 1133 transcription factor genes by genotyping 192 diverse desi and kabuli chickpea genotypes (Bajaj et al. 2016).

T-DNA insertional mutagenesis can be utilized as a tool to study functional genomics in *Arabidopsis* and other higher plants (Jung and An 2013). Agrobacterium-mediated T-DNA transformation can also provide an efficient opportunity to target candidate genes into plant cells. Random insertion of T-DNA fragments in either exon or intron results in the target gene inactivation. During *Arabidopsis* functional genomics initiative, huge number of sequence-indexed T-DNA insertion lines was obtained, which are available currently in the public domain libraries (https://www.arabidopsis.org/portals/mutants/stockcenters.jsp) of *Arabidopsis* (Alonso et al. 2003). Similarly, during International Rice Genome Sequencing project 172,500 flanking sequence tags (FSTs) were submitted in Rice Functional Genomic Express database (RiceGE, http://signal.salk.edu/cgi-bin/RiceGE), which are also available from Rice *Tos17* Insertional Mutant Database (https://tos.nias.affrc.go.jp/). These T-DNA insertion mutants are a rich source for elucidating metabolic/signaling pathways and for functional analysis of genes in plants (Gao and Zhao 2012). In addition, gene inactivation can also be done by using RNAi technology. Using knockdown approach it was confirmed that SOS2 (a serine/threonine type protein kinase) and SOS3 (a calcium binding protein) loci are present in *Arabidopsis*, rice, wheat and *Brassica* (Kumar et al. 2009; Yang et al. 2009; Kushwaha et al. 2011; Feki et al. 2014). Now it is well documented that SOS3 interacts with SOS2 after receiving cytoplasmic calcium signals produced under high Na^+ concentrations. The SOS3-SOS2 complex further activates SOS1, a Na^+/H^+ antiporter gene to maintain homeostasis (Sharma et al. 2015).

1.3.4 Genome Editing Based Approaches

Currently available tools for genome editing provide intriguing possibilities for introducing targeted mutation, INDEL and sequence modifications to a predetermined location within the genome to functionally characterize plant genes and for improvement of abiotic stress tolerance in plants (Strange and Petolino 2012). Due to low homologous recombination frequency in plants, successful gene targeting is very difficult and inefficient (Xie and Yang 2013). Most commonly used genome editing tools are Zinc finger nucleases (ZFNs), transcriptional activator-like effector nucleases (TALENs) and clustered regularly interspaced short palindromic repeat (CRISPR)-Cas9 (CRISPR-associated nuclease9) (Kumar and Jain 2015). TALENs have emerged as an alternative to ZFNs for genome editing and for introducing targeted double-strand breaks. TALENs have showed a very high success rate, but their large size may limit their delivery by recombinant adeno-associated viruses (AAV) (Gaj et al. 2013).

ZFNs are designed nucleases that induce targeted double strand breaks at specific genomic loci, thereby, allowing successful targeted mutagenesis and transgene integration in plants (Petolino et al. 2010). They are fusions of the nonspecific cleavage domain from the *Fok*I restriction endonuclease with custom-designed Cys_2-His_2 zinc-finger proteins. These chimeric nucleases produce sequence-specific DNA double-strand breaks that are repaired by error-prone non-homologous end joining to induce small alterations at targeted genomic loci (Gaj et al. 2012). They can be designed to cleave any DNA sequence and thus offer a wide range of sequences to be deleted. Using ZFNs, majority of targeted genome modifications have been performed including point mutations, deletions, insertions, inversions, duplications and translocations in several organisms and cell types (Joung and Sander 2013). The latest ground-breaking technology for genome editing is the type II clustered regularly interspaced short palindromic repeat (CRISPR)/Cas9 system from *Streptococcus pyogenes* (Bortesi and Fischer 2015). The CRISPR/Cas9 system is composed of Cas9 nuclease and customizable sgRNA which guides Cas9 to recognize target DNA and creates double strand breaks to initiate non-homologous end joining and homologous recombination repair pathways, resulting in genome modifications (Zhang et al. 2016). Since its discovery, CRISPR-Cas9 system has shown robustness and versatility in applications for genome editing in various biological contexts and has opened a new door to plant functional genomics research. This technology can be utilized for analysis of loss-of-function, gain-of-function and gene expression, along with modifications in spatio-temporal gene expression. It can also contribute in understanding gene function, gene regulatory networks and engineering abiotic stress tolerance in a variety of plants (Liu et al. 2016; Khatodia et al. 2016).

1.3.5 Metabolite Analysis

Metabolomics has now emerged as a relatively new area of functional genomics that contributes to our understanding of the complex molecular interactions in biological systems (Bino et al. 2004). Several reviews published earlier have described the role of metabolomics in functional genomics research (Hall et al. 2002; Sumner et al. 2003; Schauer and Fernie 2006; Saito and Matsuda 2010). Several reports are available on its applicability for abiotic stress tolerance in plants (Jorge et al. 2015; Nakabayashi and Saito 2015; Okazaki and Saito 2016; Sun et al. 2016). Integrated metabolomics and transcriptomics studies in model plants have significantly increased our knowledge on signal transduction pathways in different crops under stress. Recently, a report on metabolite profiling in two contrasting rice genotypes i.e., FL478 (salt-tolerant) and IR64 (salt-sensitive) found 92 primary metabolites in the leaves and roots under control and salt stress conditions (Zhao et al. 2014). In general, 6 metabolites (phenylalanine, threonine, citric acid, raffinose, melicitose and galactinol) were induced in the leaves or roots, while 11 metabolites i.e., lysine, threonine, isoleucine, proline, valine, isocitric, sucrose, lactose, sorbitol, mannitol and galactopyranoside were increased specifically in leaves or roots under stress conditions. These compounds regulating sugar and amino acid metabolism pathways will increase our understanding of the physiological mechanisms underpinning salt tolerance. Similarly, comparative proteomic analysis in the shoots of IR64 and its mutant lines resulted in identification of 34 unique proteins expressed during salt stress exposure (Ghaffari et al. 2014). Similarly, Liu et al. (2014b) detected 83 proteins in roots and 61 proteins in leaves to be differentially expressed and reported of having their significant contributions against salinity stress in rice. Protein alterations upon external stimuli are vital, and thus proteomic analysis provides deep knowledge on key aspects of plant metabolic and regulatory pathways against abiotic stress (Kim et al. 2014). These differentially expressed proteins can act as an abiotic stress tolerance marker for plants (Zhang 2014). Our understanding of metabolite adaptation to abiotic stress in plants is still incomplete. Thus, it is necessary to deepen our knowledge further with targeted comprehensive metabolomics studies with more emphasis on primary and secondary metabolic pathways.

1.4 Role of Model Species and Mutant Populations

Although functional adaptation mechanisms are highly conserved among stress susceptible genotypes, the tolerant genotypes, however, evolved additional regulatory mechanisms that enhance their ability to cope with severe abiotic stresses (Joshi et al. 2016b). Whole genome sequencing of rice and *Arabidopsis* has increased our understanding of the genes playing a crucial role in providing multiple abiotic stress tolerance (Mustafiz et al. 2011; Kumar et al. 2012; Singh et al. 2012; Tripathy et al. 2012; Kaur et al. 2014). For example, in *Arabidopsis* early stages of heat stress triggers decay of 25% of the transcriptome and is catalyzed by the 5′-3′ exonuclease XRN4. cDNA libraries prepared from 21 days old heat stressed seedlings shows 19,804 distinct loci accounting for 76% of the total *Arabidopsis* genes. Out of these, only 801 (4%) were found to be upregulated, which represents proteins involved in heat and abiotic stress response, and 4,745 (25%) were found to be down-regulated (Merret et al. 2013). Similarly, RNA-Seq and digital gene expression (DGE) analysis in *Bryum argenteum*, a desiccation-tolerant moss found in largest cold desert (Gurbantunggut desert) of China, showed 4,081 and 6,709 differentially expressed genes after 2h and 24h rehydration, respectively. Further, upon rehydration, 142 TF transcripts were found to be up-regulated, including 23 members of ERF family (Gao et al. 2015).

By using modern genomics and genetic approaches, full-length cDNA populations and BAC sequences have been transferred from stress-tolerant genotypes to stress-sensitive ones to generate stress tolerant varieties with better growth and yield (Mir et al. 2012; Akpınar et al. 2013). As wheat, barley and rye are close relatives; their syntenic relationship can be utilized for positional cloning of important stress tolerant genes (Joshi et al. 2015a; Kole et al. 2015).

Mutant phenotype selection through mutational breeding is an old technique, which has successfully contributed in generating several important varieties of cereals. Using single base mismatches, several barley and wheat mutant populations have been developed for mutation studies and several projects are running throughout the globe for developing mutant populations of their diploid progenitors (Sikora et al. 2011; Dhakarey et al. 2016). Several sets of insertion mutants are already accessible for petunia, maize, snapdragon, rice and *Arabidopsis*. However, high degree of gene duplication and tight linkage between genes act as a major limiting factor to study gene function and genetic recombination in plants (Glover et al. 2015). One possible approach is to use either homologous recombination to eliminate tandem duplications by gene replacements or to introduce point mutations using RNA-DNA hybrids (Reams et al. 2012). This can also be achieved through inserting mutated sequences to generate stop codons within the conserved regions to produce null mutations in a multigene family. However, high throughput gene silencing on double-stranded RNA through bidirectional transcription of genes is broadly accepted, as it is easy to generate transgenic plants with drastic transcriptomic alterations (Zhang et al. 2015). Recently, CRISPR/Cas9 has emerged as a powerful tool to generate knock-in mutants or knock-out mutants with frameshift mutations in plants (Liu et al. 2016).

1.5 Mapping and Map-Based Cloning

Breeding programs of important crop species like rice and wheat functioning from several decades have broadened our knowledge in the mapping of several traits related to abiotic stress tolerance. Introduction of molecular marker techniques in conventional breeding gave further extension to mapping studies and in assessment of cultivated, land race and wild genotypes (Varshney et al. 2012). These studies led to identification of germplasm rich genotypes showing extensive variation at structural and expression levels under stress conditions. These variations are useful to confirm candidate genes for stress tolerance as well as for discovering alleles for further breeding programs (Ma et al. 2012). Majority of the known abiotic stress loci have been discovered as QTL, so a particular trait mapping in different genotypes using multiple populations can locate the common loci such as drought tolerance (Jaganathan et al. 2015). More than hundred abiotic stress related traits have already been mapped only in soybean in past years (Xia et al. 2013). Similarly, availability of whole genome sequence in rice, and its strong similarity with wheat and barley genomes makes rice a potential crop for marker generation from candidate loci.

Further strategy is positional cloning of functionally correlated genes for specific trait using forward genetics approach. Positional cloning may or may not identify target gene(s) associated with a particular phenotype directly. However, through complementation analysis, the target gene can be identified (Langridge and Fleury 2011). Another variant of positional cloning is map-based cloning, where chromosomal location of a gene is identified through genetic mapping using molecular markers (Kudapa et al. 2013). With faster and more accurate next-generation sequencing (NGS) technologies as well as advanced DNA polymorphism detection techniques, map based cloning and physical mapping using BAC libraries have now become more handy for different crops such as rice (Vij and Tyagi 2007), barley (Schulte et al. 2011), soybean (Fang et al. 2013; Song et al. 2016), *Brassica* (Mun et al. 2015) and wheat (Wang et al. 2015). High-density genetic linkage maps have been integrated with sequence-based physical map, thus resulting in improved resolution and accuracy of trait-specific genes/QTLs identification (Agarwal et al. 2016).

1.6 Conclusion

Functional genomics studies have played a central role in not only providing solutions to generate new varieties through genetic transformation but also have increased our understanding of cellular metabolism operating under abiotic stress. Several functionally characterized genes when inserted into crop plants have shown increased tolerance against various environmental stresses in comparison to wild type plants. These genetically engineered plants show higher osmolyte and protein accumulation and are generally more productive in terms of agricultural yield. Further, using genomic tools, stress- and organ-specific promoters have been identified and tested thoroughly for their specificity. Also, comparative genomics studies have identified genes that throw light upon conserved evolutionary mechanisms in plants

Huge wealth of data is now available for plant signaling in response to various abiotic stresses. Additionally, several transcription and signaling factors along with their interconnections and crosstalk mechanisms have increased our understanding of the intricate network that operates under stress. Despite this, full understanding of the genes controlling signaling pathways is lacking. The filtering of the huge data using bioinformatics tools and validation of the genes using advanced genomics tools like proteomics and metabolomics can alleviate this deficit. Mining of the data systematically for functional analysis by using mutants and overexpression analysis followed by microarray analyses can reveal interactions between signaling components and downstream targeted genes. Recent technological developments in functional genomics such as RNAi technology, gene editing and next generation genomics can help us uncover the variations integrated across diverse plant genomes. This can further be applied to manipulate crop species for enhanced defense strategies using conventional, marker assisted or transgenic approaches.

Acknowledgements Research in our lab is supported by funds from the Department of Biotechnology, Council of Scientific and Industrial Research, Government of India, internal grants of ICGEB, and Bioseed Research India. RJ acknowledges the Start-Up research grant (Young Scientist) from the Science and Engineering Research Board and Dr. D. S. Kothari Fellowship from University Grants Commission, Government of India.

References

Agarwal P, Parida SK, Raghuvanshi S, Kapoor S, Khurana P, Khurana JP, Tyagi AK (2016) Rice improvement through genome-based functional analysis and molecular breeding in India. Rice 9:1

Akpınar BA, Lucas SJ, Budak H (2013) Genomics approaches for crop improvement against abiotic stress. Sci World J 2013: Article ID 361921

Allardyce JA, Rookes JE, Hussain HI, Cahill DM (2013) Transcriptional profiling of *Zea mays* roots reveals roles for jasmonic acid and terpenoids in resistance against *Phytophthora cinnamomi*. Funct Integr Genom 13:217–228

Alonso JM, Stepanova AN, Leisse TJ, Kim CJ, Chen H, Shinn P, Stevenson DK, Zimmerman J, Barajas P, Cheuk R, Gadrinab C (2003) Genome-wide insertional mutagenesis of *Arabidopsis thaliana*. Science 301:653–657

Ara H, Sinha HA (2014) Conscientiousness of mitogen activated protein kinases in acquiring tolerance for abiotic stresses in plants. Proc Ind Natl Sci Acad 80:211–219

Armbrust EV, Berges JA, Bowler C, Green BR, Martinez D, Putnam NH, Zhou S, Allen AE, Apt KE, Bechner M, Brzezinski MA (2004) The genome of the diatom *Thalassiosira pseudonana*: ecology, evolution and metabolism. Science 306:79–86

Baisakh N, Subudhi PK, Varadwaj P (2008) Primary responses to salt stress in a halophyte, smooth cordgrass (*Spartina alterniflora* Loisel.). Funct Integ Genom 8:287–300

Bajaj D, Srivastava R, Nath M, Tripathi S, Bharadwaj C, Upadhyaya HD, Tyagi AK, Parida SK (2016) EcoTILLING-based association mapping efficiently delineates functionally relevant natural allelic variants of candidate genes governing agronomic traits in chickpea. Front Plant Sci 7:450

Baloglu MC, Inal B, Kavas M, Unver T (2014) Diverse expression pattern of wheat transcription factors against abiotic stresses in wheat species. Gene 550:117–122

Bhullar NK, Gruissem W (2013) Nutritional enhancement of rice for human health: the contribution of biotechnology. Biotechnol Adv 31:50–57

Bino RJ, Hall RD, Fiehn O, Kopka J, Saito K, Draper J, Nikolau BJ, Mendes P, Roessner-Tunali U, Beale MH, Trethewey RN (2004) Potential of metabolomics as a functional genomics tool. Trends Plant Sci 9:418–425

Bohra A, Singh NP (2015) Whole genome sequences in pulse crops: a global community resource to expedite translational genomics and knowledge-based crop improvement. Biotechnol Lett 37:1529–1539

Bohra A, Sahrawat KL, Kumar S, Joshi R, Parihar AK, Singh U, Singh D, Singh NP (2015) Genetics and genomics based interventions for nutritional enhancement of grain-legume crops: status and outlook. J Appl Genet 56:151–161

Bortesi L, Fischer R (2015) The CRISPR/Cas9 system for plant genome editing and beyond. Biotechnol Adv 33:41–52

Boudsocq M, Sheen J (2013) CDPKs in immune and stress signaling. Trends Plant Sci 18:30–40

Breyne P, Zabeau M (2001) Genome-wide expression analysis of plant cell cycle modulated genes. Cur Opin Plant Biol 4:136–142

Chen L, Hao L, Parry MAJ, Phillips AL, Hu YG (2014) Progress in TILLING as a tool for functional genomics and improvement of crops. J Integr Plant Biol 56:425–443

Cheng CK, Au CH, Wilke SK, Stajich JE, Zolan ME, Pukkila PJ, Kwan HS (2013) 5'-Serial analysis of gene expression studies reveal a transcriptomic switch during fruiting body development in *Coprinopsis cinerea*. BMC Genom 14:195
Chinnusamy V, Schumaker K, Zhu JK (2004) Molecular genetic perspectives on cross-talk and specificity in abiotic stress signaling in plants. J Exp Bot 55:225–236
Cramer GR, Urano K, Delrot S, Pezzotti M, Shinozaki K (2011) Effects of abiotic stress on plants: a systems biology perspective. BMC Plant Biol 11:163
Cseri A, Cserháti M, Von Korff M, Nagy B, Horváth GV, Palágyi A, Pauk J, Dudits D, Törjék O (2011) Allele mining and haplotype discovery in barley candidate genes for drought tolerance. Euphytica 181:341–356
de Lorenzo L, Merchan F, Laporte P, Thompson R, Clarke J, Sousa C, Crespi M (2009) A novel plant leucine-rich repeat receptor kinase regulates the response of *Medicago truncatula* roots to salt stress. Plant Cell 21:668–680
Dhakarey R, Kodackattumannil Peethambaran P, Riemann M (2016) Functional analysis of jasmonates in rice through mutant approaches. Plants 5:15
Dwivedi SL, Ceccarelli S, Blair MW, Upadhyaya HD, Are AK, Ortiz R (2016) Landrace germplasm for improving yield and abiotic stress adaptation. Trends Plant Sci 21:31–42
Edwards D (2016) The impact of genomics technology on adapting plants to climate change. In: Edwards D, Batley J (eds) Plant genomics and climate change. Springer, New York, pp 173–178
Ergen NZ, Budak H (2009) Sequencing over 13000 expressed sequence tags from six subtractive cDNA libraries of wild and modern wheats following slow drought stress. Plant Cell Environ 32:220–236
Fang C, Li W, Li G, Wang Z, Zhou Z, Ma Y, Shen Y, Li C, Wu Y, Zhu B, Yang W (2013) Cloning of Ln gene through combined approach of map based cloning and association study in soybean. J Genet Genom 40:93–96
Feist AM, Palsson BO (2008) The growing scope of applications of genome-scale metabolic reconstructions using *Escherichia coli*. Nat Biotechnol 26:659–667
Feki K, Quintero FJ, Khoudi H, Leidi EO, Masmoudi K, Pardo JM, Brini F (2014) A constitutively active form of a durum wheat Na^+/H^+ antiporter SOS1 confers high salt tolerance to transgenic *Arabidopsis*. Plant Cell Rep 33:277–288
Fenart S, Chabi M, Gallina S, Huis R, Neutelings G, Riviere N, Thomasset B, Hawkins S, Lucau-Danila A (2013) Intra-platform comparison of 25-mer and 60-mer oligonucleotide Nimblegen DNA microarrays. BMC Res Notes 6:43
Ford R, Khan S, Mantri N (2015) Towards understanding the transcriptional control of abiotic stress tolerance mechanisms in food legumes. In: Pandey GK (ed) Elucidation of abiotic stress signaling in plants. Springer, New York, pp 29–43
Gahlaut V, Mathur S, Dhariwal R, Khurana JP, Tyagi AK, Balyan HS, Gupta PK (2014) A multi-step phosphorelay two-component system impacts on tolerance against dehydration stress in common wheat. Funct Integ Genom 14:707–716
Gaj T, Guo J, Kato Y, Sirk SJ, Barbas CF 3rd (2012) Targeted gene knockout by direct delivery of zinc-finger nuclease proteins. Nat Methods 9:805–807
Gaj T, Gersbach CA, Barbas CF 3rd (2013) ZFN, TALEN, and CRISPR/Cas-based methods for genome engineering. Trends Biotechnol 31:397–405
Galagan JE, Calvo SE, Borkovich KA, Selker EU, Read ND, Jaffe D, FitzHugh W, Ma LJ, Smirnov S, Purcell S, Rehman B (2003) The genome sequence of the filamentous fungus *Neurospora crassa*. Nature 422:859–868
Gao Y, Zhao Y (2012) Epigenetic suppression of T-DNA insertion mutants in *Arabidopsis*. Mol Plant 6:539–545
Gao B, Zhang D, Li X, Yang H, Zhang Y, Wood AJ (2015) De novo transcriptome characterization and gene expression profiling of the desiccation tolerant moss *Bryum argenteum* following rehydration. BMC Genom 16:416

Ghaffar MBA, Norliza MB, Pritchard J, Ford-Lloyd BV (2016) Identification of candidate genes involved in brown plant hopper resistance in rice using microarray analysis. J Trop Agric Food Sci 44:49–62
Ghaffari A, Gharechahi J, Nakhoda B, Salekdeh GH (2014) Physiology and proteome responses of two contrasting rice mutants and their wild type parent under salt stress conditions at the vegetative stage. J Plant Physiol 171:31–44
Glover NM, Daron J, Pingault L, Vandepoele K, Paux E, Feuillet C, Choulet F (2015) Small-scale gene duplications played a major role in the recent evolution of wheat chromosome 3B. Genome Biol 16:188
Gul A, Ahad A, Akhtar S, Ahmad Z, Rashid B, Husnain T (2016) Microarray: gateway to unravel the mystery of abiotic stresses in plants. Biotechnol Lett 38:527–543
Guo M, Liu JH, Ma X, Luo DX, Gong ZH, Lu MH (2016) The plant heat stress transcription factors (HSFs): structure, regulation, and function in response to abiotic stresses. Front Plant Sci 7:114
Gupta B, Huang B (2014) Mechanism of salinity tolerance in plants: physiological, biochemical, and molecular characterization. Intl J Genom 2014:701596
Gupta B, Tripathi AK, Joshi R, Pareek A, Singla-Pareek SL (2015) Designing climate smart future crops employing signal transduction components. In: Pandey GK (ed) Elucidation of abiotic stress signaling in plants: functional genomics perspectives, vol 2. Springer, New York, pp 393–414
Hall R, Beale M, Fiehn O, Hardy N, Sumner L, Bino R (2002) Plant metabolomics: the missing link in functional genomics strategies. Plant Cell 14:1437–1440
Hamilton JP, Buell CR (2012) Advances in plant genome sequencing. Plant J 70:177–190
Hasanuzzaman M, Roychowdhury R, Karmakar J, Dey N, Nahar K, Fujita M (2015) Recent advances in biotechnology and genomic approaches for abiotic stress tolerance in crop plants. In: Thangadurai D, Sangeetha J (eds) Genomics and proteomics: principles, technologies, and applications. CRC Press, Boca Raton, pp 333–366
Hirakawa H, Shirasawa K, Miyatake KO, Nunome T, Negoro S, Ohyama AK, Yamaguchi H, Sato S, Isobe S, Tabata S, Fukuoka H (2014) Draft genome sequence of eggplant (*Solanum melongena* L.): the representative solanum species indigenous to the old world. DNA Res 21:649–660
Hoang XLT, Thu NBA, Thao NP, Tran LSP (2014) Transcription factors in abiotic stress responses: their potentials in crop improvement. In: Ahmad P, Wani MR, Azooz MM, Tran LSP (eds) Improvement of crops in the era of climatic changes. Springer, New York, pp 337–366
Hruz T, Laule O, Szabo G, Wessendorp F, Bleuler S, Oertle L, Widmayer P, Gruissem W, Zimmermann P (2008) Genevestigator V3: a reference expression database for the meta-analysis of transcriptomes. Advan Bioinform 2008:42074
Jaganathan D, Thudi M, Kale S, Azam S, Roorkiwal M, Gaur PM, Kishor PK, Nguyen H, Sutton T, Varshney RK (2015) Genotyping-by-sequencing based intra-specific genetic map refines a "QTL-hotspot" region for drought tolerance in chickpea. Mol Genet Genomics 290:559–571
Jan AT, Singhal P, Haq QMR (2013) Plant abiotic stress: deciphering remedial strategies for emerging problem. J Plant Inter 8:97–108
Jorge TF, Rodrigues JA, Caldana C, Schmidt R, van Dongen JT, Thomas-Oates J, António C (2015) Mass spectrometry-based plant metabolomics: metabolite responses to abiotic stress. Mass Spectrom Rev 35:620–649
Joshi R, Chinnusamy V (2014) Antioxidant enzymes: defense against high temperature stress. In: Ahmad P (ed) Oxidative damage to plants: antioxidant networks and signaling. Elsevier, New York, pp 369–396
Joshi R, Karan R, Singla-Pareek SL, Pareek A (2012) Microarray technology. In: Gupta AK, Pareek A, Gupta SM (eds) Biotechnology in medicine and agriculture: principles and practices. IK International Publishing House Pvt. Ltd., India, pp 273–296

Joshi R, Ramanarao MV, Bedre R, Sanchez L, Pilcher W, Zandkarimi H, Baisakh N (2015a) Salt adaptation mechanisms of halophytes: improvement of salt tolerance in crop plants. In: Pandey GK (ed) Elucidation of abiotic stress signaling in plants: functional genomics perspectives, vol 2. Springer, New York, pp 243–280

Joshi R, Singh B, Bohra A, Chinnusamy V (2015b) Salt stress signaling pathways: specificity and crosstalk. In: Wani SH, Hossain MA (eds) Managing salinity tolerance in plants: molecular and genomic perspectives. CRC Press, Boca Raton, pp 51–78

Joshi R, Karan R, Singla-Pareek SL, Pareek A (2016a) Ectopic expression of Pokkali phosphoglycerate kinase-2 (OsPGK2-P) improves yield in tobacco plants under salinity stress. Plant Cell Rep 35:27–41

Joshi R, Wani SH, Singh B, Bohra A, Dar ZA, Lone AA, Pareek A, Singla-Pareek SL (2016b) Transcription factors and plant response to drought stress: current understanding and future directions. Front Plant Sci 7:1029

Joung JK, Sander JD (2013) TALENs: a widely applicable technology for targeted genome editing. Nat Rev Mol Cell Biol 14:49–55

Jung KH, An G (2013) Functional characterization of rice genes using a gene-indexed T-DNA insertional mutant population. In: Yang Y (ed) Rice protocols. Methods Mol Biol 956:57–67

Jung KH, Dardick C, Bartley LE, Cao P, Phetsom J, Canlas P, Seo YS, Shultz M, Ouyang S, Yuan Q, Frank BC (2008) Refinement of light-responsive transcript lists using rice oligonucleotide arrays: evaluation of gene-redundancy. PLoS ONE 3:e3337

Jung KH, Gho HJ, Giong HK, Chandran AK, Nguyen QN, Choi H, Zhang T, Wang W, Kim JH, Choi HK, An G (2013) Genome-wide identification and analysis of Japonica and Indica cultivar-preferred transcripts in rice using 983 Affymetrix array data. Rice 6:19

Kaur C, Ghosh A, Pareek A, Sopory SK, Singla-Pareek SL (2014) Glyoxalases and stress tolerance in plants. Biochem Soc Trans 42:485–490

Khan MS, Ahmad D, Khan MA (2015) Utilization of genes encoding osmoprotectants in transgenic plants for enhanced abiotic stress tolerance. Electr J Biotechnol 18:257–266

Khatodia S, Bhatotia K, Passricha N, Khurana SMP, Tuteja N (2016) The CRISPR/Cas genome-editing tool: application in improvement of crops. Front Plant Sci 7:506

Kim ST, Kim SG, Agrawal GK, Kikuchi S, Rakwal R (2014) Rice proteomics: a model system for crop improvement and food security. Proteomics 14:593–610

Kole C, Muthamilarasan M, Henry R, Edwards D, Sharma R, Abberton M, Batley J, Bentley A, Blakeney M, Bryant J, Cai H (2015) Application of genomics-assisted breeding for generation of climate resilient crops: progress and prospects. Front Plant Sci 6:563

Kudapa H, Ramalingam A, Nayakoti S, Chen X, Zhuang WJ, Liang X, Kahl G, Edwards D, Varshney RK (2013) Functional genomics to study stress responses in crop legumes: progress and prospects. Funct Plant Biol 14:1221–1233

Kumar V, Jain M (2015) The CRISPR-Cas system for plant genome editing: advances and opportunities. J Exp Bot 66:47–57

Kumar G, Purty RS, Sharma MP, Singla-Pareek SL, Pareek A (2009) Physiological responses among *Brassica* species under salinity stress show strong correlation with transcript abundance for SOS pathway-related genes. J Plant Physiol 166:507–520

Kumar G, Kushwaha HR, Panjabi-Sabharwal V, Kumari S, Joshi R, Karan R, Mittal S, Pareek SL, Pareek A (2012) Clustered metallothionein genes are co-regulated in rice and ectopic expression of OsMT1e-P confers multiple abiotic stress tolerance in tobacco via ROS scavenging. BMC Plant Biol 12:107

Kushwaha HR, Kumar G, Verma PK, Singla-Pareek SL, Pareek A (2011) Analysis of a salinity induced BjSOS3 protein from *Brassica* indicate it to be structurally and functionally related to its ortholog from *Arabidopsis*. Plant Physiol Biochem 49:996–1004

Kushwaha HR, Joshi R, Pareek A, Singla-Pareek SL (2016) MATH-domain family shows response towards abiotic stress in *Arabidopsis* and rice. Front Plant Sci 7:923

Langridge P, Fleury D (2011) Making the most of 'omics' for crop breeding. Trends Biotechnol 29:33–40

Le DT, Nishiyama R, Watanabe Y, Mochida K, Yamaguchi-Shinozaki K, Shinozaki K, Tran LSP (2011) Genome-wide expression profiling of soybean two-component system genes in soybean root and shoot tissues under dehydration stress. DNA Res 18:17–29

Le DT, Nishiyama R, Watanabe Y, Tanaka M, Seki M, Yamaguchi-Shinozaki K, Shinozaki K, Tran LS (2012) Differential gene expression in soybean leaf tissues at late developmental stages under drought stress revealed by genome-wide transcriptome analysis. PLoS ONE 7: e49522

Lee LS, Till BJ, Hill H, Huynh OA, Jankowicz-Cieslak J (2014) Mutation and mutation screening. In: Henry RJ, Furtado A (eds) Cereal genomics. Method Mol Biol 1099:77–95

Li F, Fan G, Wang K, Sun F, Yuan Y, Song G, Li Q, Ma Z, Lu C, Zou C, Chen W (2014a) Genome sequence of the cultivated cotton *Gossypium arboreum*. Nat Genet 46:567–572

Li J, Sun X, Yu G, Jia C, Liu J, Pan H (2014b) Generation and analysis of expressed sequence tags (ESTs) from halophyte *Atriplex canescens* to explore salt-responsive related genes. Int J Mol Sci 15:1172–11189

Liu JH, Peng T, Dai W (2014a) Critical cis-acting elements and interacting transcription factors: key players associated with abiotic stress responses in plants. Plant Mol Biol Rep 32:303–317

Liu Z, Zhang M, Kong L, Lv Y, Zou M, Lu G, Cao J, Yu X (2014b) Genome-wide identification, phylogeny, duplication, and expression analyses of two-component system genes in Chinese cabbage (*Brassica rapa* ssp. pekinensis). DNA Res 21:379–396

Liu D, Hu R, Palla KJ, Tuskan GA, Yang X (2016) Advances and perspectives on the use of CRISPR/Cas9 systems in plant genomics research. Curr Opin Plant Biol 30:70–77

Ma L, Chen C, Liu X, Jiao Y, Su N, Li L, Wang X, Cao M, Sun N, Zhang X, Bao J (2005) A microarray analysis of the rice transcriptome and its comparison to *Arabidopsis*. Genome Res 15:1274–1283

Ma Y, Qin F, Tran LS (2012) Contribution of genomics to gene discovery in plant abiotic stress responses. Mol Plant 5:1176–1178

Mahalingam R, Gomez-Buitrago A, Eckardt N, Shah N, Guevara-Garcia A, Day P, Raina R, Fedoroff NV (2003) Characterizing the stress/defense transcriptome of *Arabidopsis*. Genome Biol 4:R20

Martínez-Andújar C, Pluskota WE, Bassel GW, Asahina M, Pupel P, Nguyen TT, Takeda-Kamiya N, Toubiana D, Bai B, Górecki RJ, Fait A (2012) Mechanisms of hormonal regulation of endosperm cap-specific gene expression in tomato seeds. Plant J 71:575–586

Matsumura H, Urasaki N, Yoshida K, Krüger DH, Kahl G, Terauchi R (2012) SuperSAGE: powerful serial analysis of gene expression. In: Jin H, Gassmann W (eds) RNA abundance analysis: methods and protocols. Springer, New York, pp 1–17

Merret R, Descombin J, Juan YT, Favory JJ, Carpentier MC, Chaparro C, Charng YY, Deragon JM, Bousquet-Antonelli C (2013) XRN4 and LARP1 are required for a heat-triggered mRNA decay pathway involved in plant acclimation and survival during thermal stress. Cell Rep 5:1279–1293

Mickelbart MV, Hasegawa PM, Bailey-Serres J (2015) Genetic mechanisms of abiotic stress tolerance that translate to crop yield stability. Nat Rev Genet 16:237–251

Mir RR, Zaman-Allah M, Sreenivasulu N, Trethowan R, Varshney RK (2012) Integrated genomics, physiology and breeding approaches for improving drought tolerance in crops. Theor Appl Genet 125:625–645

Mun JH, Yu HJ, Park BS (2015) Genomic resources and physical mapping of the *B. rapa* genome. In: Wang X, Kole C (eds) The *Brassica rapa* genome. Springer, New York, pp 25–39

Mustafiz A, Singh AK, Pareek A, Sopory SK, Singla-Pareek SL (2011) Genome-wide analysis of rice and *Arabidopsis* identifies two glyoxalase genes that are highly expressed in abiotic stresses. Funct Integr Genom 11:293–305

Muthurajan R, Balasubramanian P (2009) pyramiding genes for enhancing tolerance to abiotic and biotic stresses. In: Jain SM, Brar DS (eds) Molecular techniques in crop improvement. Springer, New York, pp 163–184

Nakabayashi R, Saito K (2015) Integrated metabolomics for abiotic stress responses in plants. Curr Opin Plant Biol 24:10–16

Nakano M, Nobuta K, Vemaraju K, Tej SS, Skogen JW, Meyers BC (2006) Plant MPSS databases: signature-based transcriptional resources for analyses of mRNA and small RNA. Nucl Acid Res 34:D731–D735

Naredo MEB, Cairns J, Wang H, Atienza G, Sanciangco MD, Melgar RJ, Kumar A, Ramaiah V, Serraj R, Mc Nally KL (2009) EcoTILLING as a SNP discovery tool for drought candidate genes in *Oryza sativa* germplasm. Philippine J Crop Sci 34:10–16

Negrão SC, Pires AI, McNally KL, Oliveira MM (2011) Use of EcoTILLING to identify natural allelic variants of rice candidate genes involved in salinity tolerance. Plant Genet Resour 9:300–304

Nelson DE, Repetti PP, Adams TR, Creelman RA, Wu J, Warner DC, Anstrom DC, Bensen RJ, Castiglioni PP, Donnarummo MG, Hinchey BS (2007) Plant nuclear factor Y (NF-Y) B subunits confer drought tolerance and lead to improved corn yields on water-limited acres. Proc Natl Acad Sci USA 104:16450–16455

Nielsen KL, Høgh AL, Emmersen J (2006) DeepSAGE-digital transcriptomics with high sensitivity, simple experimental protocol and multiplexing of samples. Nucl Acid Res 34: article e133

Nishiyama R, Watanabe Y, Leyva-Gonzalez MA, Van Ha C, Fujita Y, Tanaka M, Seki M, Yamaguchi-Shinozaki K, Shinozaki K, Herrera-Estrella L, Tran LS (2013) *Arabidopsis* AHP2, AHP3 and AHP5 histidine phosphotransfer proteins function as redundant negative regulators of drought stress response. Proc Natl Acad Sci USA 110:4840–4845

Nobuta K, Venu RC, Lu C, Beló A, Vemaraju K, Kulkarni K, Wang W, Pillay M, Green PJ, Wang GL, Meyers BC (2007) An expression atlas of rice mRNAs and small RNAs. Nat Biotechnol 25:473–477

Okazaki Y, Saito K (2016) Integrated metabolomics and phytochemical genomics approaches for studies on rice. Gigascience 5:11

Pandey R, Joshi G, Bhardwaj AR, Agarwal M, Katiyar-Agarwal S (2014) A comprehensive genome-wide study on tissue-specific and abiotic stress-specific miRNAs in *Triticum aestivum*. PLoS ONE 9:e95800

Pareek A, Singh A, Kumar M, Kushwaha HR, Lynn AM, Singla-Pareek SL (2006) Whole-genome analysis of *Oryza sativa* reveals similar architecture of two-component signaling machinery with *Arabidopsis*. Plant Physiol 142:380–397

Pareek A, Sopory SK, Bohnert HK, Govindjee (eds) (2010) Abiotic stress adaptation in plants: physiolgical, molecular and genomic foundation. Springer, The Netherlands

Park S, Lee CM, Doherty CJ, Gilmour SJ, Kim Y, Thomashow MF (2015) Regulation of the *Arabidopsis* CBF regulon by a complex low-temperature regulatory network. Plant J 82:193–207

Peng X, Wu Q, Teng L, Tang F, Pi Z, Shen S (2015) Transcriptional regulation of the paper mulberry under cold stress as revealed by a comprehensive analysis of transcription factors. BMC Plant Biol 15:108

Petolino JF, Worden A, Curlee K, Connell J, Strange Moynahan TL, Larsen C, Russell S (2010) Zinc finger nuclease-mediated transgene deletion. Plant Mol Biol 73:617–628

Pitzschke A (2015) Modes of MAPK substrate recognition and control. Trends Plant Sci 20:49–55

Pitzschke A, Datta S, Persak H (2014) Salt stress in *Arabidopsis*: lipid transfer protein AZI1 and its control by mitogen-activated protein kinase MPK3. Mol Plant 7:722–738

Prasch CM, Sonnewald U (2013) Simultaneous application of heat, drought, and virus to *Arabidopsis* plants reveals significant shifts in signaling networks. Plant Physiol 162:1849–1866

Quijano CD, Brunner S, Keller B, Gruissem W, Sautter C (2015) The environment exerts a greater influence than the transgene on the transcriptome of field-grown wheat expressing the Pm3b allele. Transgenic Res 24:87–97

Rahman M, Rahmat Z, Gul M, Zafar Y (2016) Plant functional genomics: approaches and applications. In: Khan MS, Khan IA, Barh D (eds) Applied molecular biotechnology: the next generation of genetic engineering. CRC Press, Boca Raton, pp 157–186

Rasmussen S, Barah P, Suarez-Rodriguez MC, Bressendorff S, Friis P, Costantino P, Bones AM, Nielsen HB, Mundy J (2013) Transcriptome responses to combinations of stresses in *Arabidopsis*. Plant Physiol 161:1783–1794

Reams AB, Kofoid E, Kugelberg E, Roth JR (2012) Multiple pathways of duplication formation with and without recombination (RecA) in *Salmonella enterica*. Genetics 192:397–415

Rejeb IB, Pastor V, Mauch-Mani B (2014) Plant responses to simultaneous biotic and abiotic stress: molecular mechanisms. Plants 3:458–475

Richards CL, Rosas U, Banta J, Bhambhra N, Purugganan MD (2012) Genome-wide patterns of *Arabidopsis* gene expression in nature. PLoS Genet 8:e1002662

Roy SJ, Tucker EJ, Tester M (2011) Genetic analysis of abiotic stress tolerance in crops. Curr Opin Plant Biol 14:232–239

Rudd S (2003) Expressed sequence tags: alternative or complement to whole genome sequences? Trends Plant Sci 8:321–329

Russell SD, Gou X, Wong CE, Wang X, Yuan T, Wei X, Bhalla PL, Singh MB (2012) Genomic profiling of rice sperm cell transcripts reveals conserved and distinct elements in the flowering plant male germ lineage. New Phytol 195:560–573

Sahoo KK, Tripathi AK, Pareek A, Singla-Pareek SL (2013) Taming drought stress in rice through genetic engineering and transcription factors and protein kinases. Plant Stress 1:60–72

Saito K, Matsuda F (2010) Metabolomics for functional genomics, systems biology, and biotechnology. Annu Rev Plant Biol 61:463–489

Sánchez-León N, Arteaga-Vázquez M, Alvarez-Mejía C, Mendiola-Soto J, Durán-Figueroa N, Rodríguez-Leal D, Rodríguez-Arévalo I, García-Campayo V, García-Aguilar M, Olmedo-Monfil V, Arteaga-Sánchez M (2012) Transcriptional analysis of the *Arabidopsis* ovule by massively parallel signature sequencing. J Exp Bot 63:3829–3842

Saxena RK, Cui X, Thakur V, Walter B, Close TJ, Varshney RK (2011) Single feature polymorphisms (SFPs) for drought tolerance in pigeonpea (*Cajanus* spp.). Funct Integ Genom 11:651–657

Schauer N, Fernie AR (2006) Plant metabolomics: towards biological function and mechanism. Trends Plant Sci 11:508–516

Schulte D, Ariyadasa R, Shi B, Fleury D, Saski C, Atkins M, Wu CC, Graner A, Langridge P, Stein N (2011) BAC library resources for map-based cloning and physical map construction in barley (*Hordeum vulgare* L.). BMC Genom 12:247

Sharan A, Soni P, Nongpiur RC, Singla-Pareek SL, Pareek A (2017) Mapping the 'two component system' network in rice. Sci Rep 7:9287

Sharma R, Mishra M, Gupta B, Parsania C, Singla-Pareek SL, Pareek A (2015) De Novo assembly and characterization of stress transcriptome in a salinity-tolerant variety CS52 of *Brassica juncea*. PLoS ONE 10:e0126783

Sikora P, Chawade A, Larsson M, Olsson J, Olsson O (2011) Mutagenesis as a tool in plant genetics, functional genomics, and breeding. Intl J Plant Genom 2011: Article ID 314829

Singh AK, Kumar R, Pareek A, Sopory SK, Singla-Pareek SL (2012) Overexpression of rice CBS domain containing protein improves salinity, oxidative, and heavy metal tolerance in transgenic tobacco. Mol Biotechnol 52:205–216

Singh A, Kushwaha HR, Soni P, Gupta H, Singla-Pareek SL, Pareek A (2015a) Tissue specific and abiotic stress regulated transcription of histidine kinases in plants is also influenced by diurnal rhythm. Front Plant Sci 6:711

Singh B, Bohra A, Mishra S, Joshi R, Pandey S (2015b) Embracing new-generation 'omics' tools to improve drought tolerance in cereal and food-legume crops. Biol Plant 59:413–428

Sinha AK, Jaggi M, Raghuram B, Tuteja N (2011) Mitogen-activated protein kinase signaling in plants under abiotic stress. Plant Signal Behav 6:196–203

Sofi P, Trag AR (2006) Genomics in rice improvement. Asian J Biochem 1:194–210

Song Q, Jenkins J, Jia G, Hyten DL, Pantalone V, Jackson SA, Schmutz J, Cregan PB (2016) Construction of high resolution genetic linkage maps to improve the soybean genome sequence assembly Glyma1.01. BMC Genom 17:33

Sreenivasulu N, Sopory SK, Kavi Kishor PB (2007) Deciphering the regulatory mechanisms of abiotic stress tolerance in plants by genomic approaches. Gene 388:1–13

Strange TL, Petolino JF (2012) Targeting dna to a previously integrated transgenic locus using zinc finger nucleases. In: Dunwell JM, Wetten AC (eds) Transgenic plants: methods and protocols, methods in molecular biology. Springer, New York, pp 391–397

Sumner LW, Mendes P, Dixon RA (2003) Plant metabolomics: large-scale phytochemistry in the functional genomics era. Phytochem 62:817–836

Sun CX, Li MQ, Gao XX, Liu LN, Wu XF, Zhou JH (2016) Metabolic response of maize plants to multi-factorial abiotic stresses. Plant Biol 1:120–129

Sunkar R, Chinnusamy V, Zhu J, Zhu JK (2007) Small RNAs as big players in plant abiotic stress responses and nutrient deprivation. Trends Plant Sci 12:301–309

Tao X, Gu YH, Wang HY, Zheng W, Li X, Zhao CW, Zhang YZ (2012) Digital gene expression analysis based on integrated de novo transcriptome assembly of sweet potato [*Ipomoea batatas* (L.) Lam.]. PLoS ONE 7:e36234

Teotia S, Singh D, Tang X, Tang G (2016) Essential RNA-based technologies and their applications in plant functional genomics. Trends Biotechnol 34:106–123

Tripathi AK, Pareek A, Sopory SK, Singla-Pareek SL (2012) Narrowing down the targets for yield improvement in rice under normal and abiotic stress conditions *via* expression profiling of yield-related genes. Rice 5:37

Tripathi P, Rabara RC, Rushton PJ (2014) A systems biology perspective on the role of WRKY transcription factors in drought responses in plants. Planta 239:255–266

Tripathy MK, Tyagi W, Goswami M, Kaul T, Singla-Pareek SL, Deswal R, Reddy MK, Sopory SK (2012) Characterization and functional validation of tobacco PLC delta for abiotic stress tolerance. Plant Mol Biol Rep 30:488–497

Umezawa T, Fujita M, Fujita Y, Yamaguchi-Shinozaki K, Shinozaki K (2006) Engineering drought tolerance in plants: discovering and tailoring genes to unlock the future. Curr Opin Plant Biotech 17:113–122

Urano K, Kurihara Y, Seki M, Shinozaki K (2010) 'Omics' analyses of regulatory networks in plant abiotic stress responses. Curr Opin Plant Biol 13:132–138

Varshney RK, Hoisington DA, Tyagi AK (2006) Advances in cereal genomics and applications in crop breeding. Trends Biotechnol 24:490–499

Varshney RK, Ribaut JM, Buckler ES, Tuberosa R, Rafalski JA, Langridge P (2012) Can genomics boost productivity of orphan crops? Nat Biotechnol 30:1172–1176

Varshney RK, Saxena RK, Upadhyaya HD, Khan AW, Yu Y, Kim C, Rathore A, Kim D, Kim J, An S, Kumar V (2017) Whole-genome resequencing of 292 pigeonpea accessions identifies genomic regions associated with domestication and agronomic traits. Nat Genet 49:1082–1088

Vega-Sánchez ME, Gowda M, Wang GL (2007) Tag-based approaches for deep transcriptome analysis in plants. Plant Sci 173:371–380

Verdier J, Torres-Jerez I, Wang M, Andriankaja A, Allen SN, He J, Tang Y, Murray JD, Udvardi MK (2013) Establishment of the *Lotus japonicas* gene expression atlas (LjGEA) and its use to explore legume seed maturation. Plant J 74:351–362

Vij S, Tyagi AK (2007) Emerging trends in the functional genomics of the abiotic stress response in crop plants. Plant Biotechnol J 5:361–380

Vogel JT, Zarka DG, Van Buskirk HA, Fowler SG, Thomashow MF (2005) Roles of the CBF2 and ZAT12 transcription factors in configuring the low temperature transcriptome of *Arabidopsis*. Plant J 41:195–211

Wang ZL, Li PH, Fredricksen M, Gong ZZ, Kim CS, Zhang C, Bohnert HJ, Zhu JK, Bressan RA, Hasegawa PM, Zhao YX (2004) Expressed sequence tags from *Thellungiella halophila*, a new model to study plant salt-tolerance. Plant Sci 166:609–616

Wang M, Wang S, Xia G (2015) From genome to gene: a new epoch for wheat research? Trends Plant Sci 20:380–387

Wang H, Wang H, Shao H, Tang X (2016) Recent advances in utilizing transcription factors to improve plant abiotic stress tolerance by transgenic technology. Front Plant Sci 7:67

Wei CL, Ng P, Chiu KP, Wong CH, Ang CC, Lipovich L, Liu ET, Ruan Y (2004) 5′ Long serial analysis of gene expression (LongSAGE) and 3′ LongSAGE for transcriptome characterization and genome annotation. Proc Natl Acad Sci USA 101:11701–11706

Wen Y, Li X, Guo C, Ma C, Duan W, Lu W, Xiao K (2015) Characterization and expression analysis of mitogen-activated protein kinase cascade genes in wheat subjected to phosphorus and nitrogen deprivation, high salinity, and drought. J Plant Biochem Biotechnol 24:184–196

Wheeler DA, Wang L (2013) From human genome to cancer genome: the first decade. Genome Res 23:1054–1062

Wood V, Gwilliam R, Rajandream MA, Lyne M, Lyne R, Stewart A, Sgouros J, Peat N, Hayles J, Baker S, Basham D (2002) The genome sequence of *Schizosaccharomyces pombe*. Nature 415:871–880

Wu M, Chen A, Wang Z, Zhang J, Wang C, Li F, Wei P, Wang R, Luo Z, Wei C, Lin F, Yang J (2015) Plant microarray for gene expression profiling and their application. J Agric Technol 11:93–105

Xia Z, Zhai H, Lü S, Wu H, Zhang Y (2013) Recent achievement in gene cloning and functional genomics in soybean. The Sci World J 2013: Article ID 281367

Xie K, Yang Y (2013) RNA-guided genome editing in plants using a CRISPR-Cas system. Mol Plant 6:1975–1983

Yang Q, Chen ZZ, Zhou XF, Yin HB, Li X, Xin XF, Hong XH, Zhu JK, Gong Z (2009) Overexpression of SOS (Salt Overly Sensitive) genes increases salt tolerance in transgenic *Arabidopsis*. Mol Plant 2:22–31

Yang X, Yang YN, Xue LJ, Zou MJ, Liu JY, Chen F, Xue HW (2011) Rice ABI5-Like1 regulates abscisic acid and auxin responses by affecting the expression of ABRE-containing genes. Plant Physiol 5:1397–1409

Yang Z, Huang D, Tang W, Zheng Y, Liang K, Cutler AJ, Wu W (2013) Mapping of quantitative trait loci underlying cold tolerance in rice seedlings via high-throughput sequencing of pooled extremes. PLoS ONE 8:e68433

Yoshida T, Fujita Y, Maruyama K, Mogami J, Todaka D, Shinozaki K, Yamaguchi-Shinozaki KA (2014) Four *Arabidopsis* AREB/ABF transcription factors function predominantly in gene expression downstream of SnRK2 kinases in abscisic acid signaling in response to osmotic stress. Plant Cell Environ 38:35–49

Zhang J (2014) Salinity affects the proteomics of rice roots and leaves. Proteomics 14:1711–1712

Zhang X, Zhu Y, Liu X, Hong X, Xu Y, Zhu P, Shen Y, Wu H, Ji Y, Wen X, Zhang C (2015) Suppression of endogenous gene silencing by bidirectional cytoplasmic RNA decay in *Arabidopsis*. Science 348:120–123

Zhang B, Yang X, Yang C, Li M, Guo Y (2016) Exploiting the CRISPR/Cas9 system for targeted genome mutagenesis in petunia. Sci Rep 6:20315

Zhao X, Wang W, Zhang F, Deng J, Li Z, Fu B (2014) Comparative metabolite profiling of two rice genotypes with contrasting salt stress tolerance at the seedling stage. PLoS ONE 9: e108020

Chapter 2
Plant miRNAome: Cross Talk in Abiotic Stressful Times

Prashanti Patel, Karuna Yadav, T. R. Ganapathi and Suprasanna Penna

Abstract The "small RNA world" discovered by plant biologists has acquired a central, regulatory role in diverse and fundamental processes including genome stability, gene expression and defense. The microRNAs (miRNAs) are a group of small noncoding RNAs found in both animals and plants. Since their discovery in *Arabidopsis thaliana*, plant miRNAs have been identified and their target genes are characterized in various plant species. While some miRNAs are functionally conserved across plant species, studies have also shown that miRNAs respond to environmental stresses in a stress-, tissue-, and genotype-dependent manner. During abiotic stress, miRNAs function by regulating target genes within the miRNA–target gene network and by controlling signaling pathways. Both stress-induced and stress-inhibited miRNAs constitute a controlling mechanism for fine tuning the positive or negative regulators of different stress-regulated pathways. These properties suggest that miRNA-based genetic modifications have the potential to enhance abiotic stress tolerance in crops. Furthermore, consequent to stress perception, epigenetic changes facilitate miRNA regulation of several transcription factors, which are common to drought, salt and heavy metal stress. With the rapid advancement in methods of whole genome sequencing, several new and novel miRNAs are being identified and research efforts are underway to decipher the "microRNAome"—a comprehensive view of miRNA-mediated gene regulatory networks in plants. The horizon of 'regulatory RNA' field is expanding, and new developments will certainly enhance our understanding of microRNA interactome under stress conditions. In this article, we discuss the perspective of microRNA regulation of salt, drought and heavy metal stress with an emphasis on shared mechanisms and provide an understanding of their potential roles in plant adaptation to abiotic stress conditions.

Keywords Abiotic stress · Epigenetics · microRNA · miRNAome
Phytohormones signaling

P. Patel · K. Yadav · T. R. Ganapathi · S. Penna (✉)
Nuclear Agriculture and Biotechnology Division, Bhabha Atomic Research Centre,
Mumbai 400085, India
e-mail: penna888@yahoo.com

V. R. Rajpal et al. (eds.), *Genetic Enhancement of Crops for Tolerance to Abiotic Stress: Mechanisms and Approaches, Vol. I*, Sustainable Development and Biodiversity 20, https://doi.org/10.1007/978-3-319-91956-0_2

2.1 Introduction

The sessile nature of plants renders them especially vulnerable to attack by pathogens and abiotic stresses of varying kinds. There is an increased concern that unpredictable weather conditions are becoming more frequent and have an impact on agriculture (Abberton et al. 2016). This is exacerbated by the increase in human population and dwindling natural resources, thus creating an increased pressure on finding alternate and integrating technologies across different tenets of science and technology. Integral to this is the need to understand how crop plants respond to adverse environmental conditions and perform better to sustain productivity. Additionally, under natural conditions plants are rarely challenged by single stress at a time. Instead, they are exposed to multiple stresses, which elicit different responses depending upon the developmental stage, timing and type of stress and severity of individual stresses (Pandey et al. 2015). Drought, salinity, high or low temperatures and heavy metals are some of the environmental stresses which individually or together cause severe damaging effect on plant growth and yield (Ahuja et al. 2010; Zhang et al. 2011).

Under stress, plants have to carefully allocate precious resources and energy towards fending off the stress and protect themselves from opportunistic stresses that may co-occur. The ensuing remodeling of the transcriptome, proteome and metabolome is extensive and complex (Wu et al. 2003; Bonnecarrère et al. 2011; Gong et al. 2015). Plants have amassed an arsenal of counteractive strategies at all levels of physiological and molecular complexity (Choudhary et al. 2017; Suprasanna et al. 2018) (Fig. 2.1). Stress-induced gene expression can include genes encoding proteins with known structural or enzymatic functions, regulatory proteins and proteins with yet unknown functions (Bhatnagar-Mathur et al. 2008). At the genome level, dynamic epigenetic modifications control the level of expression of different regions of the chromatin in response to stress. The transcripts so produced may be subjected to alternative splicing to produce stress relevant mRNA isoform signatures. Non-coding RNA also plays a role in modifying the activity of these transcripts through silencing by cleavage or promoter methylation. Protein folding and activity can be manipulated by an array of covalent post-translational modifications and targeting to various organelles wherein they are functional. Enzymatic regulation of metabolite pathways channels flux into production of biomolecules to cope with stress. All these mechanisms ensure that the cell is altered to withstand stress structurally and physiologically (Lu and Huang 2008).

Different stresses (cold, high salt, drought, heavy metals) elicit response in plants, which share varying specificities in mode of action. Some exhibit convergence at a particular point, while at other levels specific responses are seen (Knight and Knight 2001). Generally, all stresses are first perceived, and then relayed through secondary messengers to regulators and finally to effectors, which exert protective function. For instance, both freezing and drought lead to dehydration. Consequently, cold acclimation can also be achieved by drought or ABA application (Mantyla et al. 1995). Not only ionic and osmotic imbalances but also

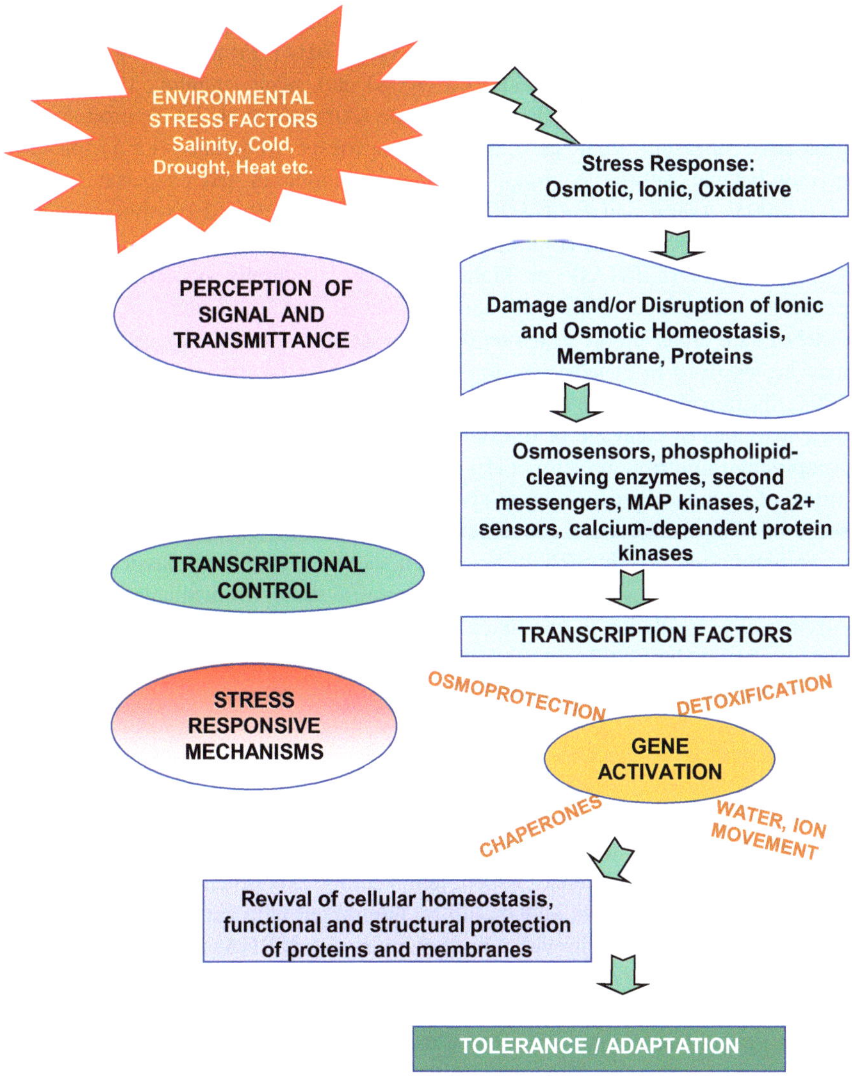

Fig. 2.1 Plant responses to abiotic stress factors. Plants exposed to various abiotic stresses (salinity, drought, extreme temperatures, toxic metals, etc.) initiate a cascade of changes in plants' functioning such as imbalanced water and nutrient uptake, stomatal closure, altered gaseous exchange, improper functioning of photosynthetic system and generation of reactive oxygen species (ROS). The ensuing effect of these induce oxidative damage to functional and structural molecules (DNA, proteins, lipids, and carbohydrates) making changes in the redox, osmotic, ionic, and energetic homeostasis of the plant. The stress signals also trigger the downstream signaling processes and gene activation through transcription factors. Defense mechanisms involve antioxidant machinery for detoxification of ROS, osmolytes production for osmotic balance and protection to structural molecules, ionic homeostasis and maintenance of redox and energetics. Source: Lokhande and Suprasanna (2012).

several metabolic and molecular disturbances are perceived under multiple abiotic stresses (Golldack et al. 2014; Nakashima et al. 2014). Principle players in these pathways are reactive oxygen species (ROS) and sugar signaling (Couée et al. 2006; Keunen et al. 2013), calcium signatures (Albrecht et al. 2003), production of protective effectors such as Late Embryogenesis Abundant (LEA) proteins, glycine-betaine, antifreeze proteins, proteinase inhibitors and chaperones (Goyal et al. 2005; Ashraf and Foolad 2007; Szalai et al. 2009) and phytohormone networks (Kohli et al. 2013) (Fig. 2.2).

MicroRNAs (miRNAs) are highly conserved naturally occurring transcripts generally undersized (20–24 nt), single-stranded and non-coding. A number of miRNAs are either up- or down-regulated by abiotic stresses, suggesting that they may be involved in gene expression during stress and variation (Sunkar and Zhu 2004; Shriram et al. 2016). miRNAs regulate all the stress-related processes through their activities of cleavage or translational inhibition of diverse gene targets, most comprising transcription factors (TF) (Sunkar et al. 2012). With the ever-increasing influx of sequencing data, newer species-specific miRNAs are being uncovered; some of whose targets are convergent with the known ones. On the other hand, some highly conserved miRNAs have been shown to have non-classical targets in different species, thus adding to the repertoire of strategies available to the plant. Additionally, regulation is fine-tuned through the deployment of bistable, coherent and incoherent loops (Flynt and Lai 2008; Meng et al. 2011; Jeong and Green 2013). Induction of specific miRNAs under different stress conditions varies between plant species and hence there is no general "medicine-for-all-ailments" approach when trying to use miRNAs as targets for genetic improvement of plant tolerance to abiotic stress (Djami-Tchatchou et al. 2017; Song et al. 2019). Also, it has been observed that there are overlapping responses to combined stress such as heat and drought, salt and drought, abiotic stress and pathogen attack, which further add to the complexity. Hence, genetic engineering experiments for improved tolerance may suffer setbacks and should be interpreted with caution.

In this chapter, we begin with a brief overview on miRNAs in epigenetic control of stress in plants. We move onto attempting to provide a portal into this complexity by integrating the known findings in miRNA regulation of salt, drought and heavy metal stress with an emphasis on shared mechanisms. In particular SnRKs, (Sucrose non-fermenting Related Kinases), TCPs (Teosinte-branched/Cycloidea/Pcf) and ABI (Abscissic Acid Insensitive) proteins are recurrent themes in the four cardinal responses to the above mentioned stressors namely, signaling by phytohormones, ROS, sugar and calcium. Majorly, the miRNAs that feature in developmental processes such as miRNAs 156, 159, 160/161, 162, 164, 165/166, 167, 168, 169, 170/171, 172, 319, 390, 393, 395, 396, 397, 398, 399 and 400 have also been found to have stress-related expression (Patel et al. 2017). The major abiotic stresses affecting plant and crop growth and development are drought, salinity, temperature (heat, cold, chilling and freezing), nutrient, high light intensity, anaerobic stress and ozone. In general, stress-responsive miRNAs are miR319a/b, miR319b.2 and miR400 (Barciszewska-Pacak et al. 2015). Interestingly, miR168 can be considered top in the hierarchy of miRNA regulation as it targets AGO1 (Argonaute), a key player in the miRNA biogenesis pathway; thus its activity modulates the fundamental originating pathway for miRNAs under stress. In *Arabidopsis*, induction of miR168, miR171, miR393 and miR396 was common to salt, drought and cold (Liu et al. 2008). The promoter regions of miRNAs contain abiotic stress response elements. These include CCAAAT (heat shock responsive element), MYB, MYC binding elements, ERD (early response to dehydration), CuREs (copper responsive elements), ERE [ethylene (ET) responsive elements] and GARE (gibberellic acid responsive elements). Through the binding of various TFs to these elements with different temporal and spatial specificities along with combinations of binding partners, the expression of the effector genes are controlled. For instance, Niu et al. (2016) observed that the freezing-induced Psu-miR475b and its targets as well as Psu-miR475b promoter activity are modulated by freezing stress and exogenous MeJA, SA or GA, suggesting cross talk between stress signaling and hormonal pathways. In a recent study, **TaeMiR408** was found to be involved in the cross talk between phosphorus starvation and salinity stress through modulation of ABA signalling genes which play a role in physiological processes of osmolyte and photosynthetic processes (Bai et al. 2018).

2.2 Micro RNAs Controlling Stress-Induced Epigenetic Changes

Information on epigenetic regulation of abiotic stress response in plants is still nascent. It is known that a high salt environment hypomethylates histones in natural mangroves while histones of drought-responsive genes are differentially acetylated and trimethylated (Golldack et al. 2011). Micro RNAs (miRNAs) regulating chromatin modifications are only just emerging. miRNA 820 is a rice specific miRNA down regulated by drought and targets DRM2 (DNA (cytosine-5)-methyltransferase). However, when miR820 is downregulated, the cognate target is also surprisingly downregulated. In addition, two different isoforms of miR820, which are 21- and 24-nucleotide forms, are known to exist (Wu et al. 2010; Jeong et al. 2011). Two siRNAs 441 and 446 are upregulated under drought stress and upregulate their target MAIF1, an F-box protein (Jeong and Green 2013). MAIF1 is the specificity factor in SCF E3 ubiquitin ligases and is strongly induced by abiotic stress, sucrose and the hormones abscisic acid (ABA), cytokinins (CKs) and auxin (AUX). However, it is proposed to negatively regulate tolerance to stress as its overexpression (OE) decreases ABA response and promotes root growth, both of which decrease tolerance to stress (Yan et al. 2011). miRNA 402 is predicted to target Demeter-Like 3 (DML3), a DNA glycosylase domain demethylase. This enzyme is responsible for activating the maternal allele of *MEDEA* (*MEA*), a Polycomb gene required for endosperm development (Choi et al. 2002). *MEA* represses transcription through chromatin modeling to prevent untimely development of the endosperm before fertilization (Kiyosue et al. 1999).

ABA and abiotic stress induce miR402 and downregulation of DML3 (Kim et al. 2010). Overexpression of miR402 encouraged seed germination under cold, drought and salt stress, but additional seed growth only in salt stress, pointing to its unknown differential interactions with the three stressors. The positive effect of miR402 on germination could occur due to the decrease in methylation-induced silencing of germination-related genes under stress. The rice TCP19, a Class I TCP TF, is induced by several abiotic stresses especially in tolerant varieties. It directly interacts with ULT1 (*Ultra Trithorax 1*), a trithorax group (trxG) factor that recruits ATX1 (*Arabidopsis homolog of TRITHORAX 1*), a histone methyltransferase and inhibits the Polycomb group of gene repression complexes. Thus, it is involved in chromatin modification under stress (Mukhopadhyay and Tyagi 2015). Such epigenetic changes upon perception of stress facilitate miRNA regulation of several effector TFs (MYBs, SPLs, NAC, AP2/ERF, HD-ZIPs, TCP, GRAS and NFYA), many of which are common to drought, salt and heavy metal stresses. Further, metabolic pathways targeted by these TFs are those involving sugar and starch synthesis and breakdown, cellulose synthesis and cell wall modeling, stress response, hormonal signal transduction and plant development as explained in the following sections.

2.3 Phytohormone Signaling

ABA is a hormone central to the cross talk among several abiotic stresses, e.g. cold, drought, salinity and heavy metal. In the signaling pathway for ABA, the ABI proteins ABI1 and ABI2 are homologs of type 2 phosphatases, which negatively regulate ABA response (Rubio et al. 2009; Lee and Luan 2012), while ABI 3, 4 and 5 are TFs positively regulating ABA response. These are very likely points of convergence among different abiotic stresses since they participate as signaling nodes integrating multiple pathways. For example, ABI5 combines sugar and ABA response (Brocard et al. 2002) and tobacco protein phosphatase 2C (PP2C) is up regulated in drought but downregulated in heat and oxidative stress (Vranová et al. 2000). The phenotypic changes associated with abiotic stress are a product of ABA interaction with other hormonal pathways in an intricate network with differential outcomes (Rowe et al. 2016; Lian et al. 2018).

The antagonistic interaction between auxin and ABA is particularly well studied in case of root architecture modeling (Rock and Sun 2005; Fukaki and Tasaka 2009). Under drought stress, normal auxin-promoted cell multiplication and lateral root (LR) meristem activation is repressed by ABA through the differential and tissue-specific induction of the transcription factor MYB86 by both hormones. Intriguingly, MYB96 too targets different genes in root and shoot: *RD22* in shoot but *Gretchen Hagen 3* (*GH3*) in the roots (Seo et al. 2009). *GH3* genes negatively regulate lateral root formation by inactivating excess auxin by conjugation (Fu et al. 2011); thus, the indirect induction of *GH3* by ABA and auxin in roots leads to lesser lateral root initiation. *GH3* genes are also stress-responsive as exemplified by rice *GH3-2*, which is induced by drought but repressed by cold. Overexpression of *GH3-2* reduced ABA levels (by reducing free auxin levels) and thus increased drought sensitivity, as ABA-activated protective mechanisms such as stomatal closure could not occur. However, this loss of free auxin also led to better ROS scavenging and increased cold responsive gene transcription, thus providing tolerance to cold (Du et al. 2012).

The auxin signaling miRNAs, miR164, 167, 160, 393 and 394 are also responsive to ABA treatment via their action on auxin responsive factors (ARFs), which control levels of free auxin and therefore lateral root development. ABI3, a B3 domain containing TF and a positive regulator of ABA signaling (Reeves et al. 2011) is induced by ARF 10/16, which are in turn targets of miR160 and are required to maintain ABA-induced seed dormancy (Liu et al. 2013). Overexpression of miR160 promoted germination through repression of ARF10 and ARF16 (Liu et al. 2007). Auxin is involved in root modeling and organ polarity both of which change under abiotic stress. Downregulation of miR167 under drought (Liu et al. 2008) de-represses the hydrolase *IAA-Alanine Resistant 3* (*IAR3*) and liberates free auxin, which induces lateral root growth for efficient water uptake (Kinoshita et al. 2012). This miRNA also targets ARF6/8 in *Arabidopsis* and thereby regulates auxin-mediated gynoecium and anther development (Wu et al. 2006). miRNA 167 is also downregulated by ABA leading to upregulation of its targets, thereby permitting auxin activity. The miR160-ABA-miR167 regulatory axis thus controls seed germination and lateral root elongation. Considering that ABI factors are important hubs of cross talk, it would not be surprising if they are miRNA targets. Recently, a degradome study in *Physcomitrella patens* revealed that the plant-specific miR536 could target the moss ABI3 (Xia et al. 2016). However, in higher plants the miR165/166–PHABULOSA (PHB) module indirectly regulates ABA signaling. PHB, a HOMEODOMAIN ZIP III (HD-ZIP III) TF directly upregulates ABI4 and a β-glucosidase (BG1). ABI4 activates downstream ABA signaling genes while BG1 generates active ABA from inactive conjugates (Yan et al. 2016). Thus, through concomitant induction of the target PHB, reduced miR165/166 levels promoted cold and drought tolerance in *Arabidopsis*. miRNA 165/166 is responsive to ROS, salt, cold, and heat, thus linking its targets, the HD-ZIP III factors to alterations in plant development and patterning under stress (Sun et al. 2015).

miRNAs 846 and 842 are alternatively spliced forms with related functions. Their levels are regulated by ABA. This hormone when exogenously applied, lowers the levels of miR846 but increases miR842 (Jia and Rock 2013a). The lectin jacalin is a target of miR846 and is induced upon exogenous application of ABA suggesting its involvement in different abiotic and biotic stresses (Jia and Rock 2013b), though its function is currently unknown. Heat stress downregulates miR400 by manipulating the processing of this miRNA from the intron of a protein coding gene. However, the primary miRNA levels remain unchanged, indicating possible post-transcriptional regulation (Barciszewska-Pacak et al. 2015).

Gibberellic acid (GA) is a growth-promoting hormone, while ABA inhibits growth and maintains seed dormancy (Achard et al. 2006). GA binding to its receptor removes DELLA repression of downstream GAMYB TFs, thereby activating a multitude of plant development programs such as flowering and anther development (Achard et al. 2004). DELLA positively regulates XERICO, a RING-H2 zinc finger TF involved in tolerance to drought and ABA biosynthesis (Ko et al. 2006; Ariizumi et al. 2013). Thus, GA-mediated degradation of DELLA ultimately leads to repression of ABA signaling. ABA induces ABI3 ultimately leading to an increase in miR159. This miRNA in turn targets MYB TFs (MYB33 and MYB101), which positively regulate ABA response and osmotic stress tolerance (Reyes and Chua 2007). The antagonistic effects of ABA and GA on post-germination growth converge at ABI4 through its contrasting stabilization by both hormones. This TF also binds to the promoters of *NCED6* (a key ABA biosynthetic gene) as well as that of *GA2ox7* (a GA catabolic gene) and induces both genes due to which ABA signaling is potentiated while GA levels decrease. ABA promotes GA degradation by *GA2ox7* or conversely GA represses ABA synthesis. Thus, ABA, ABI and GA interact at multiple levels (Shu et al. 2016).

Stress alters the membrane lipid profile as well as induces lipid-stimulated signaling through generation of lipid secondary messengers (Welti et al. 2002; Testerink and Munnik 2005). In rice, phosphatidylinositol 4, 5-bisphosphate (PIP2), phosphatidic acid (PA) and diacylglycerol pyrophosphate (DGPP) are upregulated in salt stress (Darwish et al. 2009). ABA interacts with lipid metabolism to maintain membrane integrity under stress (Golldack et al. 2014). Jasmonic acid (JA) is a small lipid-derived hormone serving as an indicator of membrane changes in stress. It responds to biotic and abiotic stress through different pathways and induces antioxidant production as well as metabolites tailored for the particular stress, such as Pathogenesis-Related (PR) proteins, proteinase inhibitors and others (Dar et al. 2015; Kazan 2015). One component of JA signaling is *Jasmonate Zim-domain* (JAZ), an inhibitory protein. JA induces DELLA (RGL3) through the COI1 and JIN1/MYC2 pathway. The accumulation of this DELLA inhibits JAZ and relieves repression of JA-responsive downstream genes. The maize JAZ14 protein connects JA, ABA and GA pathways and positively regulates ABA-promoted stress tolerance when overexpressed in *Arabidopsis* (Zhou et al. 2015). Recently, a study by Aleman et al. (2016) described ABA-mediated interaction of the ABA receptor PYL6 and transcription factor MYC2 ultimately altering interaction of the latter with JAZ gene promoters. Further, under pathogen attack, ABA is crucial for the activation of JA-mediated defense (Adie et al. 2007). The ANAC019 and ANAC055 TFs were postulated to connect JA and ABA signaling since both hormones induced them. Both TFs interact with RHA2a, a RING E3 ligase, which promotes tolerance to salt and abiotic stress in early stages of germination (Bu et al. 2009; Jiang et al. 2009). miRNA 319 targets TCP2 involved in jasmonate biosynthesis (Schommer et al. 2008). This miRNA is upregulated in a wide variety of abiotic stresses, especially arsenic and its overexpression leads to tolerance to salt and drought in creeping bentgrass (Liu and Zhang 2012; Zhou et al. 2013). Besides TCP2, it also regulates other TCP family members, which are repressors of cell division at the leaf surface (Koyama et al. 2007; Schommer et al. 2014). JA also interacts with GA and ABA in that its induction increases the former but reduces sensitivity to the latter under maternal herbivory stress in *Arabidopsis* (Singh et al. 2017).

Brassinosteroids (BR) are a class of growth-promoting plant hormones, with multifarious roles in cell elongation, photomorphogenesis, flowering, ageing, photosynthesis, defense and abiotic stress tolerance (Kagale et al. 2007). They are usually more active in young germinated shoots and regulate cell wall composition and cellular transport (Zhu et al. 2013). Thus, perturbations in cell wall integrity induce BR signaling and response to stress via modulating pectin, lignin and cellulose deposition, loosening of cell wall, and phenolic cross-linking (Rao and Dixon 2017). Exogenous application of BR-induced stress-responsive genes, ROS detoxification, protein synthesis maintenance and greater photosynthetic efficiency (Divi and Krishna 2009). One of the mechanisms by which BR induces stress tolerance is through activation of NADPH oxidase-mediated production of H_2O_2 as was shown in cold-stressed cucumber (Xia et al. 2009), which signals via the MAPK pathway to induce heat shock proteins (HSPs) and antioxidant enzymes. In rice, exogenous BR mitigated salinity-induced inhibition of germination by restoring nucleic acid and protein synthesis (Anuradha and Rao 2001). In addition, in case of cadmium (Hayat et al. 2007) and aluminum (Ali et al. 2008) toxicity, stress was considerably alleviated through BR-enhanced antioxidant production. Elucidation of the players in the BR signaling pathway is still incomplete. More recently, *Arabidopsis WRKY46*, *WRKY54* and *WRKY70* were shown to negatively affect drought tolerance and positively influence BR signaling genes (Chen et al. 2017). Also, the NAC TF *RD26*, a positive regulator of drought tolerance was found to negatively affect BR signaling by interaction with *BRI1 EMS SUPPRESSOR1* (BES1), a downstream BR signaling TF, which in turn inhibits *RD26* transcription. Thus under drought, this feedback loop operates to conserve energy by downregulating BR-induced growth, while under normal conditions *RD26* is inhibited and plant growth resumes (Ye et al. 2017). However, in light of the contrasting reports of BR-induced drought tolerance, this negative feedback loop may not be operational under all conditions (Divi et al. 2010).

A number of miRNAs orchestrate BR action. miRNA 1,848 targets the obtusifoliol 14α-demethylase gene *OsCYP51G3*, involved in demethylating precursors during phytosterol and brassinosteroid biosynthesis (Xia et al. 2015). Overexpression of miR1848 reduced *OsCYP51G3* levels and consequently BR concentration. These plants had symptoms typical of BR deficiency such as dwarfism, sterility and altered leaf morphology, as well as heightened sensitivity to salt stress. A mutant disrupted in BR signaling was rescued by overexpression of this miRNA172 (Kim et al. 2014), thus suggesting its role in inducing plant growth and transition from vegetative to reproductive stage. Similarly, overexpression of rice miR397 increased grain size and panicle branching due to downregulation of its target rice laccase. The miRNA overexpressing lines had elevated levels of BR and hence better growth, while *OsLAC* overexpression displayed symptoms characteristic of BR deficiency. Laccases are known to catalyze monolignol polymerization and lignin biosynthesis along with a host of other redox activities on different substrates, not all yet elucidated (Gavnholt and Larsen 2002). Thus, in conjunction with the known effects of BR on cell wall modification, laccase is a plausible downstream component of BR signaling (Oh et al. 2011; Zhang et al. 2013). Besides these, miRNAs 156, 159, 160 and 824 were also responsive to exogenous epibrassinolide application, concomitant with the role of BRs in plant growth and patterning, phase transition (miR156), root morphology and auxin signaling (miR160), GA signaling (miR159), stomatal number and flowering time (miR824) (Lin et al. 2013). Interestingly BIN2 kinase, a negative regulator of BR signaling, phosphorylates and stabilizes ABI5, thus mediating the antagonistic interaction between ABA and BR during seed germination and growth (Hu and Yu 2014).

OsTCP19, described above, also regulates several hormonal interactions i.e. auxin, ABA and jasmonate signaling under stress. Accordingly, *Lipooxygenase 2* (*LOX2*) (jasmonate biosynthesis) was downregulated, while ABI3 and ABI4 (positive regulators of ABA signaling) as well as IAA3 (Auxin- conjugate hydrolase) were induced in *TCP19* overexpressing rice plants. It was found to directly regulate ABI4, which increases triacylglycerol (TAG) deposition for improved stress tolerance (Mukhopadhyay and Tyagi 2015). ABI4 is a potent regulator of diverse processes, activating some (seed maturation, ABA signaling) and repressing others (photosynthesis, fatty acid biosynthesis, pigment and wax metabolism, ROS formation, plastid-nuclear signaling), through competitive binding to the overlapping CE1 and G-box promoter elements (Wind et al. 2013). Plastid-to-nuclear retrograde signaling (PNRS) is crucial to coordination of chloroplast development and functioning and import of nuclear-encoded machinery, to ensure homeostasis, especially under stress (Nott et al. 2006). It involves several essential GUN (Genomes Uncoupled) proteins whose mutants show deranged crosstalk. Part of this cross-organelle talk is negatively regulated by ABI4-mediated repression of Photosynthesis-Associated Nuclear Gene expression (PhANG) downstream of GUN1 (Wind et al. 2013).

A novel function for *Arabidopsis* miR395a was proposed wherein it targets GUN5, the H-subunit of Mg-chelatase required for PNRS and chlorophyll synthesis, in a BR-dependent manner. Exogenous BR induced miR395a, which in turn repressed GUN5. GUN5 is postulated to act with the downstream player ABI4, which in itself is downregulated by BR (Lin et al. 2013). Thus, BR-treated plants showed lower veinal chlorophyll levels and increased LR formation consistent with the role of ABI4 in repressing lateral root formation through activation of cytokinin and ABA signaling pathways and repression of polar auxin transport (Shkolnik-Inbar and Bar-Zvi 2010). A positive role has been ascribed to GUN5 and ABI4 in preventing oxidative damage to the photosynthetic apparatus during water stress. In addition, miR395 is responsive to sulfur (S) deprivation by targeting ATP Sulfurylase (APS) and Sulphate Transporter 2; 1 (SULTR2; 1) to conserve the plant's S reserves (Kawashima et al. 2009). Sulfur, a key component of primary and secondary metabolites, is required for brassinosteroid and jasmonate-sulfation by Sulfotransferase (SOT) enzymes (Kopriva et al. 2012), to fine-tune their activity. The dual regulation of S allocation between primary metabolism and sulfation-mediated regulation of hormonal activity is thus, elegantly controlled by miR395, especially under stress. Though indirect, regulation of cellular gene expression and hormonal signaling by TCP19 as well as organellar crosstalk and metabolic regulation by ABI4, are intricately orchestrated by miR395a, representing the crossroad between hormonal pathways and nutrient metabolism.

Roots are the first plant tissue to sense and respond to heavy metal stress. Under copper deficiency/excess and cadmium excess, levels of miR319a/b increase as expected as the promoter of this miRNA contains 8 CuREs. This miRNA targets TBL10 (Trichome Birefringence-Like 10), a DUF domain protein of unknown function. It is expected therefore that levels of TBL10 mRNA would decrease under copper deficiency; instead, the levels did not change in comparison to control. A probable reason for this constant expression is the post-transcriptional cleavage of excess TBL10 by miR319b.2 thereby stabilizing gene expression levels. miRNA 319b.2 does not, however, cleave TBL10 in adult plants but cleaves RAP2.12 in adult *Arabidopsis* (Barciszewska-Pacak et al. 2015). Conversely, in *Arabidopsis* seedlings, miR319b.2 cleaves TBL10 and not RAP2.12 (Sobkowiak et al. 2012). RAP2.12 is an ethylene responsive factor involved in tolerance to multiple stresses (Papdi et al. 2015), thus pointing to the involvement of ethylene in the gamut of hormone crosstalk. Levels of miR319a/b increase under salinity stress whereas miR319 b.2-derived from the same precursor as that of miR319b decreases under high salt (Barciszewska-Pacak et al. 2015). They target TCP transcription factors and hence most likely are involved in leaf development, floral organ identity and flowering time.

miRNA 171 targets the *Scarecrow-Like* (SCL) family of *Gibberellic-Acid Insensitive (GAI), Repressor of GAI (RGA)* and *Scarecrow (SCR)* (GRAS) domain containing TFs which regulate varied aspects of plant growth, root hair differentiation, light signaling, GA signaling and vegetative to floral transition (Llave et al. 2002; Wang et al. 2010; Ma et al. 2014; Fan et al. 2015). Overexpression of miR171 delays flowering; its induction under stress could therefore represent a temporal block on energy consuming processes until the stress is alleviated (Zhang and Sonnewald 2017). miRNA 393 targets a TIR1/AFB2 auxin receptor. It was found that overexpressing a cleavage resistant form of TIR1 led to increased salt tolerance in *Arabidopsis* (Chen et al. 2014). Growth Regulating Factors (GRFs) targeted by miR396 are cell cycle regulators, which control plant growth and differentiation (Liu et al. 2009). Interestingly, miR396 itself is induced by TCP4, which is in turn a target of miR319 (Schommer et al. 2014). TCP3 in the cotyledons induces miR164 which in turn targets the *NAM-ATAF-CUC* (NAC) TFs which control leaf boundary shape, organ separation and cell multiplication (Mallory et al. 2004; Koyama et al. 2010); these miRNAs contribute to control of leaf and flower shape and axillary meristem maintenance as they balance differentiation and proliferation of cell masses and hence morphogenesis. Stress induction of miR396 likely represses cell multiplication under unfavorable conditions.

2.4 Calcium Signaling

Calcium (Ca) is well known as a prominent secondary messenger in a variety of plant stress responses (Dodd et al. 2010; Kudla et al. 2010). Different stressors generate different 'Ca signatures' (Quiles-Pando et al. 2013), manifesting as unique spatio-temporal patterns of Ca concentration in the cell. Consequently, a host of Ca sensors (calmodulins, CDPKs, Ca-regulated phosphatase) as well as pumps, channels, and transporters (Batistič and Kudla 2012) work in synchrony to ensure efficacious transmission of the stress signal to the nucleus. Differential localization of Ca sensors in subcellular spaces may contribute to the specificity of calcium 'signatures' to a particular stress. Under drought and arsenic stress, inhibition of barley miR1432 led to accumulation of its target, a calmodulin-related calcium sensor protein (Ferdous et al. 2017). In citrus, miR482 targets a Calcineurin-like phosphoesterase, while a novel citrus miRNA miR049 targeted calcium dependent protein kinase 6 (Xie et al. 2017). The former miRNA is downregulated in salt but induced in drought, whereas its target is highly induced under both stresses. However, miR049 and its target are both downregulated. This is probably evidence of the existence of coherent and non-coherent miRNA-target relationships occurring in plants. The monocot-conserved miR444 family member miR444d, which is inhibited by drought but induced by powdery mildew infection, is also predicted to target a calmodulin protein (Sunkar et al. 2008).

Calcium is also essential for somatic embryogenesis through exogenous application of cytokinin. Spruce miRNAs 1160, 5638, 1315 and 5225 were downregulated while their targets, which were predicted to be CDPKs and CBLs, were induced proving the necessity of Ca signaling during callus differentiation (Li et al. 2017a, b). Interestingly, Calcineurin Binding Like 10 (CNBL10) has evolved to be a target of miR167 in apple. Though the sequence of mature miR167 is conserved among higher plants, evolutionary changes have brought about sequence variation in both apple miR167 and CNBL10, such that the two now form miRNA and cognate target (Kumar and Sarkar 2017). CNBL10 is an SOS3 (Salt Over-Sensitive 3) homolog well known for its positive role in the signaling pathways of salt and drought stress. The interaction of CNBL10 and CBL-interacting protein kinase 27 (CIPK27) leads to sequestration of Na^+ in the vacuole. In *Brassica rapa*, miRX7 was also predicted to target CNBL10 (Srivastava et al. 2017) which is upregulated by combined thiourea and NaCl treatment. Heavy metal (arsenic) stress downregulated miR1318 and upregulated its target, a Ca^{++}-ATPase. Manganese toxicity strongly induced miR1508 in nodules of *Phaseolus vulgaris* but down regulated it in roots. miRNA 1508 was predicted to target a calcium dependent protein kinase (Valdés-López et al. 2010).

2.5 Sugar Signaling

Response to abiotic stress in plants is heavily dependent on signaling by sugars and their polymers (Rosa et al. 2009; Radomiljac et al. 2013), which integrate hormonal signaling to respond to stress. For instance, starch is remobilized in a tissue-specific and differential manner under various abiotic stresses to provide sustenance as carbon and energy and thus increase plant fitness. A host of starch metabolizing enzymes with stress-specific signatures; β-amylase (BAM1), chloroplastic α-amylase (AMY3) and plastidial starch phosphorylase (PHS1) in drought and glucan water dikinase (GWD) and β-amylase (BAM3) in cold, are employed. Starch catabolism intersects with ABA signaling through the SnRK2 and ABA response element-binding factor (AREB/ABF) mediated induction of BAM1 and AMY3, in drought (Thalmann and Santelia 2017). Sugar catalytic enzymes are also involved in signaling. The tea plant hexokinases *Cs*HXK3 and *Cs*HXK 4 are induced by cold, while *Cs*HXK1 is induced under salt and drought in roots (Li et al. 2017a). *Arabidopsis* hexokinases have been shown to act as sugar sensors apart from their catalytic role and influence on glucose-mediated photosynthetic repression and hormonal pathways (Granot et al. 2013), thus serving as metabolic status indicators. Transportation of sugar is altered under stress. A class of sugar transporters SWEET (Sugar Will Eventually be Exported), were found to be expressed differentially in two cultivars of banana under cold, drought, salt and fungal attack (Miao et al. 2017), whereby they facilitate sugar transport and signaling for fruit development under stress.

SnRK1 is a serine/threonine protein kinase functioning as a critical sensor for energy and carbon in plant cells. Under stress, both metabolic pathways are compromised, leading to a reduction in the plant's ability to set seed. Starvation recruits SnRK1 to activate catabolic processes and inhibit anabolism, to liberate carbon and energy for survival. Thus, it phosphorylates and inactivates sucrose phosphate synthase, nitrate reductase and HMG-coA synthase (Halford et al. 2003). It is also a hub for phytohormone-dependent stress responses. ABA, a general growth inhibitor, mediates the interaction between SnRK1A and SKIN2 (SnRK interacting negative regulator) to prevent movement of SnRK1A to the nucleus, thus repressing SnRK1 function (Lin et al. 2014). *SnRK1* overexpressing *Arabidopsis* is hypersensitive to ABA (Jossier et al. 2009). Under drought stress therefore, ABA signaling prevents activation of hydrolases needed for nutrient mobilization from seed endosperm and arrests seed germination (Zhang and Sonnewald 2017). SnRK1 is also connected to auxin and cytokinin signaling as well as represses GA and brassinosteroid metabolism (Sharma et al. 2013). This kinase acts through both downstream bZIP TFs and miRNAs. SnRK1 activates miR319, a well-known general stress-responsive miRNA, which in turn downregulates its target TCP factors TCP2 and TCP4. The latter are repressors of cell division at leaf margins and are also stress- responsive as they regulate organelle primary metabolism and ATP generation as already described. SnRK1 induction also decreases miRNAs 159b, 161, 775 and 824a. Under starvation, the miRNA processing mutant *dcl-9* showed aberrant expression of 831 SnRK1-regulated "starvation-genes" compared to wild type, with repression of translation, organelle function, ROS signaling, protein trafficking and folding, nucleic acid metabolism, and activation of catabolic processes and chromatin remodeling (Confraria et al. 2013). This indicates that miRNAs are involved in the SnRK1-mediated signaling under stress.

Sugar signaling requires epigenetic control as the histone acetyltransferase (*hac*1) mutant) has impaired sensitivity to sugar and the GA synthesis inhibitor paclobutrazol (Heisel et al. 2013). Histone acetyltransferase 1 (HAC1) positively regulates transcription by acetylation of histones. Glucose and fructose inhibit early germination stages in seedlings, through repression of *HAC1*. However, *hac1* mutant plants are insensitive to sugar because of decreased levels of the SnRK1 complex proteins *At*PV42a and *At*PV42b, which act as the central sensor for energy and carbon status in the cell. Additionally, transcripts of genes for gibberellin biosynthesis, anthocyanin production and seed dormancy were downregulated in mutant lines along with reduced fertility, thus pointing to the essentiality of chromatin modification in sugar signaling.

The *ABA-Hypersensitive Germination 1 (AHG1)* locus encodes a protein phosphatase 2C (PP2C), which is a negative regulator in the ABA signaling pathway. The mutant *ahg1* is hypersensitive to ABA, glucose, sucrose, NaCl, KCl and mannitol, reflecting its importance in stress response during germination and post germination. Mutant seeds had longer dormancy periods and higher accumulation of ABA (Nishimura et al. 2007). As described earlier, protein phosphatases are also involved in crosstalk with calcium signaling in abiotic stress. PP2Cs are therefore versatile integrators of sugar, calcium and hormonal signaling. Also, though they are biochemically similar in action, they interact with different protein partners in various compartments, as well as express at different times, all of which confer specificity (Nishimura et al. 2007). For instance, ABI4 and ABI5 together confer sugar sensitivity during early germination. Furthermore, a link between sugar, ABA and ethylene signaling exists, as ethylene inhibits the sucrose-mediated induction of ABA during the sensitive seed germination stage (Gazzarrini and McCourt 2001). This antagonistic interaction between ABA and ET was recently shown to occur through the repression of the ET biosynthesis genes *1-Aminocyclopropane-1-Carboxylic Acid synthase* (*ACS*) by ABA (Dong et al. 2016). The PhANG *RBCS* small subunit of ribulose 1,5-bisphosphate carboxylase in addition to being light-controlled, is also sugar- and ABA-responsive. ABI4 binds to the S-box element closely associated with the light-responsive G-box of its promoter and mediates both sugar- and ABA-dependent inhibition of this gene (Acevedo-Hernández et al. 2005).

Other sucrose responsive miRNAs included miR398 and miR408. The former targets the copper-zinc superoxide dismutase CSD1 and CSD2 (Sunkar et al. 2006), a copper chaperone CCS (Beauclair et al. 2010), and in an infrequently encountered example of binding leniency, a subunit of cytochrome oxidase (COX) (Yamasaki et al. 2007). miRNA 408 targets plantacyanin and some members of the laccase family (Abdel-Ghany and Pilon 2008). Both these miRNAs are induced under copper deficiency. Negative regulation of these copper enzyme and protein targets frees up precious copper reserves for sustaining photosynthesis (Yamasaki et al. 2007). However, the sucrose induction of miR398 is both copper dependent and independent (Dugas and Bartel 2008; Ren and Tang 2012), suggesting links between copper nutritional status and sugar signaling as well as enigmatic non-canonical targets of miR398. Additionally miR319 and 160, with established functions in hormonal signaling (jasmonate and auxin response), are also induced by sucrose, thus revealing the highly connected nature of ROS, sugar and phytohormone signaling (Ren and Tang 2012).

2.6 Reactive Oxygen Species

The reactive forms of oxygen namely, superoxide anions, hydroxyl radicals, peroxyl ions and hydrogen peroxide, are essential signaling molecules for survival and defense but at the same time capable of playing havoc with the delicate machinery of the cell. The balance between the ability of plants to utilize ROS positively and its deleterious effects is dependent on the amount of ROS generated, its subcellular localization and the activity of detoxifying enzymes (Choudhary et al. 2017). A number of biotic and abiotic stressors generate "signaling ROS" commonly through sensor-mediated activation of RBOHs (Respiratory Burst Oxidase Homologs) such as apoplastic NADPH oxidase, some amine oxidases and peroxidases as well as organellar specific oxidases (Miller et al. 2008; Choudhary et al. 2017). These are activated by calcium signaling and kinase-mediated phosphorylation (Pei et al. 2000; Kobayashi et al. 2007) accompanied by organellar morphological changes wherein they connect physically to the nuclear membrane (Noctor and Foyer 2016). Redox homeostasis as one of core regulators of cellular processes has been shown to play crucial roles in mediating miRNA and hormone based regulation through possible mechanisms of post-transcriptional gene regulation or decoupling miRNA:mRNA inverse relationship (Srivastava et al. 2017). Abiotic stress-produced ROS induces glutamate dehydrogenase expression for effective channeling of excess ammonia (the by-product of proteolysis) into glutamate and further synthesis of protectants like proline (Skopelitis et al. 2006). In concordance with the role of SODs in detoxifying ROS, miR398 is downregulated by ozone, *Pseudomonas syringae* infection (Jagadeeswaran et al. 2009) and/or cold (Chen et al. 2013) but upregulated by heat where it plays a protective role (Guan et al. 2013). Regulation of miR398 targets themselves is an example of the intricate connection with other physiological processes; for instance, the diurnal, salt and ABA treatment-induced oscillation of miR398 expression (Jia et al. 2009; Siré et al. 2009), which is also species specific. Both miRNAs 398 and 408 have been reported to play conflicting roles under drought stress in *M. truncatula* and pea (Trindade et al. 2010; Jovanović et al. 2014), again probably acting in a genotype-

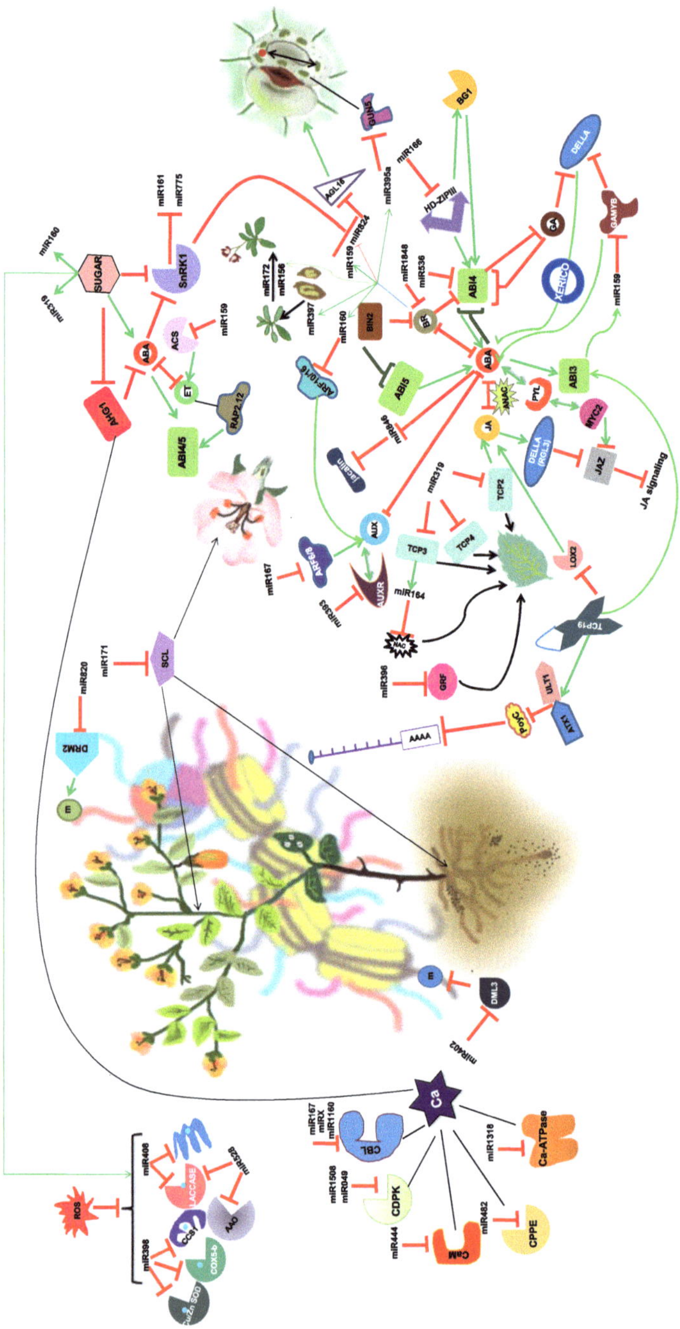
ROS
miR398
miR408
miR528
LACCASE
CCS
COX5-b
Cu/Zn SOD
AAO
CBL
miR167
miRX
miR1160
CDPK
miR1508
CaM
miR444
CPPE
miR482
Ca-ATPase
miR1318
Ca
DML3
miR402
m
DRM2
miR820
m
SCL
miR171
AAAA
PolyC
ATX1
ULT1
TCP19
GRF
miR396
NAC
miR164
AUXR
miR393
ARF6/8
miR167
AUX
TCP3
TCP4
miR319
TCP2
LOX2
JA
JA signaling
JAZ
DELLA (RGL3)
MYC2
PYL
ANAC
ABI3
ABA
ABI5
BIN2
BR
ABI4
miR536
miR1848
HD-ZIPIII
miR166
BG1
GUN5
miR395a
AGL16
miR824
miR159
SnRK1
miR161
miR775
SUGAR
miR160
miR319
ABA
ACS
miR159
ET
RAP2.12
ABI4/5
ARG1
jacalin
miR846
ARF10/16
miR160
miR397
miR172
miR156
GA
DELLA
GAMYB
XERICO
miR159

◂**Fig. 2.2** A plant miRNA-target interaction network depicting the cross-talk between the four cardinal players ROS, sugar, hormones and calcium in response to stress. The network is controlled fundamentally by chromatin modifications shown as a backdrop to the plant presented in the image. Anatomical features affected are shown as insets at appropriate places in the diagram. Single red cross bars and green arrows depict inhibition and activation respectively. Double red cross bars indicate mutual inhibitory action of hormones, and green arrows indicate either molecule-receptor binding or interaction between proteins. Simple black lines indicate signaling between different components. Black curves designate interaction between different players while single black arrows show the action of transcription factors on plant morphology. The double headed black arrow in the stomatal guard cell depicts plastid-to-nuclear retrograde signaling. Key: : Methylated histone, : Methylated DNA, : Polycomb repressor complex, : Repression, : hormone, : Double repression, : Protein stabilization, : Protein destabilization, CaM: calmodulin, CBL: Calcineurin B-like, Ca-ATPase: Calcium ATPase, CPPE: Calcineurin like phosphoesterase, : Calcium ions, : Plantacyanin, : Copper ion

and stress intensity-dependent manner. Recently, miRNA ghr-414C was shown to target the expression of the first antioxidant defense gene, iron-superoxide dismutase (*GhFSD1*) in cotton plants subjected to salt stress (Wang et al. 2019) implying miRNA mediated regulation of salt stress through ROS.

Heavy metal (aluminium, cadmium, iron, copper, mercury and arsenic) stress is a predominant generator of ROS. The legume *Medicago truncatula* is known to cope with heavy metal stress considerably better than other plants (Zhou et al. 2008). In this plant, heavy metal-responsive miRNAs were categorized into two groups. The miRNAs 171, 319, 393 and 529 comprised the first group, which was generally upregulated under heavy metal stress. Of these, miR319 (slightly induced by Cd and Hg) and miR393 (induced by Cd and Al) are also responsive to drought, salinity, ABA and cold as mentioned earlier. The miR529 is evolutionarily related to miR156 and likely targets *Squamosa promoter binding Protein-Like* (SPLs) (Morea et al. 2016), though its exact function is not elucidated. The second group of miRNAs comprising miR166 and miR398 was downregulated under heavy metal stress. miRNA 166 targets the HD-ZIPIII family of TFs, which determine leaf polarity while miR398 targets Cu/Zn SODs, which scavenge free radicals as described earlier. Thus, inhibition of these miRNAs reflects the plants' attempts to rectify leaf developmental abnormalities and mitigate oxidative stress, which are products of heavy metal stress. Heavy metals like Al bind to carboxylate and phosphate groups in the soil and reduce mineral uptake and physiological processes. Cytoplasmic calcium imbalance leads to generation of ROS. Soybean treated with Al showed upregulation of miR171, miR319, miR393, miR519 miR390, miR396, unlike in *M. truncatula* where these were downregulated, hinting at species specificity (Chen et al. 2012; Zeng et al. 2012).

Challenging *Brassica* spp with arsenic elicited widespread changes in metabolism. Photosynthesis and nitrate assimilation were impaired while ROS and lipid peroxidation increased (Jha and Dubey 2004; Requejo and Tena 2005). In rice, 36 new arsenic-responsive miRNAs were discovered with predominant roles in jasmonate and lipid biosynthesis, for the synthesis of triacylglyceride oil droplets and for epicuticular wax deposition. miRNAs 156, 166, 168, 171, 319 and 396, which are responsive to salt, drought and other abiotic stresses, were also strongly affected by arsenic (Yu et al. 2012). In rice roots, miR408, 528, 397b were upregulated, while 1316 and 390 were downregulated (Srivastava et al. 2012). In addition, miR319 was found to be responsive to arsenic toxicity. As is the case with other heavy metals, arsenic too perturbs sulfur metabolism. The miRNAs 395, 838 and 854 are predicted to target the sulfate assimilating enzymes APS and Serine Acetyltransferase (SAT) and transporter SULTR2;1. miRNAs 319 and 838 target TCP4 and a lipase leading to enhance production of jasmonates. miRNAs 164, 167 and 390 target auxin responsive factors (ARF) and NAC TFs leading to increased auxin metabolism and lateral root growth. miRNAs 159 targets GAMYB and ACC synthase altering biosynthesis of GA, ethylene and ABA, which leads to heightened defense response and changes in root architecture. Other miRNAs such as miRNAs 156, 162, 165, 169, 172, 426, 535 and 1436 target a plethora of TFs and downstream genes, leading to morphological changes and defense response. All this leads to increased SO_4^{2-} uptake imparting tolerance towards arsenic (Srivastava et al. 2012).

Under Manganese stress in *P. vulgaris*, various miRNAs were induced or inhibited, of which miR1515 (targeting HSP), miR1510/2110 (targeting NBS-LRR like proteins), miR1532 (targeting RLKs) were responsive (Valdés-López et al. 2010). Under 80 μM Cd exposure, strong upregulation of miR395 in roots, miR393 in leaf, miR156a and 167a/c in roots and leaves was observed in *Brassica napus* (Huang et al. 2010). Also levels of miR164b, which targets the NAC TF in leaves, and miR160 targeting an ARF responsive to auxin decreased under conditions of Cd exposure whereas miR394a/b/c increased in all tissues (Huang et al. 2010). miRNAs 394a targets an F-box protein and confers drought tolerance. Enhanced Cd accumulation or reduced Cd tolerance is mediated by miR390, which targets the stigma specific S-locus Receptor Kinase (SRK) (Ding et al. 2016). Interestingly, Cd excess mimics S deficiency in that miR395 is induced under both stresses. Transgenic lines overexpressing miR395 in *Arabidopsis* showed a reduction in APS (ATP Sulfurylase) activity, which catalyzes the first step of S assimilation. Micro RNA 395 also targets SULTR2;1, a leaf sulfate transporter through spatial compartmentalization. As a result, SO_4^{2-} accumulates in the leaves due to increased root translocation. This is expected as S-derived secondary metabolites such as metal chelating compounds ameliorate heavy metal toxicity (Yadav 2010; Matraszek et al. 2016).

2.7 Conclusions

Drought, salinity, high or low temperatures and heavy metals are some of the environmental stresses, which singly or together affect plant growth and productivity. Stress induced modulations include metabolic pathways of ROS and sugar signaling, calcium signatures, production of protective effector molecules and phytohormones. Studies have also shown that small RNAs (miRNAs) respond to environmental stresses in a stress-, tissue-, and genotype-dependent manner. During abiotic stress, miRNAs function by regulating target genes within the miRNA–target gene network and by controlling signaling pathways. It is of great interest in finding out the nature and integration of different signals into plant responses under a given stress condition. Hence, 'microRNA interactome' is becoming a research topic of contemporary interest. Studies are unfolding novel microRNAs, their mode of action and the targets in drought, salt and heavy metal stresses and this information has attracted the attention of plant scientists for their use in genetic engineering for higher stress tolerance. Successful examples include miR397 overexpression leading to enhanced overall grain yield in rice up to 25% and overexpression of miR156 significantly increasing the number of plant biomass by 300% in *Arabidopsis* (Zhang and Wang 2016).

The role of microRNAs in molecular cross talk across different abiotic stresses and identification of abiotic stress response elements needs further study. Characterization of novel and specific miRNAs from plants with exceptional stress tolerance can be useful to elucidate their distinctive mechanisms. Being important regulators of plant developmental pathways, miRNAs can be integrated into recapitulating stress memory. To summarize, understanding the way microRNAs regulate plant responses to abiotic stresses through different layers of associated pathways should enable manipulation of miRNA-guided gene regulation to engineer plants with improved stress tolerance.

References

Abberton M, Batley J, Bentley A, Bryant J, Cai H et al (2016) Global agricultural intensification during climate change: a role for genomics. Plant Biotechnol J 14(4):1095–1098

Abdel-Ghany SE, Pilon M (2008) MicroRNA-mediated systemic down-regulation of copper protein expression in response to low copper availability in *Arabidopsis*. J Biol Chem 283 (23):15932–15945

Acevedo-Hernández GJ, León P, Herrera-Estrella LR (2005) Sugar and ABA responsiveness of a minimal RBCS light-responsive unit is mediated by direct binding of ABI4. Plant J 43(4):506–519

Achard P, Cheng H, De Grauwe L, Decat J, Schoutteten H et al (2006) Integration of plant responses to environmentally activated phytohormonal signals. Science 311:91–94

Achard P, Herr A, Baulcombe DC, Harberd NP (2004) Modulation of floral development by a gibberellin-regulated microRNA. Development 131(14):3357–3365

Adie BA, Pérez-Pérez J, Pérez-Pérez MM, Godoy M, Sánchez-Serrano JJ et al (2007) ABA is an essential signal for plant resistance to pathogens affecting JA biosynthesis and the activation of defenses in Arabidopsis. Plant Cell 19(5):1665–1681

Ahuja I, de Vos RC, Bones AM, Hall RD (2010) Plant molecular stress responses face climate change. Trends Plant Sci 15(12):664–674

Albrecht V, Weinl S, Blazevic D, D'angelo C, Batistic O et al (2003) The calcium sensor CBL1 integrates plant responses to abiotic stresses. Plant J 36(4):457–470

Aleman F, Yazaki J, Lee M, Takahashi Y, Kim AY et al (2016) An ABA-increased interaction of the PYL6 ABA receptor with MYC2 Transcription Factor: A putative link of ABA and JA signaling. Sci Rep 6:28941

Ali B, Hasan SA, Hayat S, Hayat Q, Yadav S et al (2008) A role for brassinosteroids in the amelioration of aluminium stress through antioxidant system in mung bean (*Vigna radiata* L. Wilczek). Environ Exp Bot 62(2):153–159

Anuradha S, Rao SS (2001) Effect of brassinosteroids on salinity stress induced inhibition of seed germination and seedling growth of rice (*Oryza sativa* L.). Plant Growth Regul 33(2):151–153

Ariizumi T, Hauvermale AL, Nelson SK, Hanada A, Yamaguchi S et al (2013) Lifting DELLA repression of *Arabidopsis* seed germination by nonproteolytic gibberellin signaling. Plant Physiol 162(4):2125–2139

Ashraf M, Foolad M (2007) Roles of glycine betaine and proline in improving plant abiotic stress resistance. Environ Exp Bot 59(2):206–216

Bai Q, Wang X, Chen X, Shi G, Liu Z, Guo C, Xiao K (2018). Wheat miRNA taemir408 acts as an essential mediator in plant tolerance to pi deprivation and salt stress via modulating stress-associated physiological processes. Front Plant Sci 9:499. https://doi.org/10.3389/fpls.2018.00499

Barciszewska-Pacak M, Milanowska K, Knop K, Bielewicz D, Nuc P et al (2015) *Arabidopsis* microRNA expression regulation in a wide range of abiotic stress responses. Front Plant Sci 6

Batistič O, Kudla J (2012) Analysis of calcium signaling pathways in plants. Biochim Biophys Acta (BBA)-General Subjects 1820(8):1283–1293

Beauclair L, Yu A, Bouché N (2010) microRNA-directed cleavage and translational repression of the copper chaperone for superoxide dismutase mRNA in Arabidopsis. Plant J 62(3):454–462

Bhatnagar-Mathur P, Vadez V, Sharma KK (2008) Transgenic approaches for abiotic stress tolerance in plants: retrospect and prospects. Plant Cell Rep 27(3):411–424

Bonnecarrère V, Borsani O, Díaz P, Capdevielle F, Blanco P et al (2011) Response to photoxidative stress induced by cold in japonica rice is genotype dependent. Plant Sci 180 (5):726–732

Brocard IM, Lynch TJ, Finkelstein RR (2002) Regulation and role of the *Arabidopsis* abscisic acid-insensitive 5 gene in abscisic acid, sugar, and stress response. Plant Physiol 129(4):1533–1543

Bu Q, Li H, Zhao Q, Jiang H, Zhai Q et al (2009) The *Arabidopsis* RING finger E3 ligase RHA2a is a novel positive regulator of abscisic acid signaling during seed germination and early seedling development. Plant Physiol 150(1):463–481

Chen J, Nolan TM, Ye H, Zhang M, Tong H et al (2017) *Arabidopsis* WRKY46, WRKY54, and WRKY70 transcription factors are involved in brassinosteroid-regulated plant growth and drought responses. Plant Cell 29(6):1425–1439

Chen L, Wang T, Zhao M, Tian Q, Zhang WH et al (2012) Identification of aluminum-responsive microRNAs in *Medicago truncatula* by genome-wide high-throughput sequencing. Planta 235 (2):375–386

Chen Y, Jiang J, Song A, Chen S, Shan H et al (2013) Ambient temperature enhanced freezing tolerance of *Chrysanthemum dichrum Cd*ICE1 *Arabidopsis* via miR398. BMC Biol 11(1):121

Chen Z, Hu L, Han N, Hu J, Yang Y et al (2014) Overexpression of a miR393-resistant form of transport inhibitor response protein 1 (mTIR1) enhances salt tolerance by increased osmoregulation and Na + exclusion in *Arabidopsis thaliana*. Plant Cell Physiol 56(1):73–83

Choi Y, Gehring M, Johnson L, Hannon M, Harada JJ et al (2002) DEMETER, a DNA glycosylase domain protein, is required for endosperm gene imprinting and seed viability in *Arabidopsis*. Cell 110(1):33–42

Choudhary FK, Rivero RM, Blumwald E, Mittler R (2017) Reactive oxygen species, abiotic stress and stress combination. Plant J 90(5):856–867

Confraria A, Martinho C, Elias A, Rubio-Somoza I, Baena-González E (2013) miRNAs mediate SnRK1-dependent energy signaling in *Arabidopsis*. Front Plant Sci 4

Couée I, Sulmon C, Gouesbet G, El Amrani A (2006) Involvement of soluble sugars in reactive oxygen species balance and responses to oxidative stress in plants. J Exp Bot 57(3):449–459

Dar TA, Uddin M, Khan MM, Hakeem KR, Jaleel H (2015) Jasmonates counter plant stress: a review. Environ Exp Bot 115:49–57
Darwish E, Testerink C, Khalil M, El-Shihy O, Munnik T (2009) Phospholipid signaling responses in salt-stressed rice leaves. Plant Cell Physiol 50(5):986–997
Ding Y, Ye Y, Jiang Z, Wang Y, Zhu C (2016) MicroRNA390 is involved in cadmium tolerance and accumulation in rice. Front Plant Sci 7
Divi UK, Krishna P (2009) Brassinosteroid: a biotechnological target for enhancing crop yield and stress tolerance. N Biotechnol 26(3):131–136
Divi UK, Rahman T, Krishna P (2010) Brassinosteroid-mediated stress tolerance in *Arabidopsis* shows interactions with abscisic acid, ethylene and salicylic acid pathways. BMC Plant Biol 10 (1):151
Djami-Tchatchou AT, Sanan-Mishra N, Ntushelo K, Dubery IA (2017) Functional roles of microRNAs in Agronomically important plants—potential as targets for crop improvement and protection. Front Plant Sci 8
Dodd AN, Kudla J, Sanders D (2010) The language of calcium signaling. Annu Rev Plant Biol 61:593–620
Dong Z, Yu Y, Li S, Wang J, Tang S et al (2016) Abscisic acid antagonizes ethylene production through the ABI4-mediated transcriptional repression of ACS4 and ACS8 in *Arabidopsis*. Mol Plant 9(1):126–135
Du H, Wu N, Fu J, Wang S, Li X et al (2012) A GH3 family member, OsGH3-2, modulates auxin and abscisic acid levels and differentially affects drought and cold tolerance in rice. J Exp Bot 63(18):6467–6480
Dugas DV, Bartel B (2008) Sucrose induction of *Arabidopsis* miR398 represses two Cu/Zn superoxide dismutases. Plant Mol Biol 67(4):403–417
Fan T, Li X, Yang W, Xia K, Ouyang J (2015) Rice osa-miR171c mediates phase change from vegetative to reproductive development and shoot apical meristem maintenance by repressing four *Os*HAM transcription factors. PLoS ONE 10(5):e0125833
Ferdous J, Sanchez-Ferrero JC, Langridge P, Milne L, Chowdhury J et al (2017) Differential expression of microRNAs and potential targets under drought stress in barley. Plant Cell Environ 40(1):11–24
Flynt AS, Lai EC (2008) Biological principles of microRNA-mediated regulation: shared themes amid diversity. Nat Rev Genet 11:831
Fu J, Yu H, Li X, Xiao J, Wang S (2011) Rice GH3 gene family: regulators of growth and development. Plant Signal Behav 6(4):570–574
Fukaki H, Tasaka M (2009) Hormone interactions during lateral root formation. Plant Mol Biol 69 (4):437
Gavnholt B, Larsen K (2002) Molecular biology of plant laccases in relation to lignin formation. Physiol Plant 116(3):273–280
Gazzarrini S, McCourt P (2001) Genetic interactions between ABA, ethylene and sugar signaling pathways. Curr Opin Plant Biol 4(5):387–391
Golldack D, Li C, Mohan H, Probst N (2014) Tolerance to drought and salt stress in plants: unraveling the signaling networks. Front Plant Sci 5
Golldack D, Lüking I, Yang O (2011) Plant tolerance to drought and salinity: stress regulating transcription factors and their functional significance in the cellular transcriptional network. Plant Cell Rep 30(8):1383–1391
Gong F, Hu X, Wang W (2015) Proteomic analysis of crop plants under abiotic stress conditions: where to focus our research? Front Plant Sci 6:418
Goyal K, Walton LJ, Tunnacliffe A (2005) LEA proteins prevent protein aggregation due to water stress. Biochem J 388(1):151–157
Granot D, David-Schwartz R, Kelly G (2013) Hexose kinases and their role in sugar-sensing and plant development. Front Plant Sci 4
Guan Q, Lu X, Zeng H, Zhang Y, Zhu J (2013) Heat stress induction of miR398 triggers a regulatory loop that is critical for thermotolerance in *Arabidopsis*. Plant J 74(5):840–851

Halford NG, Hey S, Jhurreea D, Laurie S, McKibbin RS et al (2003) Metabolic signalling and carbon partitioning: role of Snf1-related (SnRK1) protein kinase. J Exp Bot 54(382):467–475
Hayat S, Ali B, Hasan SA, Ahmad A (2007) Brassinosteroid enhanced the level of antioxidants under cadmium stress in *Brassica juncea*. Environ Exp Bot 60(1):33–41
Heisel TJ, Li CY, Grey KM, Gibson SI (2013) Mutations in HISTONE ACETYLTRANSFERASE1 affect sugar response and gene expression in Arabidopsis. Front Plant Sci 4
Hu Y, Yu D (2014) BRASSINOSTEROID INSENSITIVE2 interacts with ABSCISIC ACID INSENSITIVE5 to mediate the antagonism of brassinosteroids to abscisic acid during seed germination in *Arabidopsis*. Plant Cell 26(11):4394–4408
Huang SQ, Xiang AL, Che LL, Chen S, Li H et al (2010) A set of miRNAs from *Brassica napus* in response to sulphate deficiency and cadmium stress. Plant Biotechnol J 8(8):887–899
Jagadeeswaran G, Saini A, Sunkar R (2009) Biotic and abiotic stress down-regulate miR398 expression in *Arabidopsis*. Planta 229(4):1009–1014
Jeong DH, Green PJ (2013) The role of rice microRNAs in abiotic stress responses. J Plant Biol 56 (4):187–197
Jeong DH, Park S, Zhai J, Gurazada SG, De Paoli E et al (2011) Massive analysis of rice small RNAs: mechanistic implications of regulated microRNAs and variants for differential target RNA cleavage. Plant Cell 23(12):4185–4207
Jha AB, Dubey RS (2004) Arsenic exposure alters activity behaviour of key nitrogen assimilatory enzymes in growing rice plants. Plant Growth Regul 43(3):259–268
Jia F, Rock CD (2013a) MIR846 and MIR842 comprise a cistronic MIRNA pair that is regulated by abscisic acid by alternative splicing in roots of *Arabidopsis*. Plant Mol Biol 81(4–5): 447–460
Jia F, Rock CD (2013b) Jacalin lectin At5g28520 is regulated by ABA and miR846. Plant Signal Behav 8(6):e24563
Jia X, Wang WX, Ren L, Chen QJ, Mendu V (2009) Differential and dynamic regulation of miR398 in response to ABA and salt stress in *Populus tremula* and *Arabidopsis thaliana*. Plant Mol Biol 71(1–2):51–59
Jiang H, Li H, Bu Q, Li C (2009) The RHA2a-interacting proteins ANAC019 and ANAC055 may play a dual role in regulating ABA response and jasmonate response. Plant Signal Behav 4 (5):464–466
Jossier M, Bouly JP, Meimoun P, Arjmand A, Lessard P et al (2009) SnRK1 (SNF1-related kinase 1) has a central role in sugar and ABA signaling in *Arabidopsis thaliana*. Plant J 59(2):316–328
Jovanović Ž, Stanisavljević N, Mikić A, Radović S, Maksimović V (2014) Water deficit down-regulates miR398 and miR408 in pea (*Pisum sativum* L.). Plant Physiol Biochem 83:26–31
Kagale S, Divi UK, Krochko JE, Keller WA, Krishna P (2007) Brassinosteroid confers tolerance in *Arabidopsis thaliana* and *Brassica napus* to a range of abiotic stresses. Planta 225(2): 353–364
Kawashima CG, Yoshimoto N, Maruyama-Nakashita A, Tsuchiya YN, Saito K et al (2009) Sulphur starvation induces the expression of microRNA-395 and one of its target genes but in different cell types. Plant J 57(2):313–321
Kazan K (2015) Diverse roles of jasmonates and ethylene in abiotic stress tolerance. Trends Plant Sci 20(4):219–229
Keunen EL, Peshev D, Vangronsveld J, Van Den Ende WI, Cuypers AN (2013) Plant sugars are crucial players in the oxidative challenge during abiotic stress: extending the traditional concept. Plant Cell Environ 36(7):1242–1255
Kim BH, Kwon Y, Lee BH, Nam KH (2014) Overexpression of miR172 suppresses the brassinosteroid signaling defects of *bak1* in *Arabidopsis*. Biochem Biophys Res Commun 447 (3):479–484
Kim JY, Kwak KJ, Jung HJ, Lee HJ, Kang H (2010) MicroRNA402 affects seed germination of *Arabidopsis thaliana* under stress conditions via targeting DEMETER-LIKE Protein3 mRNA. Plant Cell Physiol 51(6):1079–1083

Kinoshita N, Wang H, Kasahara H, Liu J, MacPherson C et al (2012) IAA-Ala Resistant3, an evolutionarily conserved target of miR167, mediates *Arabidopsis* root architecture changes during high osmotic stress. Plant Cell 24(9):3590–3602
Kiyosue T, Ohad N, Yadegari R, Hannon M, Dinneny J et al (1999) Control of fertilization-independent endosperm development by the MEDEA polycomb gene in *Arabidopsis*. Proc Natl Acad Sci 96(7):4186–4191
Knight H, Knight MR (2001) Abiotic stress signalling pathways: specificity and cross-talk. Trends Plant Sci 6(6):262–267
Ko JH, Yang SH, Han KH (2006) Upregulation of an Arabidopsis RING-H2 gene, XERICO, confers drought tolerance through increased abscisic acid biosynthesis. Plant J 47(3):343–355
Kobayashi M, Ohura I, Kawakita K, Yokota N, Fujiwara M et al (2007) Calcium-dependent protein kinases regulate the production of reactive oxygen species by potato NADPH oxidase. Plant Cell 19(3):1065–1080
Kohli A, Sreenivasulu N, Lakshmanan P, Kumar PP (2013) The phytohormone crosstalk paradigm takes center stage in understanding how plants respond to abiotic stresses. Plant Cell Rep 32 (7):945–957
Kopriva S, Mugford SG, Baraniecka P, Lee BR, Matthewman CA et al (2012) Control of sulfur partitioning between primary and secondary metabolism in *Arabidopsis*. Front Plant Sci 3
Koyama T, Furutani M, Tasaka M, Ohme-Takagi M (2007) TCP transcription factors control the morphology of shoot lateral organs via negative regulation of the expression of boundary-specific genes in *Arabidopsis*. Plant Cell 19(2):473–484
Koyama T, Mitsuda N, Seki M, Shinozaki K, Ohme-Takagi M (2010) TCP transcription factors regulate the activities of ASYMMETRIC LEAVES1 and miR164, as well as the auxin response, during differentiation of leaves in *Arabidopsis*. Plant Cell 22(11):3574–3588
Kudla J, Batistič O, Hashimoto K (2010) Calcium signals: the lead currency of plant information processing. Plant Cell 22(3):541–563
Kumar A, Sarkar AK (2017) Apple CALCINEURIN B-LIKE PROTEIN10 genes have evolved to be novel targets of miR167 s through sequence variation. Curr Sci 112(1):147–150
Lee SC, Luan S (2012) ABA signal transduction at the crossroad of biotic and abiotic stress responses. Plant Cell Environ 35(1):53–60
Li NN, Qian WJ, Wang L, Cao HL, Hao XY et al (2017a) Isolation and expression features of hexose kinase genes under various abiotic stresses in the tea plant (*Camellia sinensis*). J Plant Physiol 209:95–104
Li Q, Deng C, Xia Y, Kong L, Zhang H et al (2017b) Identification of novel miRNAs and miRNA expression profiling in embryogenic tissues of *Picea balfouriana* treated by 6-benzylaminopurine. PLoS ONE 12(5):e0176112
Lian C, Yao K, Duan H, Li Q, Liu C, Yin W, Xia X (2018). Exploration of ABA responsive miRNAs reveals a new hormone signaling crosstalk pathway regulating root growth of *Populus euphratica*. Int J Mol Sci 19(5):1481. https://doi.org/10.3390/ijms19051481
Lin CR, Lee KW, Chen CY, Hong YF, Chen JL (2014) SnRK1A-interacting negative regulators modulate the nutrient starvation signaling sensor SnRK1 in source-sink communication in cereal seedlings under abiotic stress. Plant Cell 26(2):808–827
Lin LL, Wu CC, Huang HC, Chen HJ, Hsieh HL et al (2013) Identification of microRNA 395a in 24-epibrassinolide-regulated root growth of *Arabidopsis thaliana* using microRNA arrays. Int J Mol Sci 14(7):14270–14286
Liu D, Song Y, Chen Z, Yu D (2009) Ectopic expression of miR396 suppresses GRF target gene expression and alters leaf growth in *Arabidopsis*. Physiol Plant 136(2):223–236
Liu HH, Tian X, Li YJ, Wu CA, Zheng CC (2008) Microarray-based analysis of stress-regulated microRNAs in *Arabidopsis thaliana*. RNA 14(5):836–843
Liu PP, Montgomery TA, Fahlgren N, Kasschau KD, Nonogaki H et al (2007) Repression of AUXIN RESPONSE FACTOR10 by microRNA160 is critical for seed germination and post-germination stages. Plant J 52(1):133–146
Liu Q, Zhang H (2012) Molecular identification and analysis of arsenite stress-responsive miRNAs in rice. J Agric Food Chem 60(26):6524–6536

Liu X, Zhang H, Zhao Y, Feng Z, Li Q et al (2013) Auxin controls seed dormancy through stimulation of abscisic acid signaling by inducing ARF-mediated ABI3 activation in *Arabidopsis*. Proc Natl Acad Sci 110(38):15485–15490

Llave C, Xie Z, Kasschau KD, Carrington JC (2002) Cleavage of scarecrow-like mRNA targets directed by a class of *Arabidopsis* miRNA. Science 297(5589):2053–2056

Lokhande VH, Suprasanna P (2012) Prospects of halophytes in understanding and managing abiotic stress tolerance. In Ahmad P, Prasad MNV (eds)Environmental adaptations and stress tolerance of plants in the era of climate change. Springer, London, pp 29–56

Lu XY, Huang XL (2008) Plant miRNAs and abiotic stress responses. Biochem Biophys Res Commun 368(3):458–462

Ma Z, Hu X, Cai W, Huang W, Zhou X et al (2014) *Arabidopsis* miR171-targeted scarecrow-like proteins bind to GT cis-elements and mediate gibberellin-regulated chlorophyll biosynthesis under light conditions. PLoS Genet 10(8):e1004519

Mallory AC, Dugas DV, Bartel DP, Bartel B (2004) MicroRNA regulation of NAC-domain targets is required for proper formation and separation of adjacent embryonic, vegetative, and floral organs. Current Biol 14(12):1035–1046

Mantyla E, Lang V, Palva ET (1995) Role of abscisic acid in drought-induced freezing tolerance, cold acclimation, and accumulation of LT178 and RAB18 proteins in *Arabidopsis thaliana*. Plant Physiol 107(1):141–148

Matraszek R, Hawrylak-Nowak B, Chwil S, Chwil M (2016) Macronutrient composition of nickel-treated wheat under different sulfur concentrations in the nutrient solution. Environ Sci Pollut Res Int 23(6):5902–5914

Meng Y, Shao C, Wang H, Chen M (2011) The regulatory activities of plant microRNAs: a more dynamic perspective. Plant Physiol 157(4):1583–1595

Miao H, Sun P, Liu Q, Miao Y, Liu J et al (2017) Genome-wide analyses of SWEET family proteins reveal involvement in fruit development and abiotic/biotic stress responses in banana. Sci Rep 7(1):3536

Miller G, Shulaev V, Mittler R (2008) Reactive oxygen signaling and abiotic stress. Physiol Plant 133(3):481–489

Morea EG, da Silva EM, e Silva GF, Valente GT, Rojas CH, Vincentz M, Nogueira FT et al (2016) Functional and evolutionary analyses of the miR156 and miR529 families in land plants. BMC Plant Biol 16(1):40

Mukhopadhyay P, Tyagi AK (2015) *OsTCP19* influences developmental and abiotic stress signaling by modulating ABI4-mediated pathways. Sci Rep 5

Nakashima K, Yamaguchi-Shinozaki K, Shinozaki K (2014) The transcriptional regulatory network in the drought response and its crosstalk in abiotic stress responses including drought, cold, and heat. Front Plant Sci 5

Nishimura N, Yoshida T, Kitahata N, Asami T, Shinozaki K et al (2007) ABA-Hypersensitive Germination1 encodes a protein phosphatase 2C, an essential component of abscisic acid signaling in Arabidopsis seed. Plant J 50(6):935–949

Niu J, Wang J, Hu H, Chen Y, An J et al (2016) Cross-talk between freezing response and signaling for regulatory transcriptions of MIR475b and its targets by miR475b promoter in *Populus suaveolens*. Sci Rep 6

Noctor G, Foyer CH (2016) Intracellular redox compartmentation and ROS-related communication in regulation and signaling. Plant Physiol 171(3):1581–1592

Nott A, Jung HS, Koussevitzky S, Chory J (2006) Plastid-to-nucleus retrograde signaling. Annu Rev Plant Biol 57:739–759

Oh MH, Sun J, Oh DH, Zielinski RE, Clouse SD et al (2011) Enhancing *Arabidopsis* leaf growth by engineering the BRASSINOSTEROID INSENSITIVE1 receptor kinase. Plant Physiol 157 (1):120–131

Pandey P, Ramegowda V, Senthil-Kumar M (2015) Shared and unique responses of plants to multiple individual stresses and stress combinations: physiological and molecular mechanisms. Front Plant Sci 6

Papdi C, Pérez-Salamó I, Joseph MP, Giuntoli B, Bögre L, Koncz C, Szabados L (2015) The low oxygen, oxidative and osmotic stress responses synergistically act through the ethylene response factor VII genes RAP 2.12, RAP 2.2 and RAP 2.3. Plant J 82(5):772–784

Patel P, Yadav K, Ganapathi TR (2017) Small and hungry: microRNAs in micronutrient homeostasis of plants. MicroRNA 6(1):22–41

Pei ZM, Murata Y, Benning G, Thomine S (2000) Calcium channels activated by hydrogen peroxide mediate abscisic acid signaling in guard cells. Nature 406(6797):731

Quiles-Pando C, Rexach J, Navarro-Gochicoa MT, Camacho-Cristóbal JJ, Herrera-Rodríguez MB et al (2013) Boron deficiency increases the levels of cytosolic Ca 2+ and expression of Ca 2+-related genes in *Arabidopsis thaliana* roots. Plant Physiol Biochem 65:55–60

Radomiljac JD, Whelan J, van der Merwe M (2013) Coordinating metabolite changes with our perception of plant abiotic stress responses: emerging views revealed by integrative-omic analyses. Metabolites 3(3):761–786

Rao X, Dixon RA (2017) Brassinosteroid mediated cell wall remodeling in grasses under abiotic stress. Front Plant Sci 8

Reeves WM, Lynch TJ, Mobin R, Finkelstein RR (2011) Direct targets of the transcription factors ABA-Insensitive (ABI) 4 and ABI5 reveal synergistic action by ABI4 and several bZIP ABA response factors. Plant Mol Biol 75(4–5):347–363

Ren L, Tang G (2012) Identification of sucrose-responsive microRNAs reveals sucrose-regulated copper accumulations in an SPL7-dependent and independent manner in *Arabidopsis thaliana*. Plant Sci 187:59–68

Requejo R, Tena M (2005) Proteome analysis of maize roots reveals that oxidative stress is a main contributing factor to plant arsenic toxicity. Phytochemistry 66(13):1519–1528

Reyes JL, Chua NH (2007) ABA induction of miR159 controls transcript levels of two MYB factors during *Arabidopsis* seed germination. Plant J 49(4):592–606

Rock CD, Sun X (2005) Crosstalk between ABA and auxin signaling pathways in roots of *Arabidopsis thaliana* (L.) Heynh. Planta 222(1):98–106

Rosa M, Prado C, Podazza G, Interdonato R, González JA et al (2009) Soluble sugars: Metabolism, sensing and abiotic stress: A complex network in the life of plants. Plant Signal Behav 4(5):388–393

Rowe JH, Topping JF, Liu J, Lindsey K (2016) Abscisic acid regulates root growth under osmotic stress conditions via an interacting hormonal network with cytokinin, ethylene and auxin. New Phytol 211(1):225–239

Rubio S, Rodrigues A, Saez A, Dizon MB, Galle A et al (2009) Triple loss of function of protein phosphatases type 2C leads to partial constitutive response to endogenous abscisic acid. Plant Physiol 150(3):1345–1355

Schommer C, Debernardi JM, Bresso EG, Rodriguez RE, Palatnik JF (2014) Repression of cell proliferation by miR319-regulated TCP4. Mol Plant 7(10):1533–1544

Schommer C, Palatnik JF, Aggarwal P, Chételat A, Cubas P et al (2008) Control of jasmonate biosynthesis and senescence by miR319 targets. PLoS Biol 6(9):e230

Seo PJ, Xiang F, Qiao M, Park JY, Lee YN et al (2009) The MYB96 transcription factor mediates abscisic acid signaling during drought stress response in *Arabidopsis*. Plant Physiol 151 (1):275–289

Sharma R, De Vleesschauwer D, Sharma MK, Ronald PC (2013) Recent advances in dissecting stress-regulatory crosstalk in rice. Mol Plant 6(2):250–260

Shkolnik-Inbar D, Bar-Zvi D (2010) ABI4 mediates abscisic acid and cytokinin inhibition of lateral root formation by reducing polar auxin transport in *Arabidopsis*. Plant Cell 22 (11):3560–3573

Shriram V, Kumar V, Devarumath RM, Khare TS, Wani SH (2016) MicroRNAs as potential targets for abiotic stress tolerance in plants. Front Plant Sci 7

Shu K, Chen Q, Wu Y, Liu R, Zhang H et al (2016) ABI4 mediates antagonistic effects of abscisic acid and gibberellins at transcript and protein levels. Plant J 85(3):348–361

Singh P, Dave A, Vaistij FE, Worrall D, Holroyd GH et al (2017) Jasmonic acid-dependent regulation of seed dormancy following maternal herbivory in Arabidopsis. New Phytol 214 (4):1702–1711

Siré C, Moreno AB, Garcia-Chapa M, López-Moya JJ, Segundo BS (2009) Diurnal oscillation in the accumulation of *Arabidopsis* microRNAs, miR167, miR168, miR171 and miR398. FEBS Lett 583(6):1039–1044

Skopelitis DS, Paranychianakis NV, Paschalidis KA, Pliakonis ED, Delis ID et al (2006) Abiotic stress generates ROS that signal expression of anionic glutamate dehydrogenases to form glutamate for proline synthesis in tobacco and grapevine. Plant Cell 18(10):2767–2781

Sobkowiak L, Karlowski W, Jarmolowski A, Szweykowska-Kulinska Z (2012) Non-canonical processing of *Arabidopsis* pri-miR319a/b/c generates additional microRNAs to target one RAP2. 12 mRNA isoform. Front Plant Sci 3

Song X, Li Y, Cao X, Qi Y (2019) MicroRNAs and their regulatory roles in Plant-Envirnoment InteractionsAnnu. Rev. Plant Biol. https://doi.org/10.1146/annurev-arplant-050718-100334

Srivastava AK, Sablok G, Hackenberg M, Deshpande U, Suprasanna P (2017) Thiourea priming enhances salt tolerance through co-ordinated regulation of microRNAs and hormones in *Brassica juncea*. Sci Rep 7

Srivastava S, Srivastava AK, Suprasanna P, D'souza SF (2012) Identification and profiling of arsenic stress-induced microRNAs in *Brassica juncea*. J Exp Bot 64(1):303–315

Sun X, Xu L, Wang Y, Yu R, Zhu X et al (2015) Identification of novel and salt-responsive miRNAs to explore miRNA-mediated regulatory network of salt stress response in radish (*Raphanus sativus* L.). BMC Genom 16(1):197

Sunkar R, Kapoor A, Zhu JK (2006) Posttranscriptional induction of two Cu/Zn superoxide dismutase genes in *Arabidopsis* is mediated by downregulation of miR398 and important for oxidative stress tolerance. Plant Cell 18(8):2051–2065

Sunkar R, Li YF, Jagadeeswaran G (2012) Functions of microRNAs in plant stress responses. Trends Plant Sci 17(4):196–203

Sunkar R, Zhou X, Zheng Y, Zhang W, Zhu JK (2008) Identification of novel and candidate miRNAs in rice by high throughput sequencing. BMC Plant Biol 8(1):25

Sunkar R, Zhu JK (2004) Novel and stress-regulated microRNAs and other small RNAs from *Arabidopsis*. Plant Cell 16(8):2001–2019

Suprasanna P, Ghuge SA, Patade VY, Mirajkar SJ, Nikalje GC (2018) Genomic roadmaps for augmenting salinity stress tolerance in crop plants. In: Kumar V, Wani SH, Penna S, Tran L-SP (eds) Salinity responses and tolerance in plants, vol. 2. Springer International, pp 189–221

Szalai G, Kellős T, Galiba G, Kocsy G (2009) Glutathione as an antioxidant and regulatory molecule in plants under abiotic stress conditions. J Plant Growth Regul 28(1):66–80

Testerink C, Munnik T (2005) Phosphatidic acid: a multifunctional stress signaling lipid in plants. Trends Plant Sci 10(8):368–375

Thalmann M, Santelia D (2017) Starch as a determinant of plant fitness under abiotic stress. New Phytol 214(3):943–951

Trindade I, Capitão C, Dalmay T, Fevereiro MP, Dos Santos DM (2010) miR398 and miR408 are up-regulated in response to water deficit in *Medicago truncatula*. Planta 231(3):705–716

Valdés-López O, Yang SS, Aparicio-Fabre R, Graham PH, Reyes JL et al (2010) MicroRNA expression profile in common bean (*Phaseolus vulgaris*) under nutrient deficiency stresses and manganese toxicity. New Phytol 187(3):805–818

Vranová E, Langebartels C, Van Montagu M, Inzé D, Van Camp W (2000) Oxidative stress, heat shock and drought differentially affect expression of a tobacco protein phosphatase 2C. J Exp Bot 51(351):1763–1764

Wang W, Liu D, Chen D, Cheng Y, Zhang X, Song L, Hu M, Dong J, Shen F (2019) MicroRNA414c affects salt tolerance of cotton by regulating reactive oxygen species metabolism under salinity stress. RNA Biol 16(3):362–375

Wang L, Mai YX, Zhang YC, Luo Q, Yang HQ (2010) MicroRNA171c-targeted SCL6-II, SCL6-III, and SCL6-IV genes regulate shoot branching in *Arabidopsis*. Mol Plant 3(5):794–806

Welti R, Li W, Li M, Sang Y, Biesiada H et al (2002) Profiling membrane lipids in plant stress responses role of phospholipase Dα in freezing-induced lipid changes in Arabidopsis. J Biol Chem 277(35):31994–32002

Wind JJ, Peviani A, Snel B, Hanson J, Smeekens SC (2013) ABI4: versatile activator and repressor. Trends Plant Sci 18(3):125–132

Wu K, Tian L, Zhou C, Brown D, Miki B (2003) Repression of gene expression by *Arabidopsis* HD2 histone deacetylases. Plant J 34:241–247

Wu L, Zhou H, Zhang Q, Zhang J, Ni F et al (2010) DNA methylation mediated by a microRNA pathway. Mol Cell 38(3):465–475

Wu MF, Tian Q, Reed JW (2006) *Arabidopsis* microRNA167 controls patterns of ARF6 and ARF8 expression, and regulates both female and male reproduction. Development 133 (21):4211–4218

Xia J, Wang X, Perroud PF, He Y, Quatrano R et al (2016) Endogenous small-noncoding RNAs and potential functions in desiccation tolerance in *Physcomitrella patens*. Sci Rep 6:30118

Xia K, Ou X, Tang H, Wang R, Wu P (2015) Rice microRNA osa-miR1848 targets the obtusifoliol 14α-demethylase gene OsCYP51G3 and mediates the biosynthesis of phytosterols and brassinosteroids during development and in response to stress. New Phytol 208(3):790–802

Xia XJ, Wang YJ, Zhou YH, Tao Y, Mao WH et al (2009) Reactive oxygen species are involved in brassinosteroid-induced stress tolerance in cucumber. Plant Physiol 150(2):801–814

Xie R, Zhang J, Ma Y, Pan X, Dong C et al (2017) Combined analysis of mRNA and miRNA identifies dehydration and salinity responsive key molecular players in citrus roots. Sci Rep 7

Yadav SK (2010) Heavy metals toxicity in plants: an overview on the role of glutathione and phytochelatins in heavy metal stress tolerance of plants. S Afr J Bot 76(2):167–179

Yamasaki H, Abdel-Ghany SE, Cohu CM, Kobayashi Y, Shikanai T et al (2007) Regulation of copper homeostasis by micro-RNA in *Arabidopsis*. J Biol Chem 282(22):16369–16378

Yan J, Zhao C, Zhou J, Yang Y, Wang P et al (2016) The miR165/166 mediated regulatory module plays critical roles in ABA homeostasis and response in *Arabidopsis thaliana*. PLoS Genet 12(11):e1006416

Yan YS, Chen XY, Yang K, Sun ZX, Fu YP et al (2011) Overexpression of an F-box protein gene reduces abiotic stress tolerance and promotes root growth in rice. Mol Plant 4(1):190–197

Ye H, Liu S, Tang B, Chen J, Xie Z et al (2017) RD26 mediates crosstalk between drought and brassinosteroid signaling pathways. Nat Commun 8

Yu LJ, Luo YF, Liao B, Xie LJ, Chen L et al (2012) Comparative transcriptome analysis of transporters, phytohormone and lipid metabolism pathways in response to arsenic stress in rice (*Oryza sativa*). New Phytol 195(1):97–112

Zeng QY, Yang CY, Ma QB, Li XP, Dong WW et al (2012) Identification of wild soybean miRNAs and their target genes responsive to aluminum stress. BMC Plant Biol 12(1):182

Zhang B, Wang Q (2016) MicroRNA, a new target for engineering new crop cultivars. Bioengineered 7(1):7–10

Zhang H, Sonnewald U (2017) Differences and commonalities of plant responses to single and combined stresses. Plant J 90(5):839–855

Zhang L, Zhao G, Jia J, Liu X, Kong X (2011) Molecular characterization of 60 isolated wheat MYB genes and analysis of their expression during abiotic stress. J Exp Bot 63(1):203–214

Zhang YC, Yu Y, Wang CY, Li ZY, Liu Q et al (2013) Overexpression of microRNA OsmiR397 improves rice yield by increasing grain size and promoting panicle branching. Nat Biotechnol 31(9):848–852

Zhou M, Li D, Li Z, Hu Q, Yang C et al (2013) Constitutive expression of a miR319 gene alters plant development and enhances salt and drought tolerance in transgenic creeping bentgrass. Plant Physiol 161(3):1375–1391

Zhou X, Yan S, Sun C, Li S, Li J et al (2015) A maize jasmonate Zim-domain protein, *Zm*JAZ14, associates with the JA, ABA, and GA signaling pathways in transgenic *Arabidopsis*. PLoS ONE 10(3):e0121824

Zhou ZS, Huang SQ, Yang ZM (2008) Bioinformatic identification and expression analysis of new microRNAs from *Medicago truncatula*. Biochem Biophys Res Commun 374(3):538–542

Zhu JY, Sae-Seaw J, Wang ZY (2013) Brassinosteroid signalling. Development 140(8):1615–1620

Chapter 3
Epigenetic Response of Plants to Abiotic Stress: Nature, Consequences and Applications in Breeding

Manoj K. Dhar, Rahul Sharma, Parivartan Vishal and Sanjana Kaul

Abstract Stress is inevitable in the life cycle of living organisms, including plants. Being sessile, plants are more prone to the deleterious effects of environmental stress. Therefore, plants have developed complex mechanisms to survive under these challenging conditions. Tolerance, avoidance, and resistance are the three major strategies followed by plants to counter the recurring biotic and abiotic stresses. These mechanisms involve genes associated with several interconnected pathways, which lead them towards better stress tolerance. Plants resort to various modifications in their morphological traits, physiology, and so forth in response to stress. Modulations in various regulatory mechanisms, including epigenetic modifications, play a pivotal role in developing stress tolerance in plants. These involve changes in either the plant homeostasis or heritable changes in gene expression pattern. The trans-generational changes are brought about, more often, by dynamic changes in epigenetic marks rather than development of stress resistant alleles via gene mutation. A large number of stress resistant transgenics have been developed over the years all over the world. However, the traditional breeding has remained indispensable. Much emphasis has been laid on identification and characterization of stress resistance genes and developing transgenic crop varieties, while the epigenomic aspects have been given less importance. The present chapter focuses on the essential components of epigenetic machinery, different epigenetic alterations involved in conversion of active euchromatin to silent heterochromatin and vice versa during stress, and integration of epigenetic data with breeding programs to devise better strategies towards development of stress resistant crops.

Keywords Epigenomics · Functional genomics · DNA methylation
Histone modifications · Small RNA

M. K. Dhar (✉) · R. Sharma · P. Vishal · S. Kaul
Genome Research Laboratory, School of Biotechnology, University of Jammu, Jammu 180006, India
e-mail: manojkdhar@rediffmail.com

V. R. Rajpal et al. (eds.), *Genetic Enhancement of Crops for Tolerance to Abiotic Stress: Mechanisms and Approaches, Vol. I*, Sustainable Development and Biodiversity 20, https://doi.org/10.1007/978-3-319-91956-0_3

3.1 Introduction

The environment is a dynamic entity and is thus under a continuous state of change. Unlike animals, plants being sessile are not able to escape from the environmental vagaries. However, they have developed various mechanisms to resist the changes and survive under harsh conditions. Environmental stress can be biotic (pathogen, pest) or abiotic (heat, salinity or drought), but both limit the plant productivity. Among different types of stresses, water scarcity is the most prominent stress responsible for huge losses in crop yield worldwide. Temperature fluctuation due to global warming have already led to increased episodes of drought in many drought prone areas. Global increase in desertification, decrease in total agricultural land and drought like conditions due to water scarcity and salinity pose an urgency to develop high yielding drought tolerant varieties of crops.

Water acts as the most important component of plant transport system playing a pivotal role in the transport of metabolites from one part of the plant to the other. Water acts as universal solvent helping in general metabolism of the plant. However, water scarcity poses many threats to normal growth and development of the plant. Drought can be defined as a type of abiotic stress where the plant does not receive adequate quantity of water required for optimal growth and productivity (Diekman et al. 2012). Drought affects plants by reducing their leaf size, disturbing plant water equilibrium and reducing water use efficiency. Overall, drought has negative effect on the growth, development and yield of the plants. Water scarcity is a complex stress affecting many interrelated biochemical and molecular pathways and thus making it very difficult to pinpoint the exact mechanism. One of the major issues is the silencing of genes involved in cell division and protein synthesis during drought. This leads to cell dormancy and/or slow cell division, resulting in impaired growth and low yield. Water scarcity further leads to osmotic imbalances resulting in decreased turgor pressure affecting cell elongation along with various physiological and morphological impairments (Fig. 3.1).

The impact and intensity of the stress are the major factors prompting response from the living organisms. When the effect exceeds the tolerance limit, the organisms exhibit direct and indirect response to stress. Plants have evolved various strategies to overcome the effect of drought stress. These include but are not limited to; escapism, avoidance and tolerance. Many plants under drought stress show early germination and early flowering to escape the harsh effects of drought stress. Some plants lessen the effect of stress by avoidance strategies, such as; plants develop deep roots to increase root zone effect, show leaf rolling to decrease total surface area, leading to stomatal closure to reduce water loss by transpiration. Plants can also show resistance or tolerance to the stress condition, which involves up-regulation and down-regulation of various stress responsive genes (Fig. 3.2). Drought tolerance is a quantitative trait involving many genes and linkage of these genes with other desirable and undesirable genes makes it difficult for the crop breeders to develop efficient drought tolerant varieties with high yield potential (Varshney et al. 2011; Bhardwaj et al. 2015). Effect of and response to stress is

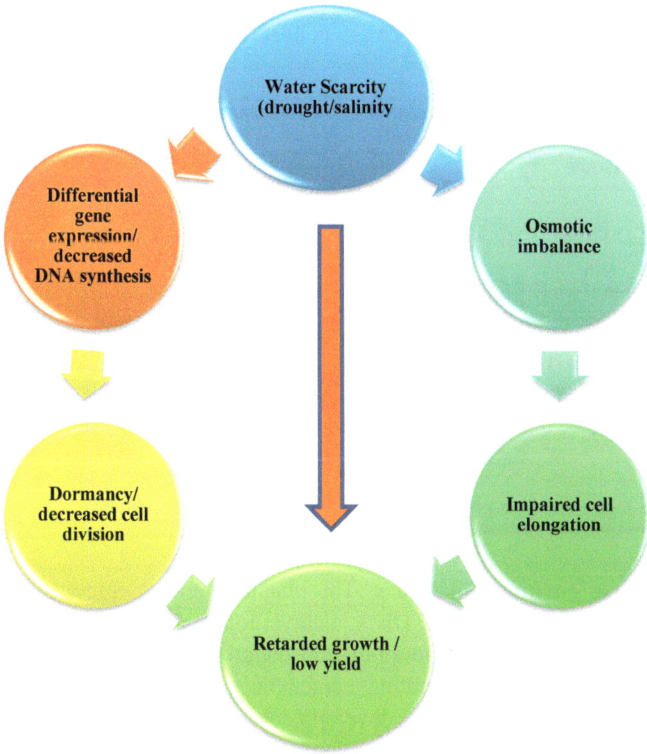

Fig. 3.1 Effect of water scarcity on the growth and yield

further dependent on the time of stress induction during the life span of the plant. The stress response of the plant, at morphological and molecular levels, differs during germination and flowering. These temporal and spatial regulation strategies employed by the plants further make stress response a complex process to understand. A vast number of drought stress responsive genes have been identified in plants. Many studies amply describe their expression levels, however, the factors controlling the expression of these genes have not received due attention. These genes are involved in cross-talk with one another and function in a well-knit network. Breeding, mass screening, marker-assisted selection, transgene transfer or engineering for drought resistance are the few strategies, which have been commonly employed to identify, characterize and deploy these genes for developing stress tolerant varieties.

The environmental stress is responsible for generation of many genomic, transcriptional and proteomic variants, ultimately leading to phenotypic variants. Apart from genomic variants, environmental stress is also responsible for generating new epimutants via epigenetic changes, which show modulated gene expression to produce phenotypic variants beneficial for the plant including conferring protection

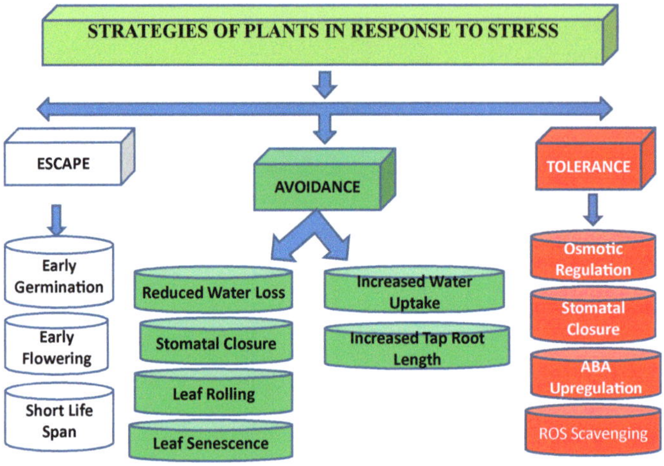

Fig. 3.2 Strategies employed by the plants to cope harsh stress conditions

against adverse environmental cues. The chapter reviews epigenetic changes in plants under stress mainly via DNA methylation and histone post-translational modifications. Available tools to study and quantify epigenetic changes are highlighted. The potential of integrating epigenetic data with breeding programs to devise better strategies towards development of stress resistant crops is also discussed.

3.2 Epigenetic Changes

Very often the variation observed in different individuals of a population is due to various genetic factors, however, sometimes the variation is epigenetic i.e. it does not result from DNA sequence polymorphism. The term epigenetics was coined by Waddington (1953), who described it as the role of gene interactions and environmental factors in introducing a particular phenotype. However, presently epigenetics refers to heritable/non-inheritable changes, meiotic or mitotic in terms of chromatin structure, cytosine methylation of DNA and modifications in histone proteins, thereby generating various global and locus-specific epialleles (Manning et al. 2006). Post-translational modifications and accessibility of chromatin also affect the gene expression, which leads to variant phenotypes. The study encompassing the above factors/phenomena, affecting the phenotype of an organism is called as epigenomics (Lane et al. 2014). Despite being epigenetic marks, many chromatin modifications are not heritable (Springer 2013). Therefore, epigenetic variation stems from various chromatin marks such as methylation of cytosine residues in the DNA, histone tail modifications and small RNAs, etc. There is voluminous literature available on DNA methylation as it constitutes the most studied chromatin modification, particularly with regard to its stability and heritability. Many studies have shown suppression of individual genes, repetitive DNA sequences and transposable elements due to methylation. However, in some cases there are occasional changes in the methylation status indicating the semistable heritability of methylation. In recent years, more emphasis has been laid in exploring the epigenetic modifications since they are potential source of hitherto un-assessed heritable variations, which can be exploited for predicting phenotypic variations and devising newer strategies for efficient breeding. In this direction, many studies have been conducted not only for characterization of epigenetic variation but also to understand the association between epigenetic and genetic changes at specific loci.

Epigenetic marks, present on the genome of any organism such as plant, animal or microbe, define that organism by regulating its gene expression. Merely observing epigenetic marks may not be of much use; therefore, it would be meaningful if the factors responsible for relative modification of epigenetic marks were studied in detail.

First reported case of epigenetic mutation was occurrence of asymmetric flowers showing peloric variation in *Linaria* (Jeggo and Holliday 1986). Later, methylation mediated *Lcyc* gene silencing was suggested as the possible reason for generation of flower variants. Cubas et al. (1999) confirmed the inheritance of *Lcyc* gene methylation to next generation. Thus, methylation marks were the first and most important epimarks to be studied.

As discussed above, epigenetic machinery involves DNA methylation, histone post-translational modifications and smRNA. Apart from stress response, epigenetic modifications are also involved in maintaining genome stability, suppressing transposon activity and controlling gene expression by manipulations at post-transcriptional and post-translational levels. A variety of epigenetic marks present on amino terminal of histone tails due to post-translational modifications along with other histone variants, constitute highly complex histone code (Chinnusamy and Zhu 2009). Methylation, acetylation, phosphorylation, sumoylation, biotinylation, ubiquitination, carbonylation, glycosylation and ADP ribosylation are the various types of post-translational modifications that occur on histone tails (Zentner and Henikoff 2013). Apart from these, chromatin remodeling is another area of interest for studying epigenetic modifications though, these marks show short-term transgenerational inheritance, which in turn affects the DNA accessibility of transcriptional factors (Bonasio et al. 2010).

3.3 Stress Perception and Epigenetic Response

Both genetic and epigenetic changes are responsible for producing phenotypic variants but mutation rates are generally much lower than spontaneous DNA methylation events; thus, epialleles play a very important role in phenotypic diversity (Schmitz et al. 2011). Genetic variations along with heritable or metastable epigenetic variations have the potential to drive natural variation. An epiallele can be termed as a heritable modification without a change in the DNA sequence. Epialleles have been classified as obligatory, facilitated or pure (Richards 2006) based on the relative influence of genetics and epigenetics upon chromatin state. Epialleles are considered obligatory when the chromatin state is directly correlated with a genetic change. Facilitated epialleles represent a condition when a genetic change results in a poised allelic state, which can exist in an active or silenced form. In pure epialleles, no genetic changes influence the chromatin state. All these types of epialleles have been reported to exist in plant populations.

Biotic and abiotic stresses affect the plants in term of health, yield and quality; however, accumulation of transgenerational epigenetic changes can help them to tolerate the stress in future (Slaughter et al. 2012). Epigenetic factors act as a link between stress perceptions and implementation of appropriate response. Plants react to abiotic stresses such as high temperature, high salinity and drought by modifying their growth and developmental processes to lessen the extent of damage caused by these stresses. Although, most of the modifications are transient but the extent of induction depends upon the severity of the stress exposure. Many modifications reverse with the onset of favorable conditions, however, plants have the capability of memorizing certain changes at transcriptional and post-transcriptional levels and transmit them faithfully to next generation. All these modifications occur at epigenetic level without disturbing the sequence of the genome. In *Populus trichocarpa*, drought-induced genome wide increase in cytosine methylation helps to cope with harsh effects of abiotic stress (Liang et al. 2014). Post-translational modifications of histone acting primarily as a response to stress perception may not be truly epigenetic in nature as its transgenerational transference is obscure.

3.4 DNA Methylation and Machinery

DNA methylation was discovered in early twentieth century (Johnson and Coghill 1925) and has emerged as one of the most important heritable and reversible epigenetic marks responsible for modification of cytosine residue at C-5 position in all viable organisms. DNA methylation studies require prior sequence information for developing methylation maps of different plants. The studies have identified methylation as an important epimark, being present on every genome (Lane et al. 2014) though the pattern of methylation differs among genomes (West et al. 2014).

DNA methylation is one of the important epigenetic modifications that has been widely studied in plants (Niederhuth and Schmitz 2014). *Arabidopsis thaliana* being the model organism for studies in plants has been greatly exploited for genetic and epigenetic studies. The answer to mystery of phenotypic variations arising from similar DNA sequences lies in the pattern of heritable epimarks, which are responsible for altered gene expression resulting into characteristic phenotype. DNA methylation has distinct pattern in plants as compared to animals. The methylation is known to occur at CG along with CHG and CHH residues. In case of mammals, it occurs on symmetric dinucleotides-CG, whereas in plants, both symmetric (CG, CHG) and asymmetric CHH residues are involved; H stands for either of adenine, cytosine, or thymine nucleotides. DNA methylation can occur without any preference for genic or intergenic region.

The complete pattern of arrangement of nucleotide methylation in an organism's genome is called as methylome. Methylome studies carried out in different plants have brought to light the role of methylation. *Arabidopsis thaliana* genome shows 24% CG methylation (Cokus et al. 2008), *Glycine max* has 51% (Schmitz et al. 2013) and *Oryza sativa* genome has 59% CG methylation (Feng et al. 2010). Methylome analysis of *Brassica oleracea* revealed that 54.9% of whole genome CGs are methylated, whereas symmetric CHGs count 9.4% and asymmetric CHHs count 2.4% only (Parkin et al. 2014). CG methylation has usually negative effects on gene expression and hence is considered as a repressive epigenetic mark (Suzuki and Bird 2008).The establishment of methylation at these residues occurs through different modes. All three types of methylation marks are known to produce silencing effects at the target region, but many a times can show position effect, silencing the neighboring sequences (Law and Jacobsen 2010). CG methylation plays a very important role in maintaining the genome stability, by silencing repetitive elements. The centromeric regions of the chromosomes are rich in repetitive elements and hence show high CG methylation, which ultimately leads to heterochromatinization of the region. However, with regard to promoters, methylation is negatively correlated with transcriptional activity of the genes.

DNA methylation can be spontaneous or induced. Environmental changes are known to induce significant changes in DNA methylation leading to new variations. Methyltransferase1 (MET1) is considered responsible for maintaining methylation marks at CG residues in *A. thaliana* (Lister et al. 2008). Histone specific post-translational modifications also influence the methylation status of DNA sequences. H3K9 methylation controls DNA methylation at CHG and CHH residues but a group of different methyltransferases known as chloromethyltransferases (CMT) namely, CMT2 and CMT3 are known to maintain the methylation at these residues (Stroud et al. 2014).

Compared to higher plant DNA, which is up to 50% methylated, mammalian DNA is methylated from 2 to 8% only (Zhu 2009). Domains Rearranged Methylase 1 and 2 (DRM1 & DRM2) help in asymmetric DNA methylation in plants, following RNA-dependent DNA methylation pathway (RdDM), whereas symmetrical DNA sequences CpG and CpHpG maintain their methylation status by MET1 and CMT3. DNA Methyltransferase 1 (DNMT1) found in metazoans maintains the 5mC status along with DRM1 performing de novo CG sites methylation. MET1 produces methylation patterns that are transmitted through mitosis as well as meiosis (Saze 2008).

Chromatin remodeling is another method of indirectly maintaining the methylation status. Decrease in DNA Methylation 1 (DDM1) is a chromatin remodeling factor maintaining the CG and CHG methylation status in *Arabidopsis* (Brzeski and Jerzmanowski 2003). Fujimoto et al. (2008) induced whole genome hypomethylation by inserting DDM1 RNAi construct in *B. oleracea*.

DNA demethylases also manipulate DNA methylation status of plants by removing methylation marks. DNA glycosylases family includes several enzymes such as Demeter (DME), Repressor of Silencing1 (ROS1) and Demeter-Like (DML) proteins. DML2 and DML3 are responsible for localized as well as global demethylation (Saze et al. 2012). ROS1, a DNA demethylase, interacts with ROS3, which possesses a motif capable of recognizing and binding small RNA, which in turn guides ROS1 and ROS3 to specific loci on genome for demethylation (Zheng et al. 2008). It is hypothesized that this pathway can be used to demethylate specific gene loci by targeting loci-specific small RNA to ROS3, which in turn can act as a link between Sm RNA and ROS1 for site-specific demethylation.

Massively parallel sequencing of mononucleosomes for genome-wide nucleosome position analysis suggested that nucleosome positioning influences genome-wide methylation patterns. DNA methyltranferases targeting the nucleosome-bound DNA suggest that nucleosome and DNA methyltranferases share a conserved relationship. Of many studies conducted on various plant species, the distribution of compact methylation of cytosine was found at repeats and transposons with an average of 90% sequence methylation, creating a transcriptionally repressed chromatin state (Wang et al. 2009). Comparative studies of methylome and transcriptome in *Arabidopsis* revealed that coding region exhibited high methylation whereas high and low expressed genes showed a significantly lower methylation level (Lister et al. 2008). RNA pol II processivity has been dependent on nucleosome positioning as RNA pol II has been found relatively more on exons than introns suggesting that methylation has an important role in defining and evolving the exons. Exons possess more methylation marks than introns and this pattern can be used to define intron-exon boundary, moreover linker region of the nucleosome possesses less methylation as compared to sequence wrapping histone octamer (Chodavarapu et al. 2010).

Increase in methylation of coding regions of genes, due to activity of small interfering RNA (siRNA), is responsible for repression inside promoters (Lauria and Rossi 2011). Whereas the methylation of flanking regions of the gene body i.e. promoter, 5' UTR and 3' UTR and nearby coding region has negative effect on gene expression and can play a significant role in tissue-specific expression and biotic stress response (Dowen et al. 2012). Plants have shown alternate splicing in response to stress (Ali and Reddy 2008).

It can be concluded that CG methylation is the most important type of methylation having varied effects on the expression of the gene depending upon the localization of CG. CG methylation of promoter region has been negatively correlated with gene expression, however, CG methylation when present in gene body has positive effects on transcription levels of the gene, indicating the complex nature of methylation in the context of its localization (Ball et al. 2009). In *Arabidopsis thaliana*, levels of gene body methylation have been found as a factor for maintaining transcriptional elongation. In other words, proper transcription requires hypermethylation at gene body (Zilberman et al. 2007). However, a comparative transcriptome and methylome study in *Brassica oleracea* showed certain exceptions, where highly methylated genes showed transcription and low methylated genes were non-expressing. However, a general trend of moderate level of conserved gene body methylation was associated with expression (Parkin et al. 2014).

Epialleles providing stability have been documented in *Brassica napus* as well as many plant systems (Long et al. 2011). Epialleles are mostly mitotically stable; however, many instances of meiotically stable epialleles have also been documented. Epigenetic recombinant inbred lines (epiRILs) are derived from two parents having negligible difference in DNA sequences but contrasting DNA methylation profile. epiRILs have been found associated with stability and heritability of some traits such as flowering time and plant height over eight generations in *Arabidopsis* (Johannes et al. 2009).

3.5 Methodologies for Studying Methylation

DNA methylation patterns in the genome of an organism play important role in regulation of gene expression. Over the last many years, different methods have been used to understand the methylation patterns. Primarily, restriction digestion methods using restriction enzymes, differing in the tendencies to distinguish the recognition site on the basis of methylation sensitivities, have been used to profile genome-wide distribution of DNA methylation (Schmitz and Zhang 2011). The drawback of this method is the limitation of number of isoschizomers differing in their methylation sensitivities so that they could be efficiently used in the process. Moreover, the efficacy of the restriction digestion is also variable. The distribution of the methylated marks in the genome of an organism determines the size of generated fragments. The terminal sequences explain the methylation sensitive restriction of the fragment but the target sites, which may have escaped from the restriction, present in the fragment, make it difficult to estimate exact quantification of methylation, leading to a poor resolution (Laird 2010).

Methylation-sensitive amplification polymorphism (MSAP) is a modification of Amplified Fragment Length Polymorphism (AFLP) technique, which was used for the first time to detect the differential methylation pattern of fungi during dimorphic transition (Reyna-Lopez et al. 1997). Later, isoschizomers, *Msp*I and *Hpa*II have been used commonly; however, *Mse*I was used in original protocol. *Msp*I and *Hpa*II recognize the sequence CCGG but *HpaII* is sensitive to methylation at any cytosine residue, whether internal or external, and cleaves hemimethylated sequences only. *Msp*I on the other hand is sensitive to external methylation irrespective of its semi or full methylation status. Thus, MSAP can be used to assay full methylation of internal cytosines and hemi methylation of external cytosines. Earlier, radiolabeled primers were used to carry out MSAP, but now the technique has been modified with the introduction of fluorescently tagged primers (Yang et al. 2011).

Sodium bisulphite conversion is yet another technique to identify methylated residues after sequencing. In this technique, unmethylated cytosines are converted into uracils, which are further converted into thymine during the PCR amplification process but methylated cytosines remain unchanged. Therefore, one can distinguish between modified and unmodified cytosine residues after sequencing the PCR amplicons. The converted sequences are compared to reference amplicon sequences to identify the cytosines, which have modified during the process. Methylation specific primers are designed for the process to amplify the epigenetically modified stretch of DNA (Ji et al. 2014).

Combining the sodium bisulphite conversion with high throughput sequencing, one can go up to single base resolution. Now-a-days, whole-genome bisulfite sequencing (WGBS) is employed to check the methylation status at whole genome level and understand the utilization of methylation by the genome to regulate its activities. The converted cytosines can be determined only after comparing the sequences with reference genome in case of WGBS process (Lister et al. 2008).

CpG islands are the regulatory sequences which are mostly present upstream of the genes and act as the site of methylation. Antibody specific for 5-Methylcytosine (5mC) is used to precipitate methylated cytosines and the process is called as immunoprecipitation. Coupling immunoprecipitated methylated DNA with gene microarray has laid the foundation of a high throughput technique called as MeDIP-chip (Zhang et al. 2006). Various steps involved in the process include; (i) isolation of DNA, which is further, purified and sonicated into fragments of 300–1000 base pairs, (ii) denaturation of DNA fragments and (iii) treatment with the antibodies raised specifically to detect immunoprecipitated 5-methylcytidine. The methylated DNA is separated from antibody, digested with endopeptidase K followed by phenol/chloroform extraction to purify the DNA. This methylated DNA can then be used with microarray.This technique has better resolution than methylation-sensitive restriction digestion methods.

To obtain the whole methylome, affinity enrichment of methylated regions of the genome is achieved by immunoprecipitation with antibodies specific to methylated loci followed by hybridization to a microarray (ChIP-chip) or sequencing (ChIP-seq). Affinity enrichment approach suffers from the drawback of detecting only CG region methylation, which is otherwise considered as most important. Although single base pair resolution is possible with bisulphite conversion, methylome studies are still benefited by exploring differentially methylated regions, which scale from a few to several hundred base pairs (Bock 2012). Methylation dependent immunoprecipitation forms the basis of ChIP-chip and ChIP-seq techniques, which are simple, efficient and cost-effective, however, they suffer from a few drawbacks; the heavy dependence of the approaches on distribution of 5mC in the genome, and the antibody's specificity.

Polymorphic epimutation at a single base level has not been found as effective as single nucleotide polymorphism in affecting the phenotype of the organism. However, larger regions of DNA methylation have been shown to result in phenotypic variations. Several differential methylation sites (DMS) constitute a differential methylated region (DMR) and identifying them is a challenging task. It requires a lot of time, if done manually. Many statistical tools like Bayesian statistics employing hidden Markov's model have been used to convert many DMS into one DMR. Once identified, these DMRs can be used to identify the regions of the genome directly responsible for variations, as DMRs are associated with phenotypic changes (Robinson et al. 2014). Studies in maize have shown that differentially methylated regions in the genome contain spontaneous epimutations, which produce phenotypic variations while associated with DMRs. Although, spontaneous methylation changes are associated with DMRs, however, specific epimutation conferring a phenotypic change has not been yet reported (Eichten et al. 2013). WGBS techniques are adding several novel epimutation marks. Studies of heritability, stability and expressibility of these epimarks can provide new insights in epigenetics.

3.6 Histone Modifications and Machinery

Histones are the important nucleoproteins in eukaryotes and are involved in DNA packaging and gene regulation. Histones belong to a conserved family of small basic proteins ranging from 11 to 21.5 kDa (McGhee and Felsenfeld 1980). Tightly packaged region of the DNA forms heterochromatin and loosely packaged regions form euchromatin. The packaging is based on the post-translational modifications on amino- and carboxy- terminals of histones, thereby, influencing gene activity (Rando 2012). The post-translational modifications present on the histone proteins are in the form of epimarks and a specific pattern of these epimarks on histone proteins constitutes the histone code. Understanding histone code can solve many mysteries of regulation of gene expression. Of the many post-translational modifications, methylation and acetylation of lysine residues of the histone H3 play very important roles in regulating various gene activities (Margueron et al. 2005). Histone acetylation levels, known to regulate the gene expression in case of biotic and abiotic stresses, are dependent on activity of histone acetyltransferases and histone deacetylases. Histone acetyltransferases are known to transfer acetyl group to lysine residues of the N terminal of histone tails and show positive correlation with gene activity (Zentner and Henikoff 2013).

Lysine residues carry positive charge, which can be neutralized by the transferred acetyl group, thereby relaxing the chromatin as it reduces the affinity between oppositely charged components of nucleosomal complex. Thus, acetylation marks are associated with active transcription. Histone deacetylases are known to remove the acetyl group from the histone lysine residues, thereby creating a strong affinity between components of nucleosomal complex. Deacetylation helps in the condensation of the chromatin; therefore, it is generally associated with gene silencing. Jasmonic acid (JA) regulates activity of two histone deacetylases; HDA6 and HDA19 in *Arabidopsis* in response to abiotic and biotic stress, respectively (Keqiang et al. 2008). JA-regulated pathogenesis-related genes, respond to biotic stress through HDA19 activity (Wu et al. 2008). Acetylation marks have been positively correlated with gene expression. Acetylation marks on H3K9 in the promoter region of stress responsive genes positively correlate with gene expression

(Misook et al. 2011) Similar studies in *Arabidopsis* have revealed association of H3K23 and K3K27 acetylation with coding regions of stress responsive genes (Kim et al. 2008). An overall increased acetylation has been observed in histone H3 and H4 of tobacco and *Arabidopsis*. This increase in acetylation has been correlated with upregulation of abiotic stress related genes (Sokol et al. 2007).

Methylation is one of the most important epigenetic marks on histone residues governing the heterochromatinization and expression of genes. Lysine residues are vulnerable to methylation and can get mono-, di- or tri- methylated. Each post-translational modification has its specific effect. Genome annotations of regulatory element, intron-exon border, cell type specific region and specific functional regions can be performed with the help of chromatin signatures. Histone methyltransferases (HMT) and histone demethylases (HDM) are the enzymes known to methylate and demethylate the N terminal region of the histone tails, respectively. Apart from DDM, histone methyltransferases SUVH4/KYP (SU (VAR) 3-9 HOMOLOG 4/KRYPTONITE) and SUVH5/6, or SRA-domain methylcytosine-binding protein Variant In Methylation 1/2/3(VIM1/2/3) are also involved in the maintenance of DNA methylation (Dhar et al. 2014; Saze 2008).

Histone H3K9 methylation has been known for the heterochromatinization (Gendrel et al. 2002). In *Arabidopsis*, vernalization leads to suppression of *Flowering Locus C* (*FLC*) gene activity by heavy accumulation of silencing markers; H3K9me2 and H3K27me3 marks in the promoter of *FLC* gene (Kim and Sung 2012). H3K4 and H3K36 trimethylation has been found associated with genes suggesting active transcription whereas H3K9 and H3K27 methylation marks are associated with repressed genes and are also regarded as heterochromatin marks (Dhar et al. 2009; Zentner and Henikoff 2013).

Aurora and NIMA (Never in mitosis gene A) kinase catalyze phosphorylation of histone H3 at serine 10 residue and a haspin like protein is known to phosphorylate histone H3 at threonine 3 residue (Houben et al. 2007). Histone H3 (ser)10ph/(ser) 28ph and H2B(K)143 mono ubiquitinization has been positively correlated with transcription (Khorasanizadeh 2004). In *Arabidopsis*, H2B mono-ubiquitinization led to elevated plant tolerance against necrotic fungal attack (Dhawan et al. 2009).

For mapping global histone modifications and determining transcription factor binding, ChIP followed by sequencing has been a standard procedure (Mikkelsen et al. 2007). New variants of ChIP namely, Nano-ChIP-seq and ChIP-exo are now-a-days used for better resolution and sensitivity (Terooatea et al. 2016). Some modifications like acetylation and in some cases phosphorylation and ubiquitinization are associated with gene upregulation (Zhang et al. 2007a) whereas biotinylation and sumoylation are associated with down regulation of genes (Chen et al. 2010). Depending upon the position, amino acid residue and number of methyl groups attached, methylation has different effects on gene expression. Contrary to H3K27me3 and H3K9me2, which act as main gene silencing markers, H3K4me3 and H3K36me2/me3 have been found in actively transcribing regions (Wang et al. 2009). H3K27me1/me2 and H4K20me1 have been found enriched in constitutively heterochromatinized and transposon regions (Roudier et al. 2011).

Gene expression can be altered in response to stress by ATP-dependent chromatin remodeling dynamics while altering the binding interaction between histone and DNA sequences (Gutzat and Scheid 2012). SWI/SNF ATPase, a type of chromatin remodeling protein finds its immense role in development processes and regulation of a variety of stress-related genes (Bezhani et al. 2007; Walley et al. 2008; Narlikar et al. 2013). Prolonged heat stress reversibly reduces nucleosome loading to a significant extent at Transposable element (TE) loci (Pecinka et al. 2010). Studies in *Arabidopsis* have revealed stress-responsive growth arrest duc to activity of a chromatin remodeling gene named as *AtCHR12* (Mlynarova et al. 2007).

3.7 Role of Small RNAs

Small RNAs (smRNA) play a major role in gene silencing at transcriptional level by DNA methylation and at post-transcriptional levels by RNA degradation. They belong to a non-coding class of RNA, ranging in size from 20 to 27 nucleotides. They play an important role in the regulation of gene expression by employing epigenetic mechanism during gametogenesis, fertilization and zygote development (Slotkin et al. 2009; Bourchis and Voinnet 2010).

siRNAs are the short interfering RNAs which are derived from the transcripts generated by a plant specific polymerase known as Pol IV. In *Arabidopsis thaliana*, a 24-nucleotide siRNA guides the most important pathway of non-CG methylation called as RNA-directed DNA methylation pathway (RdDM) (Zhang et al. 2007b). siRNA finds its role in maintenance of heterochromatic loci at repetitive sequences and controlling the activity of transposons by silencing those (Tran et al. 2005). Activity of RNAi has been negatively associated with gene expression (Volpe et al. 2002). The RdDM is a primary and most utilized mode of non-CG methylation but it requires the activity of a distinct group of methyltransferase family DRM1/DRM2 (Law and Jacobsen 2010).

3.8 Epigenetics and Development of New Varieties

As stated earlier, the epigenetic response of a plant occurs at varied levels. The response could range from methylation of a single nucleotide to a differential methylated region, from a stress responsive gene to whole genome methylation, from histone H3 post-translational modification to change in the histone code. The response to abiotic stress is a complex phenomenon involving many cross-talking pathways. Though plants simultaneously upregulate and downregulate many genes in response to stress, analysis of the overall results can reveal whether the plant opts an avoidance or tolerance mechanism against stress response.

Global demethylation is the most common and instant strategy employed by plants in response to stress. Abiotic stress has been found to globally demethylating the genome in maize (Steward et al. 2002) and rice (Wang et al. 2011). In contrast, *Mesembryanthemum crystallinum* plants on exposure to salt stress, showed a two-fold increase in CpHpG methylation (Dyachenko et al. 2006). According to Tsaftaris and Dickinson (2000), many plants show increasing methylation in response to biotic or abiotic stress leading to overall degeneration of genome activity, however, with the onset of favourable conditions the genome shows hypomethylation for optimum expression. However, Suji and Joel (2010) showed that drought stress resulted in hypomethylation of drought susceptible varieties whereas drought tolerant rice varieties were marked with genome hypermethylation. Hence, it can be concluded that different plants employ different strategies to survive under stress conditions. Studies in *Brassica juncea* revealed that drought stress led to hypomethylation of drought responsive genes, however, hypermethylated regions increased in number in case of genes involved in basic cellular activities. It follows that the plant shuts basic activities to a certain extent and activates the defense mechanism to survive under abiotic stress condition (Sharma et al. 2017).

Tissue culture-induced stress response has been studied in maize by MeDIP chip technique. The analysis revealed that 67% of the total 479 differential methylated regions were hypomethylated. In comparison to 75% of the hypomethylated regions, just 47% of the hypermethylated regions were observed in multiple tissue culture regenerants confirming the stability of the hypomethylated DMRs (Ji et al. 2015). Comparison of DMR variation under natural conditions and under tissue culture stress-induced DMR changes suggested that many loci were common. Therefore, it can be concluded that certain loci are more prone to stress-mediated methylation. Studies have confirmed the correlation between promoter hypomethylated changes and expression of certain genes (Stelpflug et al. 2014).

Spatial epigenetic studies in *Arabidopsis* by immunolocalization revealed increase in the level of transcriptionally active marks like H3K9ac and H3K4me3 with increasing occupancy of RNA pol II under drought stress, however, rehydration led to decrease in the level of these marks. H3K4me3 marks were decreased but maintained at low levels on the genome indicating their role in epigenetic stress memory (Kim et al. 2012). Similar results were obtained during abiotic stress studies in *Zea mays*, which revealed a rise in total acetylation levels in the roots of maize seedlings as compared to control seedlings. H3K9 acetylation marks were evenly distributed in the nucleus (Zhao et al. 2014a, b). The elevation in the levels of acetylation positively correlates with increased transcription levels of drought-responsive genes.

Stress-responsive genes include functional, structural and regulatory genes (Osakabe et al. 2014). Stress shows a global effect on regulation of gene expression but stress-responsive genes are affected the most and contribute to stress tolerance (Fig. 3.3). Stress-responsive genes can express as transcription factors, members of signal transduction pathways, transporters etc., playing important roles in stress response. Genes expressing after stress stimuli can be categorized as early responsive (expressing instantly after stress stimuli) and responsive (expressing at any point of time during stress).

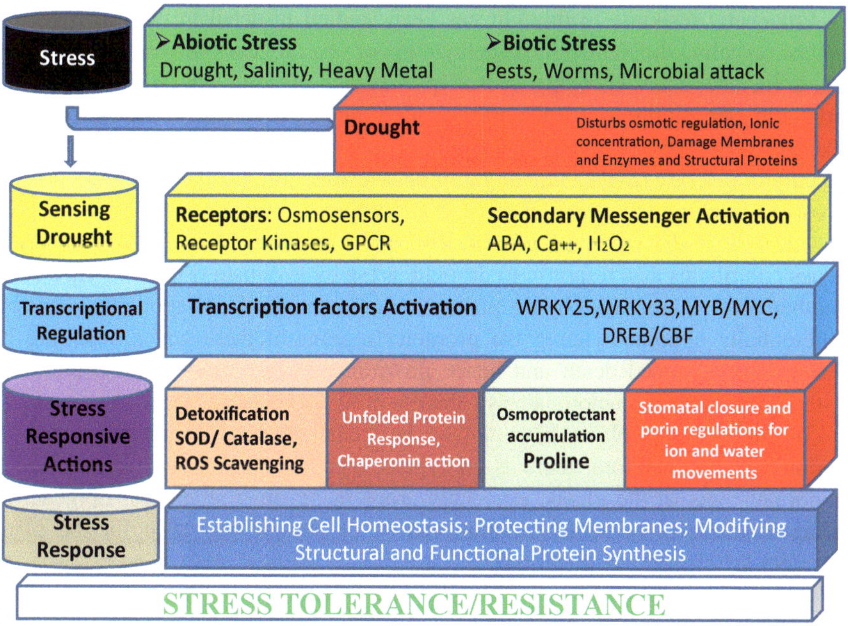

Fig. 3.3 Regulatory pathways imparting stress tolerance to the plants at different levels of stress sensing and stress response

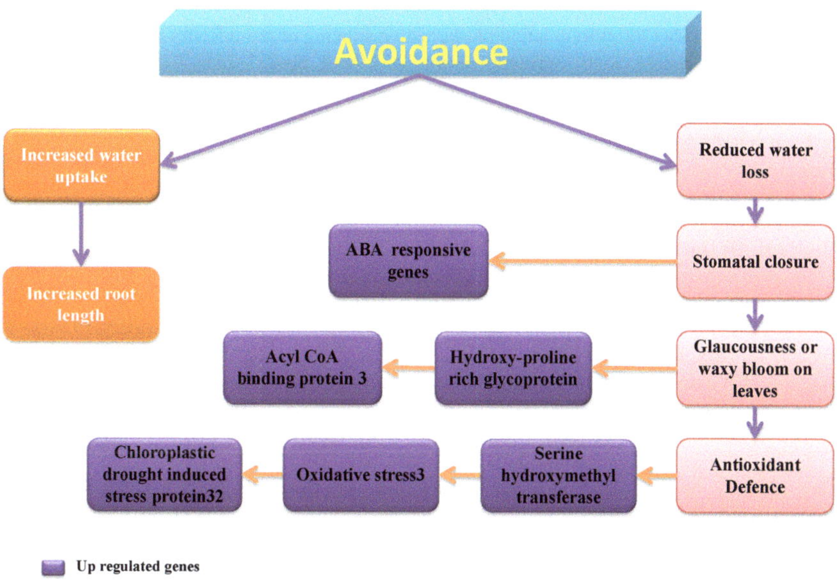

Fig. 3.4 Drought stress response in *Brassica juncea* RH30 studied by MeDIP chip

Epigenetic response to drought has been studied in detail in case of *Brassica juncea* revealing differential hypomethylation of several genes (Kosuke et al. 2017). These include, genes encoding hydroxyproline rich glycoprotein (HPRG) involved in lignin deposition to reduce transpiration rates, Acyl CoA binding protein3 (ACBP3) involved in maintaining cuticle and stomatal closure, Serine hydroxymethyltransferase (SHMT) involved in photorespiration, Oxidative stress3 (OS-3) involved in chromatin remodelling and Chloroplastic drought-induced stress protein 32 (CDSP-32) involved in oxidative stress response. Activation of all these genes contributes as a response to drought stress by avoidism (Fig. 3.4). However, another strategy employed by the plant was upregulation of anti-apoptotic genes by differentially hypomethylating the promoter regions of these genes, resulting in delayed programmed death and hence providing an opportunity to the plant, to withstand such harsh abiotic stress (Sharma et al. 2017).

Since transgenic plants have not so far gained general acceptability due to various issues, therefore, there is an urgent need to improve the agronomically important traits of the plants, especially crops using other strategies. Epigenetics being a source of heritable variation can be of tremendous value for the plant breeders to select novel phenotypes.

Acknowledgements Authors are thankful to DBT, Govt. of India for financial assistance.

References

Ali GS, Reddy ASN (2008) Regulation of alternative splicing of pre-mRNAs by stresses. Nuclear pre-mRNA. Process plants. Springer, Berlin Heidelberg, pp 257–275

Ball MP, Li J, Gao Y, Lee J, Leproust E, Park IH, Xie B, Daley GQ, Church G (2009) Targeted and genome-scale strategies reveal gene-body methylation signatures in human cells. Nat Biotechnol 27:361–368

Bezhani S, Winter C, Hershman S, Wagner JD, Kennedy JF, Kwon CS, Pfluger J, Su Y, Wagner D (2007) Unique, shared, and redundant roles for the *Arabidopsis* SWI/SNF chromatin remodeling ATPases BRAHMA and SPLAYED. Plant Cell 19:403–416

Bhardwaj AR, Joshi G, Kukreja B, Malik V, Arora P, Pandey R, Shukla RN et al (2015) Global insights into high temperature and drought stress regulated genes by RNA-Seq in economically important oilseed crop *Brassica juncea*. BMC Plant Biol 15:1–15

Bock C (2012) Analysing and interpreting DNA methylation data. Nat Rev Genetics 13:705–719

Bonasio R, Tu S, Reinberg D (2010) Molecular signals of epigenetic states. Science 330:612–616

Bourchis D, Voinnet O (2010) A small-RNA perspective on gametogenesis, fertilization, and early zygotic development. Science 330:617–622

Brzeski J, Jerzmanowski A (2003) Deficient in DNA methylation 1 (DDM1) defines a novel family of chromatin-remodeling factors. J Biol Chem 278:823–828

Chen M, Lv S, Meng Y (2010) Epigenetic performers in plants. Dev Growth Differ 52:555–566

Chinnusamy V, Zhu JK (2009) Epigenetic regulation of stress responses in plants. Curr Opin Plant Biol 12:133–139

Chodavarapu RK, Feng S, Bernatavichute YV, Chen PY, Stroud H, Yu Y, Hetzel JA et al (2010) Relationship between nucleosome positioning and DNA methylation. Nature 466:388–392

Cokus SJ, Feng S, Zhang X, Chen Z, Merriman B, Haudenschild CD, Pradhan S et al (2008) Shotgun bisulphite sequencing of the *Arabidopsis* genome reveals DNA methylation patterning. Nature 452:215–219

Cubas P, Vincent C, Coen E (1999) An epigenetic mutation responsible for natural variation in floral symmetry. Nature 401:157–161

Diekman J, Petracek M, Heard JE (2012) Drought tolerance through biotechnology: improving translation from the laboratory to farmers' fields. Curr Opin Biotechnol 23:243–250

Dhar MK, Vishal P, Sharma R, Kaul S (2014) Epigenetic dynamics: role of epimarks and underlying machinery in plants exposed to abiotic stress. Int J Genomics 2014:187146. https://doi.org/10.1155/2014/187146

Dhar PK, Thwin CS, Tun K, Tsumoto Y, Maurer-Stroh S, Eisenhaber F, Surana U (2009) Synthesizing non-natural parts from natural genomic template. J Biol Eng 3(3):2. https://doi.org/10.1186/1754-1611-3-2

Dhawan R, Luo H, Foerster AM, Abu Qamar S, Du HN, Briggs SD, Scheid OM, Mengiste T (2009) HISTONE MONOUBIQUITINATION1 interacts with a subunit of the mediator complex and regulates defense against necrotrophic fungal pathogens in *Arabidopsis*. Plant Cell 21:1000–1019

Dowen RH, Pelizzola M, Schmitz RJ, Lister R, Dowen JM, Nery JR, Dixon JE, Ecker JR (2012) Widespread dynamic DNA methylation in response to biotic stress. Proc Natl Acad Sci USA 109:E2183–E2191

Dyachenko OV, Zakharchenko NS, Shevchuk TV, Bohnert HJ, Cushman JC, Buryanov YI (2006) Effect of hypermethylation of CCWGG sequences in DNA of *Mesembryanthemum crystallinum* plants on their adaptation to salt stress. Biochem (Mosc) 71:461–465

Eichten SR, Briskine R, Song J, Li Q, Swanson-Wagner R, Hermanson PJ, Waters AJ et al (2013) Epigenetic and genetic influences on DNA methylation variation in maize populations. Plant Cell 25:2783–2797

Feng S, Cokus SJ, Zhang X, Chen PY, Bostick M, Goll MG, Hetzel J et al (2010) Conservation and divergence of methylation patterning in plants and animals. Proc Natl Acad Sci 107:8689–8694

Fujimoto R, Sasaki T, Inoue H, Nishio T (2008) Hypomethylation and transcriptional reactivation of retrotransposon-like sequences in ddm1 transgenic plants of *Brassica rapa*. Plant Mol Biol 66:463–473

Gendrel A, Lippman Z, Yordan C, Colot V, Martienssen RA (2002) Dependence of heterochromatic histone H3 methylation patterns on the *Arabidopsis* gene DDM1. Science 297:1871–1873

Gutzat R, Scheid OM (2012) Epigenetic responses to stress: triple defense? Curr Opin Plant Biol 15:568–573

Houben A, Demidov D, Caperta AD, Karimi R, Agueci F, Vlasenko L (2007) Phosphorylation of histone H3 in plants–a dynamic affair. Biochim Biophys Acta 1769(5–6):308–315

Jeggo PA, Holliday R (1986) Azacytidine-induced reactivation of a DNA repair gene in Chinese hamster ovary cells. Mol Cell Biol 6:2944–2949

Ji L, Neumann DA, Schmitz RJ (2015) Crop epigenomics: identifying, unlocking and harnessing cryptic variation in crop genomes. Mol Plant 8(6):860–870

Ji L, Sasaki T, Sun X, Ma P, Lewis ZA, Schmitz RJ (2014) Methylated DNA is over-represented in whole-genome bisulfite sequencing data. Front Genet 5:341

Johannes F, Porcher E, Teixeira FK, Saliba-Colombani V, Simon M, Agier N, Bulski A, Albuisson J et al (2009) Assessing the impact of transgenerational epigenetic variation on complex traits. PLoS Genet 5:e1000530

Johnson TB, Coghill RD (1925) Researches on pyrimidines. C111. The discovery of 5-methyl-cytosine in tuberculinic acid, the nucleic acid of the tubercle Bacillus1. J American Chem Soc 47:2838–2844

Khorasanizadeh S (2004) The nucleosome: from genomic organization to genomic regulation. Cell 116:259–272
Keqiang Wu, Zhang L, Zhou C, Chun-Wei Y, Chaikam V (2008) HDA6 is required for jasmonate response, senescence and flowering in Arabidopsis. J Exp Bot 59(2):225–234
Kim K-C, Lai Z, Fan B, Chen Z (2008) Arabidopsis WRKY38 and WRKY62 transcription factors interact with histone deacetylase 19 in basal defense. Plant Cell 20:2357–2371
Kim DH, Sung S (2012) Environmentaly coordinated epigenetic silencing of FLC by protein and long noncoding RNA components. Curr Opin Plant Biol 15:51–56
Kim JM, To TK, Ishida J, Matsui A, Kimura H, Seki M (2012) Transition of chromatin status during the process of recovery from drought stress in *Arabidopsis thaliana*. Plant Cell Physiol 53:847–856
Kosuke N, Kawagishi Y, Kawabe A, Sato M, Masuta Y, Kato A, Ito H (2017) Epigenetic regulation of a heat-activated retrotransposon in cruciferous vegetables. Epigenomes 1(1):7
Laird PW (2010) Principles and challenges of genome wide DNA methylation analysis. Nat Rev Genet 11:191–203
Lane AK, Niederhuth CE, Ji L, Schmitz RJ (2014) pENCODE: a plant encyclopedia of DNA elements. Annu Rev Genet 48:49–70
Lauria M, Rossi V (2011) Epigenetic control of gene regulation in plants. Biochimica et Biophysica Acta (BBA) 1809:369–378
Law JA, Jacobsen SE (2010) Establishing, maintaining and modifying DNA methylation patterns in plants and animals. Nat Rev Genet 11:204–220
Liang D, Zhang Z, Wu H, Huang C, Shuai P, Ye CY, Tang S et al (2014) Single-base-resolution methylomes of *Populus trichocarpa* reveal the association between DNA methylation and drought stress. BMC Genet 15:S9
Lister R, O'Malley RC, Tonti-Filippini J, Gregory BD, Berry CC, Millar AH, Ecker JR (2008) Highly integrated single-base resolution maps of the epigenome in *Arabidopsis*. Cell 133:523–536
Long Y, Xia W, Li R, Wang J, Shao M, Feng J, King GJ et al (2011) Epigenetic QTL mapping in *Brassica napus*. Genetics 189:1093–1102
Manning K, Tor M, Poole M, Hong Y, Thompson AJ, King GJ, Giovannoni JJ, Seymour GB (2006) A naturally occurring epigenetic mutation in a gene encoding an SBP-box transcription factor inhibits tomato fruit ripening. Nat Genet 38:948–952
Margueron R, Trojer P, Reinberg D (2005) The key to development: interpreting the histone code? Curr Opin Genet Dev 15:163–176
McGhee JD, Felsenfeld G (1980) Nucleosome structure. Ann Rev Biochem 49:1115–1156
Mikkelsen TS, Ku M, Jaffe DB, Issac B, Lieberman E, Giannoukos G, Alvarez P et al (2007) Genome-wide maps of chromatin state in pluripotent and lineage-committed cells. Nature 448:553–560
Misook H, Danny W-KN, Wen-Hsiung L, Chen ZJ (2011) Coordinated histone modifications are associated with gene expression variation within and between species. Genome Res 21(4):590–598
Mlynarova L, Nap JP, Bisseling T (2007) The SWI/SNF chromatin remodelling gene AtCHR12 mediates temporary growth arrest in *Arabidopsis thaliana* upon perceiving environmental stress. Plant J 51:874–885
Narlikar GJ, Sundaramoorthy R, Owen-Hughes T (2013) Mechanisms and functions of ATP-dependent chromatin-remodeling enzymes. Cell 154:490–503
Niederhuth CE, Schmitz RJ (2014) Covering your bases: inheritance of DNA methylation in plant genomes. Mol Plant 7:472–480
Osakabe Y, Osakabe K, Shinozaki K, Tran LSP (2014) Response of plants to water stress. Front Plant Sci 5:86
Parkin IAP, Koh C, Tang H, Robinson SJ, Kagale S, Clarke WE, Town CD et al (2014) Transcriptome and methylome profiling reveals relics of genome dominance in the mesopolyploid *Brassica oleracea*. Genome Biol 15:R77

Pecinka A, Dinh HQ, Baubec T, Rosa M, Lettner N, Scheid OM (2010) Epigenetic regulation of repetitive elements is attenuated by prolonged heat stress in *Arabidopsis*. Plant Cell 22:3118–3129

Rando OJ (2012) Combinatorial complexity in chromatin structure and function: revisiting the histone code. Curr Opin Genet Dev 22:148–155

Reyna-Lopez GE, Simpson J, Ruiz-Herrera J (1997) Differences in DNA methylation patterns are detectable during the dimorphic transition of fungi by amplification of restriction polymorphisms. Mol Gen Genet 253:703–710

Richards EJ (2006) Inherited epigenetic variation-revisiting soft inheritance. Nat Rev Genet 7:395–401

Robinson MD, Kahraman A, Law CW, Lindsay H, Nowicka M, Weber LM, Zhou X (2014) Statistical methods for detecting differentially methylated loci and regions. Front Genet 5:324

Roudier F, Ahmed I, Sarazin A, Mary-Huard T, Cortijo S, Bouyer D et al (2011) Integrative epigenomic mapping defines four main chromatin states in *Arabidopsis*. EMBOJ 30(10):1928–1938

Saze H (2008) Epigenetic memory transmission through mitosis and meiosis in plants. Semin Cell Dev Biol 19:527–536

Saze H, Tsugane K, Kanno T, Nishimura T (2012) DNA methylation in plants: relationship to small RNAs and histone modifications, and functions in transposon inactivation. Plant Cell Physiol 53:766–784

Schmitz RJ, He Y, Valdes-Lopez O, Khan SM, Joshi T, Urich MA, Nery JR et al (2013) Epigenome-wide inheritance of cytosine methylation variants in a recombinant inbred population. Genome Res 23:1663–1674

Schmitz RJ, Schultz MD, Lewsey MG, O'Malley RC, Urich MA, Libiger O, Schork NJ et al (2011) Transgenerational epigenetic instability is a source of novel methylation variants. Science 334:369–373

Schmitz RJ, Zhang X (2011) High-throughput approaches for plant epigenomic studies. Curr Opin Plant Biol 14:130–136

Sharma R, Vishal P, Kaul S, Dhar MK (2017) Epiallelic changes in known stress-responsive genes under extreme drought conditions in *Brassica juncea* (L.) Czern. Plant Cell Rep 36(1):203–217

Slaughter A, Daniel X, Flors V, Luna E, Hohn B, Mauch-Mani B (2012) Descendants of primed *Arabidopsis* plants exhibit resistance to biotic stress. Plant Physiol 158:835–843

Slotkin RK, Vaughn M, Borges F, Tanurdžić M, Becker JD, Feijó JA, Martienssen RA (2009) Epigenetic reprogramming and small RNA silencing of transposable elements in pollen. Cell 136:461–472

Sokol A, Kwiatkowska A, Jerzmanowski A, Prymakowska- Bosak M (2007) Up-regulation of stress-inducible genes in tobacco and *Arabidopsis* cells in response to abiotic stresses and ABA treatment correlates with dynamic changes in histone H3 and H4 modifications. Planta 227:245–254

Springer NM (2013) Epigenetics and crop improvement. Trends Genet 29:241–247

Stelpflug SC, Eichten SR, Hermanson PJ, Springer NM, Kaeppler SM (2014) Consistent and heritable alterations of DNA methylation are induced by tissue culture in maize. Genetics 198:209–218

Steward N, Ito M, Yamaguchi Y, Koizumi N, Sano H (2002) Periodic DNA methylation in maize nucleosomes and demethylation by environmental stress. J Biol Chem 277:37741–37746

Stroud H, Do T, Du J, Zhong X, Feng S, Johnson L, Patel DJ et al (2014) Non-CG methylation patterns shape the epigenetic landscape in *Arabidopsis*. Nat Struct Mol Biol 21:64–72

Suji KK, Joel AJ (2010) An epigenetic change in rice cultivars under water stress conditions. Elect J Plant Breed 1:1142–1143

Suzuki MM, Bird A (2008) DNA methylation landscapes: provocative insights from epigenomics. Nat Rev Genet 9:465–476

Terooatea TW, Pozner A, Buck-Koehntop BA (2016) PAtCh-Cap:input strategy for improving analysis of ChIP-exo data sets and beyond. Nucleic Acids Res 44(21):e159

Tran RK, Zilberman D, de Bustos C, Ditt RF, Henikoff JG, Lindroth AM, Delrow J, Boyle T, Kwong S, Bryson TD, Jacobsen SE, Henikoff S (2005) Chromatin and siRNA pathways cooperate to maintain DNA methylation of small transposable elements in Arabidopsis. Genome Biol 6:R90

Tsaftaris AS, Dickinson AN (2000) DNA methylation and plant breeding. Plant Breed Rev 18:87–176

Varshney RK, Bansal KC, Aggarwal PK, Datta SK, Craufurd PQ (2011) Agricultural biotechnology for crop improvement in a variable climate: hope or hype? Trends Plant Sci 16:363–371

Volpe TA, Kidner C, Hall IM, Teng G, Grewal SIS, Martienssen A (2002) Regulation of heterochromatic silencing and histone H3 lysine-9 methylation by RNAi. Science 297:1833–1837

Waddington CH (1953) Epigenetics and evolution. Symp Soc Exp Biol 7:186–199

Walley JW, Rowe HC, Xiao Y, Chehab EW, Kliebenstein DJ, Wagner D, Dehesh K (2008) The chromatin remodeler SPLAYED regulates specific stress signaling pathways. PLoS Pathog 4: e1000237–e1000237

Wang W, Zhao X, Pan Y, Zhu L, Fu B, Li Z (2011) DNA methylation changes detected by methylation sensitive amplified polymorphism in two contrasting rice genotypes under salt stress. J Genet Genomics 38:419–424

Wang X, Elling AA, Li X, Li N, Peng Z, He G, Sun H et al (2009) Genome-wide and organ-specific landscapes of epigenetic modifiations and their relationships to mRNA and small RNA transcriptomes in maize. Plant Cell 21:1053–1069

West PT, Li Q, Ji L, Eichten SR, Song J, Vaughn MW, Schmitz RJ et al (2014) Genomic distribution of H3K9me2 and DNA methylation in a maize genome. PLoS ONE 9:e105267

Wu K, Zhang L, Zhou C, Yu CW, Chaikam V (2008) HDA6 is required for jasmonate response, senescence and flowering in *Arabidopsis*. J Exper Bot 59:225–234

Yang C, Zhang M, Niu W, Yang R, Zhang Y, Qiu Z, Sun B et al (2011) Analysis of DNA methylation in various swine tissues. PLoS ONE 6:e16229

Zentner GE, Henikoff S (2013) Regulation of nucleosome dynamics by histone modifications. Nat Struct Mol Biol 20:259–266

Zhang X, Clarenz O, Cokus S, Bernatavichute YV, Pellegrini M, Goodrich J, Jacobsen SE (2007a) Whole-genome analysis of histone H3 lysine 27 trimethylation in *Arabidopsis*. PLoSBiol 5: e129

Zhang X, Henderson IR, Lu C, Green PJ, Jacobsen SE (2007b) Role of RNA polymerase IV in plant small RNA metabolism. Proc Natl Acad Sci 104:4536–4541

Zhang X, Yazaki J, Sundaresan A, Cokus S, Chan SWL, Chen H, Henderson IR et al (2006) Genome-wide high-resolution mapping and functional analysis of DNA methylation in *Arabidopsis*. Cell 126:1189–1201

Zhao L, Wang P, Hou H, Zhang H, Wang Y, Yan S, Huang Y et al (2014a) Transcriptional regulation of cell cycle genes in response to abiotic stresses correlates with dynamic changes in histone modifications in maize. PLoS ONE 9:e106070

Zhao L, Wang P, Yan S, Gao F, Li H, Hou H, Zhang Q et al (2014b) Promoter-associated histone acetylation is involved in the osmotic stress-induced transcriptional regulation of the maize ZmDREB2A gene. Physiol Plant 151:459–467

Zheng X, Pontes O, Zhu J, Miki D, Zhang F, Li WX, Lida K et al (2008) T ROS3 is an RNA-binding protein required for DNA demethylation in *Arabidopsis*. Nature 455:1259–1262

Zhu JK (2009) Active DNA demethylation mediated by DNA glycosylases. Ann Rev Genet 43:143

Zilberman D, Gehring M, Tran RK, Ballinger T, Henikoff S (2007) Genome-wide analysis of *Arabidopsis thaliana* DNA methylation uncovers an interdependence between methylation and transcription. Nat Genet 39:61–69

Chapter 4
Effect of Drought Stress and Utility of Transcriptomics in Identification of Drought Tolerance Mechanisms in Maize

Nidhi Singh, Shikha Mittal and Nepolean Thirunavukkarasu

Abstract Maize crop encounters a number of abiotic and biotic stresses which reduce the production and the productivity. Abiotic stresses such as drought are unpredicted environmental disturbances during the crop growth which often lead to reduced crop yield or complete crop loss in some cases. Drought occurring at flowering leads to greater yield losses than when it occurs at other developmental stages. Plant responses at various levels such as morphological, physiological, biochemical and molecular changes to cope up with the stress. It is very important to understand the genes involved in drought tolerance as well and their interactions to breed tolerant hybrids in maize. Transcriptome profiling is useful to understand the whole spectrum of genes expressed under drought condition. The assay will be useful to decipher the genes involved in specific pathways and with the help of in silico analyses, interactions of target genes can be studied. Several transcriptome studies have been carried out in maize in different stages and in tissues under drought stress. Genes involved in detoxification, stomatal regulation, photosynthesis, hormone signaling, root architecture and sugar metabolism pathways are considered as important to achieve drought tolerance. The genes identified through gene expression assays could be used as candidate genes in selection programmes to develop drought tolerant hybrids in maize.

Keywords Drought · Candidate genes · Plant responses · Transcriptomes
Tolerant mechanisms

N. Singh · S. Mittal · N. Thirunavukkarasu (✉)
Division of Genetics, ICAR-Indian Agricultural Research Institute,
New Delhi 110012, India
e-mail: tnepolean@gmail.com; tnepolean@yahoo.com

V. R. Rajpal et al. (eds.), *Genetic Enhancement of Crops for Tolerance to Abiotic Stress: Mechanisms and Approaches, Vol. I*, Sustainable Development and Biodiversity 20, https://doi.org/10.1007/978-3-319-91956-0_4

4.1 Introduction

The changing trends in environmental temperature, precipitation and sea levels are adversely affecting the crops' production worldwide. Though various biotic and abiotic stresses affect agricultural crops; drought, cold, flood and heat have been the most devastating leading to huge yield losses. Drought is an important stress that relentlessly affects the agricultural output worldwide, especially in arid and semi-arid regions (Farooq et al. 2012). Drought is a climatic glitch, represented by deprived moisture as a consequence of sub-normal rainfall, unreliable rainfall circulation, higher water need or an array of all three factors.

Present challenge for researchers is to overcome the problem of water scarcity. It is the biggest threat our agriculture is now tackling, which needs continuous efforts of scientists. It is estimated that more than one third of the arable land of the world is facing the problem of water scarcity. Other types of abiotic stresses often accompany drought stress, thus making it more complex to study (Barnabás et al. 2008; Farooq et al. 2009; Zlatev and Lidon 2012).

Though plants can tolerate drought up to some extent, however, the degree of tolerance varies from species to species (Rampino et al. 2006). They cope up with drought by adopting any of the three strategies; drought escape, drought endurance and drought avoidance to complete their life cycle. Different levels of complex interactions among stress factors and integration of morphological, physiological, biochemical and molecular processes influence plant developmental stages (Farooq et al. 2009; Zlatev and Lidon 2012). To understand the mechanism of drought tolerance, it is important to understand the changes in plants that occur in response to drought stress. The primary responses in plants against drought include reduced leaf water potential and turgor loss, stomatal closure, cessation of cell enlargement and growth, and reduction in water content (Farooq et al. 2009). The changes in gaseous exchange occur as a result in reduction in photosynthetic process and organic solute synthesis. This ultimately affects photosynthesis, respiration, translocation, ion uptake, growth factors, carbohydrate and nutrient metabolism, plant growth and cell elongation. Further, hyped intensity results in photosynthetic arrest, metabolic imbalance and eventually the death of the plant (Farooq et al. 2009; Jaleel et al. 2008). Drought also modifies CO_2 conductance and thus adds to photosynthetic imbalance by histological and leaf anatomical changes. In the following sections, we have reviewed the morphological, physiological and biochemical responses of plant to drought stress followed by responses at molecular and transcriptome levels.

4.2 Morphological Response

Plant responses vary with growth stage, exposure period, stress intensity and level of tolerance (Jaleel et al. 2008). In the subsequent subsections, the morphological changes adapted by plants in response to drought as well as to withstand drought stress conditions have been discussed.

4.2.1 Plant Growth

Growth is defined as an irreversible proliferation in plant mass resulting from both cell division (especially in meristems) and cell extension. As a result of complexity in cell growth and differentiation under drought stress, the morphological transition occurs that leads to poor growth in plants. The reduced growth is considered as an adaptive modification in plants to help them avoid energy loss under unfavorable conditions. Hydrostatic pressure is very essential for cell growth and expansion. This is the reason that cell expansion is very sensitive to water stress, which directly or physically reduces growth as a result of low hydrostatic pressure. Weak photosynthetic activity affects the plant growth, which in turn is controlled by water supply. Therefore, plant suffers a reduction in photosynthesis under poor water supply. Limitation posed in by reduced photosynthesis components results in reduced growth of the plant to conserve the stored energy.

A. **Effect on vegetative growth**: The early phase is one of the most susceptible phases in the life cycle of plants under limited moisture conditions as drought affects both elongation and expansion of cells due to low hydrostatic pressure (Kusaka et al. 2005; Shao et al. 2008). In maize, for example, elongation of stem gets reduced under drought stress during vegetative stage. The water stress condition also affects the rate of tiller appearance that in turn reduces the plant grain yield. Limited supply of moisture reduces leaf expansion rate. Constricted moisture during vegetative growth shrinks the leaf area of the plant considerably and therefore carbon usage gets reduced throughout the growing season. Denmead and Shaw (1960) reported that extended drought during vegetative stage affects the length of the internodes by affecting cell size development and assimilate storage.
B. **Effect on reproductive growth**: Flowering, silking, pollination and grain formation are the important stages of plant development. Among cereal grasses, maize is most sensitive crop to drought stress at flowering stage. The flowering interval in maize is very short and pollen remains viable for a very short time period. It has been reported that per day delay between pollen shed and silk emergence reduces sexual fertilization and increases bareness and yield loss (Sangoi and Salvador 1997). The delicate period lies from one week before silking to two weeks after silking with probable chances of ovules, kernels and ears abortion (Uhart and Andrade 1995). There is a delayed silking under moisture stress so pollen is shed much before the stigmas are formed (Herrero and Johnson 1981). In maize, the anthesis-silking interval (ASI) increases in response to the drought condition. Extended dry conditions reduce ear growth and silk appearance thus escalating ASI. Increased ASI is thought to be a cause of yield loss as it is highly correlated with kernel set (Byrne et al. 1995).

4.2.2 *Root*

One of the important components of water potential is the matrix potential, which is defined as the energy required by the plant to extract soil water. In low moisture conditions, this force is greatly enhanced and shows a high matrix potential while in dry conditions, it increases further which ultimately results in stress of plant. Another difficulty faced under drought stress is accumulation of solutes in the interior cells of the roots, which leads to reverse cell osmosis. The probable outcome of reverse osmosis is membrane collapse and finally the death of the root cells, which adversely affects the water intake capacity of the plant. Roots being the first to sense the drought conditions are highly influenced by drought than any other aerial part of the plant. Early stages of plant development are highly controlled by a well-developed shallow root system (Johansen et al. 1994). An increase in fresh weight of the roots under drought stress has been reported probably due to better water utilization than shoots. The best symptom for morphological identification of drought tolerant crop is maximum root fresh weight. In many experiments, the reduction of shoot to root ratio as a result of dehydration stress is very well documented. Under water limited conditions, there is a high root to shoot ratio (Wu and Cosgrove 2000) due to better availability of food assimilates to roots. In maize, drought at seedling stage increases the root growth and thus adapting plant to drought stress by making the apical cell walls of the root expansible. Sacks et al. (1997) reported that meristematic cells elongate with reduced cell division per unit length of tissues and cell under drought stress condition.

4.2.3 *Leaf Area*

Leaf area is a typical trait for plant photosynthesis and transpiration. Photosynthesis along with cell-growth are among the primary processes affected by drought stress (Chaves and Chaves 1991). These processes help plant to attain optimum leaf area for photosynthesis and dry matter establishment. Drought considerably reduces the number of leaves per plant, leaf size and longevity. Restricted photosynthetic area may suppress the leaf expansion due to reduced leaf region (Rucker et al. 1995).

4.2.4 *Fresh and Dry Mass*

Unpleasant drought conditions may slow the rate of fresh and dry biomass formation (Farooq et al. 2009). Plant yield under drought stress is strongly associated with the processes of dry mass partitioning and biomass distribution (Kage et al. 2004). Process of dry mass accumulation is affected by the water stress at different stages of plant growth. The allocation of dry matter between root and shoot and further partitioning of above ground dry matter into vegetative and reproductive organ are vital for crop yield under stress condition.

4.2.5 *Yield*

The stage and the duration of stress drastically influences the grain yield. Stress during the early vegetative state has little impact on yield reduction while greatest harm is done when drought stress continues until post vegetative or the reproductive stage of plant growth. Rolling of leaf is most immediate response of plant to drought stress condition at the early vegetative growth. Leaf rolling reduces the rate of photosynthesis hence negatively influencing the yield. If stress progresses to the reproductive stage of the plant, it affects the silk extension and ultimately viability of the pollen grains. If the stress continues further to post-flowering stages, yield is reduced due to reduction in kernel rows and kernel numbers. Another factor affecting the grain yield is evapotranspiration. Evapotranspiration is the loss of water from soil by evaporation as well as loss of water through transpiration. This inadequate availability of water affects the nutrient availability, uptake and transport. In maize, most sensitive stage affecting crop yield is the three-week period of silking, and drought stress at this stage results in kernel abortion, and further continuation of drought stress reduces the seed size.

4.3 Physiological Responses

4.3.1 *Photosynthesis*

Photosynthesis is one of the main physiological responses of plant negatively affected by drought stress. Drought badly affects photosystem-II than photosystem-I. Photosynthetic rate gets adversely influenced by limited CO_2 supply and metabolic processes under stress. Leaf potential becomes low under water stress and in response to reduced leaf turgor stoma closes. Enzymatic activities slow down under drought stress due to diminished supply of CO_2 to RUBisco that dissipates the energy in photosynthetic apparatus causing down regulation of photosynthesis. Photosynthesis promptly depends on relative water content and leaf potential both of which at low concentration slows the rate of photosynthesis. The major effect of drought is decreased CO_2 availability through limited diffusion through stomata and mesophyll (Flexas et al. 2004, 2007). This decrease in mesophyll conductance is linked to physical interaction or alterations in the structure of the intercellular spaces due to leaf shrinkage (Lawlor and Cornic 2002) or to alterations in the biochemistry (bicarbonate to CO_2 conversion) and/or membrane permeability (aquaporins). This pattern of metabolic changes supports the assertion by (Cornic 2000) that stomatal closing is the principle cause of decrease in photosynthetic rate under mild drought.

4.3.2 Respiration

During respiration process, plants catabolize food for ATP production and other useful metabolites. Enormous research has been conducted with relevance to photosynthesis but very less work has been done to find the effect of stress on respiration. Under drought stress, some studies reported a significant reduction in respiration rate, some showed no changes at all while some reports concluded to have increased respiratory rate under water stress condition. Hence, a unanimous conclusion has not been reached.

4.3.3 Transpiration

Transpiration is a process of evaporation of water from the aerial parts of the plants. It occurs largely when the stomata remains open for gaseous exchange. Thus, the degree of stomatal opening regulates the rate of transpiration. Other factors affecting rate of transpiration are linked to hydration level, humidity, temperature, leaf number and leaf moisture. Roots withdraw water from the soil and draw it up to stomatal openings. As water moves all the way through the system, vital nutrients are transported to different areas of the plant. The stoma releases waste products such as oxygen into the environment and brings in carbon dioxide. In addition, transpiration maintains turgor in plants leading to maintenance of water in cells. Drought often limits the growth of root and shoots, which makes the plant stunted under plant stress. Reduction in growth is followed by complete or partial stomatal closure resulting in reductions in transpiration and CO_2 uptake for photosynthesis. Therefore, stomatal closure under severe drought condition influences the photosynthesis as well as transpiration rate. The water loss by a plant depends on plant dimensions and the quantity of water absorbed in the roots. Transpiration cannot persist if its water uptake efficiency is not in equilibrium with soil water. When roots are unsuccessful in absorbing water to keep up with the rate of transpiration, turgor pressure drops and due to reduction in turgor, stomata close to minimize further water loss. If the loss in hydrostatic pressure stretched through the plant, the plant wilts and dies from lack of nutrients.

4.3.4 Pigments

Photosynthetic pigments are present in chloroplasts and are mainly involved in the process of photosynthesis by trapping sunlight and reducing power production in plants. Soil dryness mainly affects chlorophyll 'a' and 'b' activity (Farooq et al. 2009) whilst carotenoids still help plants to survive under drought condition. Ratio of chlorophyll 'a' and 'b' to carotenoids changes in response to drought stress

(Anjum et al. 2003; Farooq et al. 2009). Drought induced photosynthesis limitation has been reported in many studies (Anjum et al. 2003; Lawson et al. 2003) because of stomatal and non-stomatal limitations (Farooq et al. 2009). Carotenoids act as antioxidant defense system that helps to overcome the oxidative damage generated by increased drought stress. β-carotene of all green plants is absolutely bound to the core complexes of PS-I and PS-II. It plays a unique role in protecting photochemical processes and sustaining them (Havaux 1998). Drought has the ability to decrease the concentration of chlorophylls and carotenoids (Havaux 1998; Poormohammad Kiani et al. 2008), mainly with the generation of reactive oxygen species (ROS) in the thylakoids (Ramachandra Reddy et al. 2004).

4.4 Biochemical Responses

4.4.1 ROS and Antioxidative Enzymes

The production of ROS is one of the earliest responses in any type of abiotic stress. Decreased metabolic machinery has been known to trigger the accumulation of free radicals under desiccation. A drop in rate of photosynthesis and limited CO_2 fixation give rise to a number of ROS such as H_2O_2, O_2 and OH^-. These ROS are essential when present in minimal amount but can become deterrent when present in large amounts causing oxidative damage to the plants under water stress (Arora et al. 2002). Many studies on maize have reported increased ROS under drought stress condition. Photorespiration being a wasteful process is the main source of ROS accumulation accounting for approximately 70% of the total hydrogen peroxide production. To minimize the ROS level and fight the oxidative stress caused by them, plants express antioxidative enzymes to strengthen their antioxidative defense system. The antioxidant defense system is comprised of various enzymes such as catalase, superoxide dismutase, ascorbate peroxidase, peroxidase and helps the plant to eradicate excess ROS and minimize the damage caused by them (Li et al. 2013). The equilibrium between ROS production and antioxidative defense system decides the stress responsive pathway of the plant and thus the ability of antioxidative defense system of the plant is directly correlated with the drought resistance of the plant (Anjum et al. 2011). Chugh et al. (2011) reported increased activities of catalase, peroxidase and ascorbate peroxidase in a drought tolerant variety of maize. Polyethylene glycol (PEG)-induced water stress is thought to be relieved by increased ROS, (abscisic acid) ABA accumulation and antioxidative enzymatic activity. In plant cells, different mechanisms are available to prevent the production of toxic molecules but oxidative damage remains an expected problem as it causes perturbations in metabolism (Ramachandra Reddy et al. 2004). In maize, glutathione reductase (GR) and dehydroascorbate reductase (DHAR) were solely located in mesophyll cells whereas most of the superoxide dismutase (SOD) and ascorbate peroxidase (APX) were located in mesophyll and bundle sheath cells. Kingston-Smith and Foyer (2000) suggested that the oxidative damage under stressful conditions in C4 plants remain confined to bundle sheath cells because of inadequate antioxidant protection in this tissue.

4.4.2 Lipid Peroxidation

Lipids are chief components of the membrane system and thus maintain the integrity of the system. Increased ROS production in water stress condition damages the membrane integrity of the cell by the lipid oxidation (Liljenberg 1992). The concentration of malondialdehyde (MDA) content increases which is responsible for damages in the membrane system by altering its fluidity, protein cross-linking, transport etc. (Sharma et al. 2012). Ge et al. (2006) conducted a systematic study to know the effect of drought on antioxidative and lipid peroxidation system of maize plant. He found a significant increase in ROS scavenging enzymes with increase in stress severity along with increased MDA content. Increased MDA, indicator of lipid peroxidation, was also reported by Yin et al. (2012) in two different types of maize plant along with other biochemical changes. The alteration in membrane lipids has become a major biomarker of plant under the stress condition.

4.4.3 Osmolytes Accumulation

Regulating water potential in water stressed condition can be a rescue mechanism for plants facing stress. Presence of water ion/channel proteins and osmolytes has been reported to regulate the osmotic adjustments under drought stress (Ingram and Bartels 1996). Osmolyte accumulations result in reduction of osmotic potential and thus maintain cell turgor pressure and water uptaking capacity to sustain the plant's physiological processes. In support of this, accumulation of sugars such as raffinose family oligosaccharides (RFO), fructose and trehalose have been reported in drought stressed condition (Wanek and Richter 1997). Trehalose, a non-reducing saccharide when present in definite amount acts as a stabilizer of protein and cell membranes (Paul et al. 2008). Proline has been considered as one of the most important osmolytes that accumulates in plants in response to different environmental stresses including water stress. An investigation on importance of osmolytes accumulation under drought stress concluded that osmolytes are beneficial for plant when occur in root tips as they allow deeper root development and increased access to the water deep inside the soil (Serraj and Sinclair 2002).

4.4.4 Carbohydrates Biosynthesis

Association of soluble sugars with drought tolerance is highly reported. Alteration in carbohydrate content is particularly important because of their proximal association with plant's physiological processes. Sugar accumulation upon drought exposure results in osmoregulation as well as induction of sugar-related signaling pathways including mitogen-activated protein kinase (MAPK), $Ca2^{+}$ and calmodulins in plants (Kaur et al. 2007). Trehalose is the minimally needed simplest sugar which acts as an osmoprotectant while other soluble sugars, chiefly sucrose, have shown to increase in drought stressed condition. Sucrose being a compatible solute acts as an osmolyte and maintains plant's water potential. Sugar accumulation is also important in maintaining other processes. Phosphofructokinase, an important enzyme for glycolytic pathway usually degrades in dehydration condition. *In vitro* studies have shown the involvement of sucrose, maltose and trehalose in enzymatic stabilization under dehydration (Carpenter et al. 1987).

4.5 Molecular Responses

4.5.1 Transcriptional Factors (TFs)

Under abiotic stress, plants often influence the expression of numerous transcriptional regulators (TFs) which in turn up-regulate an array of downstream genes for survival and stress adaptation. Several families of TFs and *cis*-elements have shown to play significant roles in promoter region of stress-related genes and thus control the expression or suppression of these genes. So far, at molecular level, studies focused on identifying plant response to the drought stress condition involving initiation of stress-responsive and stress tolerating genes. ABA stimulation in plant controls stomatal closure to regulate transpiration and stress responsive transcriptional factors under drought conditions (Cutler et al. 2010). Till date, more than 7% of the coding sequences regulating plant responses to environment have been explained (Udvardi et al. 2007). Probably these TFs are thought to regulate the plant late phase under dehydration stress while some may regulate other drought responsive signaling pathways for activating drought responsive genes for tolerance (Kilian et al. 2012).

Nuclear factor Y is a ubiquitous ABA-dependent TF that has been reported to be strongly expressed under drought in maize crop at both transcriptional and post transcriptional level (Nelson et al. 2007; Li et al. 2008a, b). In maize, TF ZmNF-YB2 is shown to have an equal role as AtNF-YB1 in *Arabidopsis* in conferring improved performance under drought conditions (Nelson et al. 2007). A TF belonging to Abscisic acid Stress Ripening protein (ASR) family, ZmASR-1 protein influences branched chain amino-acid biosynthesis and maintains kernel yield in maize under water deficit conditions (Virlouvet et al. 2011). An another group of TF family, bZIP plays a vital role in ABA signaling along with other functions in plant growth and abiotic stresses. *ZmbZIP72*, a bZIP transcription factor gene in maize was found to be over-expressed in various organs by drought, salinity and ABA in seedling stages. Similarly, AP2, ERF, dehydration-responsive element-binding protein (DREB), Cys2His2 Zinc Finger (C_2H_2 ZF) TFs, MYB, bHLH are important plant stress-responsive TFs which have been shown to express or hold an important role in plant stress tolerance mechanism (Ying et al. 2012).

4.5.2 Hormonal Regulation and Signaling

Phytohormones regulate the very aspect of plant growth and development and enable plants to cope with various environmental conditions. They initiate specific signaling pathways to induce responsive gene expressions in stress condition. ABA is the key phytohormone governing plant responses in drought and other abiotic stress conditions. Importance of other phytohormones such as, cytokinins, brassinosteroids, auxins, jasmonate etc., in abiotic stress tolerance is also discovered.

ABA accumulation is very rapid in any stress condition and triggers downstream stress-responsive signaling that helps the plant to survive the stressed condition. Most of the TFs work in an ABA-dependent manner while studies suggested the presence of both ABA-dependent and ABA-independent regulatory systems (Shinozaki and Yamaguchi-Shinozaki 1996). In drought stress condition, ABA accumulation in the shoot induces stomatal closure to reduce water loss from the plant. Equilibrium between ABA biosynthesis and ABA catabolism is critical for plant survival.

Cytokinins, known for their role in cell division, growth and differentiation, decrease under drought stress, which makes shoots more responsive to ABA and ultimately resulting in stomatal closure (Goicoechea et al. 1997). Though little research has been done on the role of auxins in drought condition but a drop in indole-3-acetic acid (IAA) content under drought stress and changes in other genes of IAA biosynthesis pathway and signaling in rice implied its role in drought condition (Du et al. 2013). IAA functions antagonistic to ethylene in ABA regulation and so shut down the ethylene-initiated ABA signaling in plants (Sakamoto et al. 2008). Under drought stress, low level of auxin and increased production of ABA appears to provide drought tolerance in plants.

Salicylic acid is a hormone-like substance, which is important in improving drought tolerance ability in plants. Okuma et al. (2014) investigated salicylic acid accumulating *Arabidopsis* mutant and confirmed that these mutants were more tolerant to drought stress than the wild type by inhibiting light-induced stomatal opening. Jasmonic acid (JA) is also a signaling molecule affecting plants response at molecular level. It imparts drought tolerance by lowering oxidative stress and by enhancing expression of antioxidative enzymes. JA and ABA cross talks in signaling pathways and their interaction helps to regulate the plant signaling cascades in drought conditions.

4.6 Transcriptomes

Nearly, every cell of every organism is composed entirely of the same genome and has same set of genes. Thus, disparity in response of plant in different environmental conditions is entirely because of the differential expression of genes in different stages of cell development. The transcriptome consists of all RNA, including, rRNA, tRNA, mRNA, and non-coding RNAs expressed in one or a population of cells at a given moment. Decoding different transcriptomes associated with different cells at different times gives a more clear view and deeper insights into specific responses of cells. With the comparative analysis of transcripts of an organism in a particular condition, researchers can determine when genes express or switch off.

4.6.1 Role of Transcriptome in Maize for Drought Stress Tolerance

Accessibility of transcriptome and whole-genome sequences in public databases and with the upgradation of bioinformatics tools, detection of genetic variation in genotypes and within genotypes has become easier and more cost-effective. Maize (*Zea mays* spp. *mays* L.) is very sensitive to water constraints, particularly during flowering, pollination and embryo development. Therefore, it is important to locate candidate genes and unravel molecular mechanisms in response to drought in maize to accelerate its genetic improvement through marker-assisted selection. A general idea of identification and exploitation of gene for crop improvement has been explained in Fig. 4.1.

The progress in transcriptome analysis techniques, sequencing and bioinformatics, the genetic basis of drought tolerance in maize has been further improved. Gene expression studies in maize in response to water stress have been investigated in roots (Poroyko et al. 2007), seedlings (Zheng et al. 2004), and developing ear and tassel (Zhuang et al. 2007). Different types of transcriptomic techniques are now available such as array-based, whole-genome-based and candidate-based to understand the gene expression.

4.6.1.1 Array-Based Transcriptome

In mid 1970s, the base for the development of the novel techniques of microarray was formulated when it became possible to monitor the level of expression of nucleic acid by fluorescent labeling. Microarray technology exploits the basic fundamental characteristic of nucleic acid to anneal with its complementary nucleic acid sequence by hydrogen bonds formation. In this technique, spotted samples (cDNA, DNA and oligosaccharides) with known identities are arrested on a solid

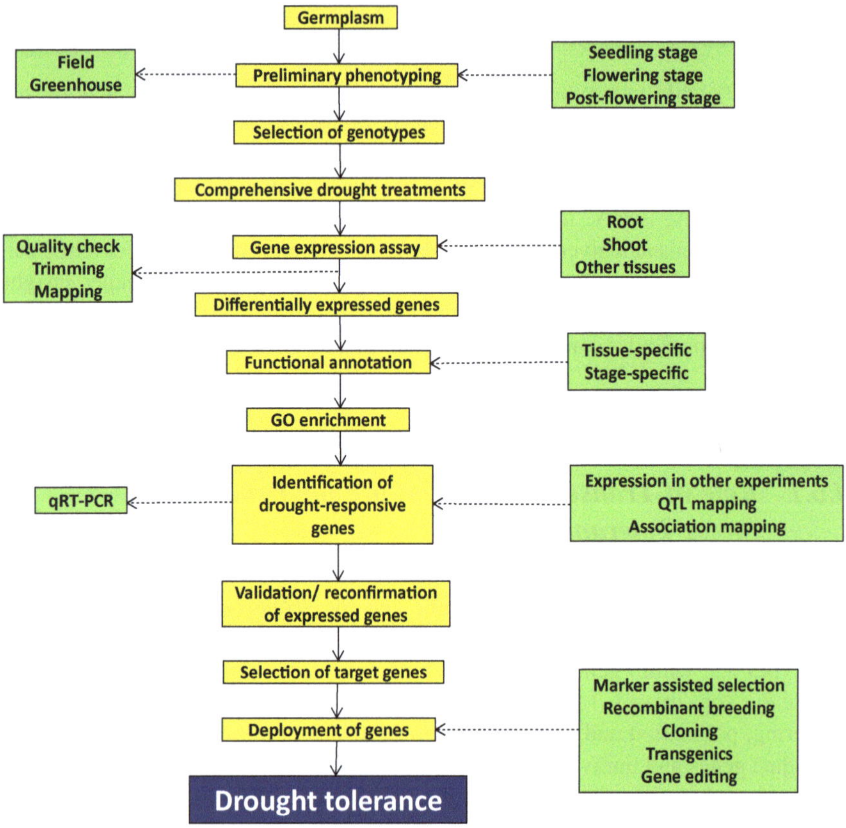

Fig. 4.1 Identification of drought tolerant genes through transcriptome approach

support like glass, silicon, and/or nylon membranes. Each spot represents a single gene, and thus a parallel gene expression for thousands of genes becomes possible at the same time.

Microarray has been successfully employed to maize crop under a range of abiotic stresses for locating potential candidate genes. A cost effective oligonucleotide microarray was developed for the maize community for gene expression analysis in maize. It consists of a total of 5,745,270 mer oligonucleotides representing 25,969 ESTs assemblies, 20,206 singleton ESTs (detected only in a single cDNA library), 9,707 assembled maize sequences, 804 non-redundant repeat elements, 467 organelle sequences, 288 maize community favorites and 11 transgenes. Replicated baseline expression profiles have been generated for 18 tissues and deposited in a database (www.maizearray.org). Advanced and commercial alternative to the public 70-mer array was developed by affymetrix known as the GeneChip Maize Genome Array. This array contains 17,555 probe sets, spanning 14,850 maize transcripts representing 13,339 maize genes. These arrays have 25-mer probes.

A recent advance includes whole genome transcript profiling with a 100 K Maize Affymetrix Gene Chip Array, which contains 100,000 probe sets to detect transcripts from *Zea mays* (Xu et al. 2009). Using microarray chip experiments, gene expression profile under drought stress have been studied in different maize parts including roots, leaves and kernels (Zheng et al. 2004; Hayano-Kanashiro et al. 2009; Marino et al. 2009; Luo et al. 2010; Humbert et al. 2013).

4.6.1.2 Whole Genome Transcriptome

Though microarray studies are relatively inexpensive and the data can be easily generated and analyzed but the detection is limited only to the sequences and homologues on the array. Next generation sequencing (NGS) of RNA (known as RNA-seq) has revolutionized transcriptomic studies by providing scope of multi-dimensional examination of whole cellular transcriptome much more efficiently, allowing identification of novel transcripts (Wang et al. 2009).

High-throughput RNA sequencing (RNA-seq) identifies the abundance of RNA and promises a comprehensive picture of the transcriptome, allowing for the full annotation and quantification of all genes and their isoforms across samples. This technology is extensively applied to identify novel transcripts, study gene expression differences, gene fusion events, alternative splicing and RNA editing.

Several studies have exploited RNA-seq to study transcriptome of many plant species including sorghum (Johnson et al. 2014), tea plant (Liu et al. 2016), maize (Song et al. 2017), lentil (Singh et al. 2017), *Arabidopsis* (Filichkin et al. 2010) and rice (Lu et al. 2010; Zhang et al. 2010). Recently, RNA-seq has become popular to study maize transcriptome and thus so a detailed transcriptome of leaf, root, reproductive leaf meristem and inflorescence has been developed in maize using RNA-seq (Li et al. 2010; Eveland et al. 2010; Opitz et al. 2016; Song et al. 2017). Many comparative studies have been made to test the effectiveness of microarray and RNA-seq in providing the genome-wide expressions in maize (Sekhon et al. 2013). RNA-seq provided extended coverage of the genome along with clarity in expression patterns among paralogs. In yet another study by Hansey et al. (2012), whole seedlings of 21 maize inbred lines were sequenced from diverse North American and exotic germplasm. Kakumanu et al. (2012) used RNA-seq to analyze drought-stressed and well-watered fertilized ovary and basal leaf meristem tissue of maize. The study showed more number of drought responsive genes in ovary (1500) than leaf meristem.

4.6.1.3 Candidate Gene-Based Transcriptome Analysis

Candidate gene is a gene governing a particular trait in an organism at any said environment or condition. Candidate gene approach is based on three successive steps. First is to identify a potential candidate gene based on the physiological, biological and functional importance of the gene in question to that condition or environment or based on linkage data of the locus under study. This is limited to existing knowledge of genes. In the second step, a molecular polymorphism or genetic variant is revealed to calculate statistical co-relation between candidate gene polymorphism and phenotypic variation or the candidate gene can be co-localized on a genetic linkage map to look for the linkage between candidate gene and loci being characterized. Detecting polymorphism in laboratory often involves sequencing of the case and control ones. The third step tests the validity of association and segregation from correlative experiments (Kwon and Goate 2000).

There exist a number of ways to detect candidate genes such as prior knowledge of the biological pathways, linkage studies, expression studies, and quantitative trait locus (QTL) analysis and genome wide association studies (GWAS). Genome wide association mapping and QTL mapping are the genomic tools, which identify a region that may be on or near to a potential candidate gene. As the identified suspected potential candidate gene is believed to have a role in the said biological pathway of the desired trait, finding an association by GWAS studies confirms its role in that pathway (Korte and Farlow 2013).

4.6.2 Important Gene Families Identified Using Transcriptomes and Their Role in Stress Tolerance

Harb et al. (2010) made comparisons between moderate and progressive microarray data that showed specific association of cell wall expansion genes under moderate stress while same genes were shown to be down-regulated in the progressive drought condition. The quantification of expansin genes i.e., *EXPA3*, *EXPA4*, *EXPA8*, *EXPA10*, and *EXPANSIN-LIKE B1* was done where most of the genes were found to be expressed in moderate drought stress.

DREB TFs belongs to AP2/ERF superfamily and has have been identified to be one of the main transcription factors to be involved in improving drought tolerance. DREB binds to dehydration responsive element (DRE) in the promoter region of many drought and/or cold stress-inducible genes (Liu et al. 1998). Over-expression of isoforms of DREB, (DREB2A-CA) protein in transgenic plant imparts significant drought and heat tolerance (Sakuma et al. 2006). Liu et al. (2013) cloned 18 *ZmDREB* genes of maize B73 genome and analyzed phylogenetic relationships and synteny with rice, maize and sorghum. They explored a significant link between genetic variation between *ZmDREB2.7* and drought tolerance at seedling stage. Further analysis revealed that the DNA polymorphisms in the promoter region of *ZmDREB2.7* was associated with different levels of drought tolerance among maize varieties.

Humbert et al. (2013) reported molecular responses in maize to drought and nitrogen stresses individually as well as in combination by customized Affymetrix maize microarray. Their study concluded effects of mild and severe drought stress on plant's photosynthetic machinery, Calvin cycle, sucrose and starch metabolism.

Table 4.1 List of important genes identified to provide drought tolerance in maize

Drought related function	Gene ID	Chr	Gene start	Gene end	Function	References
Detoxification	GRMZM2G025992	7	171775019	171778224	Oxygen radical detoxification	McKersie et al. (1996), Castillejo et al. (2008)
	GRMZM2G054559	3	12195404	12200349	Phospholipid hydrolysis	Zhu 2002
	GRMZM2G059991	6	136070517	136074741	Oxygen radical detoxification	McKersie et al. (1996), Castillejo et al. (2008)
	GRMZM2G066120	1	37470728	37476121	Reactive oxygen species homeostasis	Zhu (2002)
	GRMZM2G071021	3	221771183	221775333	Reactive oxygen species homeostasis	Miao et al. (2006), Chen et al. (2012)
	GRMZM2G125268	4	165996358	165999622	Reactive oxygen species homeostasis	Miao et al. (2006), Chen et al. (2012)
	GRMZM2G140667	2	219258176	219261097	Reactive oxygen species homeostasis	Badawi et al. (2004)
	GRMZM2G172322	1	12985602	12991971	H_2O_2 metabolism	Galle et al. (2013)
	GRMZM2G305066	8	152510200	152511639	Reactive oxygen species homeostasis	Zhu (2002)
	GRMZM5G884600	10	138607002	138608876	ROS homeostasis	Miao et al. (2006)
Stomatal closure	GRMZM2G068330	4	11278503	11281332	ABA-dependent pathway	Davletova et al. (2005)
	GRMZM2G069365	4	160153804	160155930	ABA-dependent pathway	Davletova et al. (2005)
	GRMZM2G071112	7	112658777	112661470	ABA-dependent pathway	Davletova et al. (2005)
	GRMZM2G089619	2	50140925	50142374	ABA-dependent pathway	Davletova et al. (2005)
	GRMZM2G111136	5	88828896	88831814	ABA-dependent pathway	Thirunavukkarasu et al. (2014)
	GRMZM2G122479	6	139464390	139470075	Ion homeostasis-dependent pathway	Laporte et al. (2002)
	GRMZM2G172586	10	2639262	2640147	ABA-dependent pathway	Davletova et al. (2005)
	GRMZM2G328438	8	73654879	73656447	ABA-dependent pathway	Davletova et al. (2005)

(continued)

Table 4.1 (continued)

Drought related function	Gene ID	Chr	Gene start	Gene end	Function	References
	GRMZM2G330848	7	175907236	175908949	ABA-dependent pathway	Iuchi et al. (2001), Thompson et al. (2007)
	GRMZM2G389379	2	188271896	188273136	ABA-dependent pathway	Davletova et al. (2005)
	GRMZM2G407181	1	174550907	174553815	ABA-dependent pathway	Iuchi et al. (2001), Thompson et al. (2007)
	GRMZM2G408158	2	235244890	235246909	ABA-dependent pathway	Iuchi et al. (2001), Thompson et al. (2007)
	GRMZM2G417229	5	201282108	201284278	ABA-dependent pathway	Davletova et al. (2005)
	GRMZM2G417954	7	5981141	5983425	ABA-dependent pathway	Iuchi et al. (2001), Thompson et al. (2007)
	GRMZM2G470974	10	2632775	2633842	ABA-dependent pathway	Davletova et al. (2005)
	GRMZM5G858784	3	87358369	87360132	ABA-dependent pathway	Iuchi et al. (2001), Thompson et al. (2007)
	AC197099.3_FGT005	1	96504420	96506132	ABA-dependent pathway	Abe et al. (1997)
	AF466202.2_FGP001	10	138428306	138432705	Ion homeostasis	Laporte et al. (2002)
	GRMZM2G570020	1	166137697	166137798	ABA-dependent pathway	Abe et al. (1997), Seo et al. (2011)
	GRMZM2G008250	1	174845979	174849344	ABA-dependent pathway	Li et al. (2008a, b)
	GRMZM2G009275	4	119500675	119501178	ABA-dependent pathway	Abe et al. (1997), Seo et al. (2011)
	GRMZM5G822829	10	138462252	138463015	ABA-dependent pathway	Abe et al. (1997), Seo et al. (2011)
Photosynthesis	GRMZM2G012397	7	5134217	5135120	Photosystem I reaction center	Zhang et al. (2012)
	GRMZM2G024150	5	3530018	3531150	Oxidation reduction process	Kimata and Hase (1989)
	GRMZM2G033885	7	157314547	157315990	Chlorophyll a/b-binding protein, photosystem II	Ashraf (1994)
	GRMZM2G077755	5	207185179	207187500	Chlorophyll a/b-binding protein, photosystem II	Ashraf (1994)

(continued)

Table 4.1 (continued)

Drought related function	Gene ID	Chr	Gene start	Gene end	Function	References
	GRMZM2G078409	2	24166809	24167435	Electron transfer	Efeoglu et al. (2009)
	GRMZM2G122337	6	1340417	1341388	Oxidation reduction process	Kimata and Hase (1989)
	GRMZM2G178693	2	28493204	28495467	Water transport	Jang et al. (2004), Alexandersson et al. (2005)
	GRMZM2G092125	2	169186912	169190423	Water transport	Jang et al. (2004), Alexandersson et al. (2005)
	GRMZM2G081192	4	143891075	143894123	Water transport	Jang et al. (2004), Alexandersson et al. (2005)
	GRMZM2G154628	5	195239679	195242694	Water transport	Jang et al. (2004), Alexandersson et al. (2005)
	GRMZM2G014914	7	41435006	41438812	Water transport	Jang et al. (2004), Alexandersson et al. (2005)
Hormone signalling	GRMZM2G056120	3	196638145	196644110	ABA-inducible TFs triggering stomatal closure	Furihata et al. (2006), Kim et al. (2012)
	GRMZM2G057935	1	277059620	277064623	Signalling network	Sheehan et al. (2004)
	GRMZM2G066867	5	18469442	18472522	ABA signaling network	Schafleitner et al. (2007), Mao et al. (2010)
	GRMZM2G073750	3	123881899	123888098	ABA-inducible TFs triggering stomatal closure	Furihata et al. (2006), Kim et al. (2012)
	GRMZM2G110908	10	148736929	148738463	ABA signaling network	Schafleitner et al. (2007), Mao et al. (2010)
	GRMZM2G112240	4	170944444	170947965	ABA signaling network	Zheng et al. (2010)
	GRMZM2G116557	8	159641546	159646951	ABA-inducible TFs triggering stomatal closure	Furihata et al. (2006), Kim et al. (2012)

(continued)

Table 4.1 (continued)

Drought related function	Gene ID	Chr	Gene start	Gene end	Function	References
	GRMZM2G137413	4	118717709	118723692	ABA-inducible TFs triggering stomatal closure	Furihata et al. (2006), Kim et al. (2012)
	GRMZM2G142768	4	171402585	171405007	ABA-inducible TFs triggering stomatal closure	Furihata et al. (2006), Kim et al. (2012)
	GRMZM2G165433	9	11782450	11785572	ABA signaling network	Schafleitner et al. (2007), Mao et al. (2010)
	GRMZM2G179121	1	158940439	158942519	ABA-inducible TFs triggering stomatal closure	Furihata et al. (2006), Kim et al. (2012)
	GRMZM2G015605	10	87283919	87284844	Auxin transport	Hund et al. (2009)
Root development	GRMZM2G028648	6	115935425	115937455	Auxin transport	Hund et al. (2009)
	GRMZM2G090576	5	20815035	20817354	Auxin transport	Hund et al. (2009)
	GRMZM2G091819	10	16522572	16525775	Auxin biosynthesis	Overvoorde et al. (2010)
	GRMZM2G104400	8	102056994	102060989	Auxin transport	Hund et al. (2009)
	GRMZM2G371345	10	46790874	46793155	Auxin transport	Overvoorde et al. (2010)
	GRMZM2G014844	10	34733949	34737952	Carbohydrate metabolism	Krasensky and Jonak (2012)
Sucrose metabolism	GRMZM2G016890	10	34240659	34246077	Starch biosynthesis	Hurkman et al. (2003)
	GRMZM2G058310	7	155396510	155399710	Starch degradation	Rizhsky et al. (2004)
	GRMZM2G073054	2	59216710	59226937	Starch degradation	Rizhsky et al. (2004)
	GRMZM2G130043	4	172635729	172706662	Hydrolysis of sucrose	Ruan et al. (2010)
	GRMZM2G152908	9	122479052	122485725	Sucrose metabolism	Gonzalez et al. (1995)
	GRMZM2G175218	3	1535920	1540195	Sucrose metabolism	Gonzalez et al. (1995)
	GRMZM2G175423	1	197301249	197304338	Cellulose hydrolysis	Mei et al. (2009)
	GRMZM2G318780	1	56787028	56794227	Starch biosynthesis	Ober et al. (1991)
	GRMZM2G394450	2	23504839	23509410	Cellulose hydrolysis	Mei et al. (2009)

The genes involved in photosynthesis and Calvin cycle were severely down-regulated while that of later two (sucrose and starch metabolism) were found to be up-regulated. Genes involved in amino acid biosynthesis mainly for asparagines and proline were also over-expressed in this study.

NGS provides more lucid view into the DNA variation, polymorphism detection, marker development and gene expression analysis (Barabaschi et al. 2011; Mastrangelo et al. 2012). Xu et al. (2014) studied transcriptome of maize reference genome B73 by RNA-seq and compared gene expression in fertilized ovaries and basal leaf meristem tissues collected under drought-treated and well-watered conditions. The study identified 6,385,011 SNPs from 15 maize inbreds and B73 reference genome. Several genes such as ADP-glucose pyrophosphorylase (GRMZM2G163437), glucosyltransferase (GRMZM2G179063), putative calmodulin-binding protein (GRMZM2G466563), leucine-rich repeat receptor-like protein kinase family protein (GRMZM2G428554) were identified to involve in drought tolerance (Table 4.1) (Xu et al. 2014).

4.7 Conclusions

Drought stress is one of the major abiotic stresses that affects the crop growth of maize and leads to low yield. Drought affects all developmental stages and plants respond at different levels; morphological, physiological, biochemical and molecular. At morphological level, drought stress responses include reduced plant growth, high root to shoot ratio, reduced number of leaves per plant, reduced leaf size and longevity, low grain yield etc. Physiological responses include decrease in respiratory rate, photosynthetic rate as well as transpiration rate due to stomatal closure. At biochemical level, ROS production, osmolyte accumulation and biosynthesis of carbohydrates are the major responses. At molecular stage, transcription factors and phytohormones play major role in regulation of drought tolerance. The progress in transcriptomic approaches for understanding the gene expression identified various drought-related transcription factor gene families from both ABA-dependent and ABA-independent pathways. These genes and pathways would be helpful for the development of drought tolerant maize hybrids.

References

Abe H, Yamaguchi-Shinozaki K, Urao T et al (1997) Role of *Arabidopsis* MYC and MYB homologs in drought- and abscisic acid-regulated gene expression. Plant Cell 9:1859–1868

Alexandersson E, Fraysse L, Sjövall-Larsen S et al (2005) Whole gene family expression and drought stress regulation of aquaporins. Plant Mol Biol 59:469–484

Anjum S, Xie X, Wang L (2011) Morphological, physiological and biochemical responses of plants to drought stress. Afr J Agric Res 6:2026–2032

Anjum F, Yaseen M, Rasool E et al (2003) Water stress in barley (*Hordeum vulgare* L.). I. Effect on morphological characters. Pak J Agric Sci 40:43–44
Arora A, Sairam RK, Srivastava GC (2002) Oxidative stress and antioxidative system in plants. Curr Sci 82:1227–1238
Ashraf M (1994) Breeding for salinity tolerance in plants. CRC Crit Rev Plant Sci 13:17–42
Badawi GH, Kawano N, Yamauchi Y et al (2004) Over-expression of ascorbate peroxidase in tobacco chloroplasts enhances the tolerance to salt stress and water deficit. Physiol Plant 121:231–238
Barabaschi D, Guerra D, Lacrima K et al (2011) Emerging knowledge from genome sequencing of crop species. Mol Biotechnol 50(3):250–266
Barnabás B, Jäger K, Fehér A (2008) The effect of drought and heat stress on reproductive processes in cereals. Plant Cell Environ 31:11–38
Byrne PF, Bolanos J, Edmeades GO, Eaton DL (1995) Gains from selection under drought versus multilocation testing in related tropical maize populations. Crop Sci 35:63–69
Carpenter JF, Crowe LM, Crowe JH (1987) Stabilization of phosphofructokinase with sugars during freeze-drying: characterization of enhanced protection in the presence of divalent cations. BBA Gen Subj 923:109–115
Castillejo MÁ, Maldonado AM, Ogueta S, Jorrín JV (2008) Proteomic analysis of responses to drought stress in sunflower (*Helianthus annuus*) leaves by 2DE gel electrophoresis and mass spectrometry. Open Proteom J 1:59–71
Chaves MM, Chaves MM (1991) Effects of water deficits on carbon assimilation. J Exp Bot 42:1–16
Chen J-H, Jiang H-W, Hsieh E-J et al (2012) Drought and salt stress tolerance of an *Arabidopsis* glutathione S-transferase U17 knockout mutant are attributed to the combined effect of glutathione and abscisic acid. Plant Physiol 158:340–351
Chugh V, Kaur N, Gupta AK (2011) Evaluation of oxidative stress tolerance in maize (*Zea mays* L.) seedlings in response to drought. Indian J Biochem Biophys 48:47–53
Cornic G (2000) Drought stress inhibits photosynthesis by decreasing stomatal aperture—not by affecting ATP synthesis. Trends Plant Sci 5:187–188
Cutler SR, Rodriguez PL, Finkelstein RR, Abrams SR (2010) Abscisic acid: emergence of a core signaling network. Annu Rev Plant Biol 61:651–679
Davletova S, Schlauch K, Coutu J, Mittler R (2005) The zinc-finger protein zat12 plays a central role in reactive oxygen and abiotic stress signaling in *Arabidopsis*. Plant Physiol 139:847–856
Denmead OT, Shaw RH (1960) The effects of soil moisture stress at different stages of growth on the development and yield of corn. Agron J 52:272–274
Du H, Liu H, Xiong L (2013) Endogenous auxin and jasmonic acid levels are differentially modulated by abiotic stresses in rice. Front Plant Sci 4:397
Efeoglu B, Ekmekci Y, Cicek N (2009) Physiological responses of three maize cultivars to drought stress and recovery. S Afr J Bot 75:34–42
Eveland AL, Satoh-Nagasawa N, Goldshmidt A et al (2010) Digital gene expression signatures for maize development. Plant Physiol 154:1024–1039
Farooq M, Wahid A, Kobayashi N et al (2009) Plant drought stress: effects, mechanisms and management. Agron Sustain Dev 29:185–212
Farooq M, Hussain M, Wahid A, Siddique KHM (2012) Drought stress in plants: an overview. In: Aroca R (ed) Plant responses to drought stress. Springer, Berlin, pp 1–33
Filichkin SA, Priest HD, Givan SA et al (2010) Genome-wide mapping of alternative splicing in *Arabidopsis thaliana*. Genome Res 20:45–58
Flexas J, Bota J, Loreto F et al (2004) Diffusive and metabolic limitations to photosynthesis under drought and salinity in c(3) plants. Plant Biol 6:269–279
Flexas J, Diaz-Espejo A, Galmés J et al (2007) Rapid variations of mesophyll conductance in response to changes in CO_2 concentration around leaves. Plant Cell Environ 30:1284–1298
Furihata T, Maruyama K, Fujita Y et al (2006) Abscisic acid-dependent multisite phosphorylation regulates the activity of a transcription activator AREB1. Proc Natl Acad Sci USA 103:1988–1993

Galle A, Csiszar J, Benyo D et al (2013) Isohydric and anisohydric strategies of wheat genotypes under osmotic stress: biosynthesis and function of ABA in stress responses. J Plant Physiol 170:1389–1399

Ge TD, Sui FG, Bai LP et al (2006) Effects of water stress on the protective enzyme activities and lipid peroxidation in roots and leaves of summer maize. Agric Sci China 5:291–298

Goicoechea N, Antolin MC, Sanchez-Diaz M (1997) Gas exchange is related to the hormone balance in mycorrhizal or nitrogen-fixing alfalfa subjected to drought. Physiol Plant 100:989–997

Gonzalez EM, Gordon AJ, James CL, Arrese-lgor C (1995) The role of sucrose synthase in the response of soybean nodules to drought. J Exp Bot 46:1515–1523

Hansey CN, Vaillancourt B, Sekhon RS et al (2012) Maize (*Zea mays* L.) genome diversity as revealed by rna-sequencing. PLoS One. https://doi.org/10.1371/journal.pone.0033071

Harb A, Krishnan A, Ambavaram MMR, Pereira A (2010) Molecular and physiological analysis of drought stress in *Arabidopsis* reveals early responses leading to acclimation in plant growth. Plant Physiol 154:1254–1271

Havaux M (1998) Carotenoids as membrane stabilizers in chloroplasts. Trends Plant Sci 3:147–151

Hayano-Kanashiro C, Calderón-Vásquez C, Ibarra-Laclette E et al (2009) Analysis of gene expression and physiological responses in three Mexican maize landraces under drought stress and recovery irrigation. PLoS One. https://doi.org/10.1371/journal.pone.0007531

Herrero M, Johnson R (1981) Drought stress and its effects on maize reproductive systems. Crop Sci 21:105–110

Humbert S, Subedi S, Cohn J et al (2013) Genome-wide expression profiling of maize in response to individual and combined water and nitrogen stresses. BMC Genom 14:3

Hund A, Trachsel S, Stamp P (2009) Growth of axile and lateral roots of maize: I. Development of a phenotying platform. Plant Soil 325:335–349

Hurkman WJ, McCue KF, Altenbach SB et al (2003) Effect of temperature on expression of genes encoding enzymes for starch biosynthesis in developing wheat endosperm. Plant Sci 164:873–881

Ingram J, Bartels D (1996) The molecular basis of dehydration tolerance in plants. Annu Rev Plant Physiol Plant Mol Biol 47:377–403

Iuchi S, Kobayashi M, Taji T et al (2001) Regulation of drought tolerance by gene manipulation of 9-cis-epoxycarotenoid dioxygenase, a key enzyme in abscisic acid biosynthesis in *Arabidopsis*. Plant J 27:325–333

Jaleel CA, Gopi R, Gomathinayagam M, Panneerselvam R (2008) Effects of calcium chloride on metabolism of salt-stressed *Dioscorea rotundata*. Acta Biol Cracov Ser Bot 50:63–67

Jang JY, Kim DG, Kim YO et al (2004) An expression analysis of a gene family encoding plasma membrane aquaporins in response to abiotic stresses in *Arabidopsis thaliana*. Plant Mol Biol 54:713–725

Johansen C, Baldev B, Brouwer J (1994) Biotic and abiotic stresses constraining productivity of cool season food legumes in Asia, Africa and Oceania. Seas Food Legum 19:175–194

Johnson SM, Lim F-L, Finkler A et al (2014) Transcriptomic analysis of *Sorghum bicolor* responding to combined heat and drought stress. BMC Genom 15:456

Kage H, Kochler M, Stützel H (2004) Root growth and dry matter partitioning of cauliflower under drought stress conditions: measurement and simulation. Eur J Agron 20:379–394

Kakumanu A, Ambavaram MMR, Klumas C et al (2012) Effects of drought on gene expression in maize reproductive and leaf meristem tissue revealed by RNA-Seq. Plant Physiol 160:846–867

Kaur K, Gupta AK, Kaur N (2007) Effect of water deficit on carbohydrate status and enzymes of carbohydrate metabolism in seedlings of wheat cultivars. Indian J Biochem Biophys 44:223–230

Kilian J, Peschke F, Berendzen KW et al (2012) Prerequisites, performance and profits of transcriptional profiling the abiotic stress response. Biochim Biophys Acta Gene Regul Mech 1819:166–175

Kim MJ, Park M-J, Seo PJ et al (2012) Controlled nuclear import of the transcription factor NTL6 reveals a cytoplasmic role of SnRK2.8 in the drought-stress response. Biochem J 448:353–363
Kimata Y, Hase T (1989) Localization of ferredoxin isoproteins in mesophyll and bundle sheath cells in maize leaf. Plant Physiol 89:1193–1197
Kingston-Smith AH, Foyer CH (2000) Bundle sheath proteins are more sensitive to oxidative damage than those of the mesophyll in maize leaves exposed to paraquat or low temperatures. J Exp Bot 51:123–130
Korte A, Farlow A (2013) The advantages and limitations of trait analysis with GWAS: a review. Plant Methods 9:29
Krasensky J, Jonak C (2012) Drought, salt, and temperature stress-induced metabolic rearrangements and regulatory networks. J Exp Bot 63:1593–1608
Kusaka M, Ohta M, Fujimura T (2005) Contribution of inorganic components to osmotic adjustment and leaf folding for drought tolerance in pearl millet. Physiol Plant 125:474–489
Kwon JM, Goate AM (2000) The candidate gene approach. Alcohol Res Health 24:164–168
Laporte MM, Shen B, Tarczynski MC (2002) Engineering for drought avoidance: expression of maize NADP-malic enzyme in tobacco results in altered stomatal function. J Exp Bot 53:699–705
Lawlor DW, Cornic G (2002) Photosynthetic carbon assimilation and associated metabolism in relation to water deficits in higher plants. Plant Cell Environ 25:275–294
Lawson T, Oxborough K, Morison JIL, Baker NR (2003) The responses of guard and mesophyll cell photosynthesis to CO_2, O_2, light, and water stress in a range of species are similar. J Exp Bot 54:1743–1752
Li B, Wei A, Song C et al (2008a) Heterologous expression of the TsVP gene improves the drought resistance of maize. Plant Biotechnol J 6:146–159
Li W-X, Oono Y, Zhu J et al (2008b) The *Arabidopsis* NFYA5 transcription factor is regulated transcriptionally and post-transcriptionally to promote drought resistance. Plant Cell 20:2238–2251
Li P, Ponnala L, Gandotra N et al (2010) The developmental dynamics of the maize leaf transcriptome. Nat Genet 42:1060–1067
Li Z, Shi P, Peng Y (2013) Improved drought tolerance through drought preconditioning associated with changes in antioxidant enzyme activities, gene expression and osmoregulatory solutes accumulation in white clover (*Trifolium repens* L.). Plant Omics 6:481–489
Liljenberg CS (1992) The effects of water deficit stress on plant membrane lipids. Prog Lipid Res 31:335–343
Liu Q, Kasuga M, Sakuma Y et al (1998) Two transcription factors, DREB1 and DREB2, with an EREBP/AP2 DNA binding domain separate two cellular signal transduction pathways in drought- and low-temperature-responsive gene expression, respectively, in *Arabidopsis*. Plant Cell 10:1391–1406
Liu S, Wang X, Wang H et al (2013) Genome-Wide analysis of ZmDREB genes and their association with natural variation in drought tolerance at seedling stage of *Zea mays* L. PLoS Genet. https://doi.org/10.1371/journal.pgen.1003790
Liu SC, Jin JQ, Ma JQ et al (2016) Transcriptomic analysis of tea plant responding to drought stress and recovery. PLoS ONE. https://doi.org/10.1371/journal.pone.0147306
Lu T, Lu G, Fan D et al (2010) Function annotation of the rice transcriptome at single-nucleotide resolution by RNA-seq. Genome Res 20:1238–1249
Luo M, Liu J, Lee RD et al (2010) Monitoring the expression of maize genes in developing kernels under drought stress using oligo-microarray. J Integr Plant Biol 52:1059–1074
Mao X, Zhang H, Tian S et al (2010) TaSnRK2.4, an SNF1-type serine/threonine protein kinase of wheat (*Triticum aestivum* L.), confers enhanced multistress tolerance in *Arabidopsis*. J Exp Bot 61:683–696
Marino R, Ponnaiah M, Krajewski P et al (2009) Addressing drought tolerance in maize by transcriptional profiling and mapping. Mol Genet Genomics 281:163–179
Mastrangelo AM, Mazzucotelli E, Guerra D et al (2012) Improvement of drought resistance in crops: from conventional breeding to genomic selection. In: Venkateswarlu B, Shanker AK,

Shanker C, Maheswari M (eds) Crop Stress and its management: perspective and strategies. Springer, Dordrecht, pp 225–259
McKersie BD, Bowley SR, Harjanto E, Leprince O (1996) Water-deficit tolerance and field performance of transgenic alfalfa overexpressing superoxide dismutase. Plant Physiol 111:1177–1181
Mei C, Park SH, Sabzikar R et al (2009) Green tissue-specific production of a microbial endo-cellulase in maize (*Zea mays* L.) endoplasmic-reticulum and mitochondria converts cellulose into fermentable sugars. J Chem Technol Biotechnol 84:689–695
Miao Y, Lv D, Wang P et al (2006) An Arabidopsis glutathione peroxidase functions as both a redox transducer and a scavenger in abscisic acid and drought stress responses. Plant Cell 18:2749–2766
Nelson DE, Repetti PP, Adams TR et al (2007) Plant nuclear factor Y (NF-Y) B subunits confer drought tolerance and lead to improved corn yields on water-limited acres. Proc Natl Acad Sci USA 104:16450–16455
Ober ES, Setter TL, Madison JT et al (1991) Influence of water deficit on maize endosperm development: enzyme activities and RNA transcripts of starch and zein synthesis, abscisic acid, and cell division. Plant Physiol 97:154–164
Okuma E, Nozawa R, Murata Y, Miura K (2014) Accumulation of endogenous salicylic acid confers drought tolerance to *Arabidopsis*. Plant Signal Behav 9:e280851–e280854. https://doi.org/10.4161/psb.28085
Opitz N, Marcon C, Paschold A et al (2016) Extensive tissue-specific transcriptomic plasticity in maize primary roots upon water deficit. J Exp Bot 67:1095–1107
Overvoorde P, Fukaki H, Beeckman T (2010) Auxin control of root development. Cold Spring Harb Perspect Biol. https://doi.org/10.1101/cshperspect.a001537
Paul MJ, Primavesi LF, Jhurreea D, Zhang Y (2008) Trehalose metabolism and signaling. Annu Rev Plant Biol 59:417–441
Poormohammad Kiani S, Maury P, Sarrafi A, Grieu P (2008) QTL analysis of chlorophyll fluorescence parameters in sunflower (*Helianthus annuus* L.) under well-watered and water-stressed conditions. Plant Sci 175:565–573
Poroyko V, Spollen WG, Hejlek LG et al (2007) Comparing regional transcript profiles from maize primary roots under well-watered and low water potential conditions. J Exp Bot 58 (2):279–289
Ramachandra Reddy A, Chaitanya KV, Jutur PP, Sumithra K (2004) Differential antioxidative responses to water stress among five mulberry (*Morus alba* L.) cultivars. Environ Exp Bot 52:33–42
Rampino P, Pataleo S, Gerardi C et al (2006) Drought stress response in wheat: physiological and molecular analysis of resistant and sensitive genotypes. Plant Cell Environ 29:2143–2152
Rizhsky L, Davletova S, Liang H, Mittler R (2004) The zinc finger protein Zat12 is required for cytosolic ascorbate peroxidase 1 expression during oxidative stress in *Arabidopsis*. J Biol Chem 279:11736–11743
Ruan YL, Jin Y, Yang YJ et al (2010) Sugar input, metabolism, and signaling mediated by invertase: roles in development, yield potential, and response to drought and heat. Mol Plant 3:942–955
Rucker KS, Kvien CK, Holbrook CC, Hook JE (1995) Identification of peanut genotypes with improved drought avoidance traits. Peanut Sci 24:14–18
Sacks MM, Silk WK, Burman P (1997) Effect of water stress on cortical cell division rates within the apical meristem of primary roots of maize. Plant Physiol 114:519–527
Sakamoto M, Munemura I, Tomita R, Kobayashi K (2008) Involvement of hydrogen peroxide in leaf abscission signaling, revealed by analysis with an *in vitro* abscission system in Capsicum plants. Plant J 56:13–27
Sakuma Y, Maruyama K, Qin F et al (2006) Dual function of an *Arabidopsis* transcription factor DREB2A in water-stress-responsive and heat-stress-responsive gene expression. Proc Natl Acad Sci USA 103:18822–18827

Sangoi L, Salvador R (1997) Dry matter production and partitioning of maize hybrids and dwarf lines at four plant populations. Cienc Rural 27:1–6

Schafleitner R, Gutierrez Rosales RO, Gaudin A et al (2007) Capturing candidate drought tolerance traits in two native Andean potato clones by transcription profiling of field grown plants under water stress. Plant Physiol Biochem 45:673–690

Sekhon RS, Briskine R, Hirsch CN et al (2013) Maize gene atlas developed by rna sequencing and comparative evaluation of transcriptomes based on rna sequencing and microarrays. PLoS ONE. https://doi.org/10.1371/journal.pone.0061005

Seo JS, Joo J, Kim MJ et al (2011) OsbHLH148, a basic helix-loop-helix protein, interacts with OsJAZ proteins in a jasmonate signaling pathway leading to drought tolerance in rice. Plant J 65:907–921

Serraj R, Sinclair TR (2002) Osmolyte accumulation: Can it really help increase crop yield under drought conditions? Plant Cell Environ 25:333–341

Shao HB, Chu LY, Shao MA et al (2008) Higher plant antioxidants and redox signaling under environmental stresses. C R Biol 331:433–441

Sharma P, Jha AB, Dubey RS, Pessarakli M (2012) Reactive oxygen species, oxidative damage, and antioxidative defense mechanism in plants under stressful conditions. J Bot 2012:1–26

Sheehan MJ, Farmer PR, Brutnell TP (2004) Structure and expression of maize phytochrome family homeologs. Genetics 167:1395–1405

Shinozaki K, Yamaguchi-Shinozaki K (1996) Molecular responses to drought and cold stress. Curr Opin Biotechnol 7:161–167

Singh D, Singh CK, Taunk J et al (2017) Transcriptome analysis of lentil (*Lens culinaris* Medikus) in response to seedling drought stress. BMC Genom 18:206

Song K, Kim HC, Shin S et al (2017) Transcriptome analysis of flowering time genes under drought stress in maize leaves. Front Plant Sci 8:1–12

Thirunavukkarasu N, Hossain F, Arora K et al (2014) Functional mechanisms of drought tolerance in subtropical maize (*Zea mays* L.) identified using genome-wide association mapping. BMC Genom 15(1182):1–12

Thompson AJ, Mulholland BJ, Jackson AC et al (2007) Regulation and manipulation of ABA biosynthesis in roots. Plant Cell Environ 30:67–78

Udvardi MK, Kakar K, Wandrey M et al (2007) Legume transcription factors: global regulators of plant development and response to the environment. Plant Physiol 144:538–549

Uhart SA, Andrade FH (1995) Nitrogen deficiency in maize: I. Effects on crop growth, development, dry matter partitioning, and kernel set. Crop Sci 35:1376–1383

Virlouvet L, Jacquemot M-P, Gerentes D et al (2011) The ZmASR1 protein influences branched-chain amino acid biosynthesis and maintains kernel yield in maize under water-limited conditions. Plant Physiol 157:917–936

Wanek W, Richter A (1997) Biosynthesis and accumulation of D-ononitol in *Vigna umbellata* in response to drought stress. Physiol Plant 101:416–424

Wang Z, Gerstein M, Snyder M (2009) RNA-Seq: a revolutionary tool for transcriptomics. Nat Rev Genet 10:57–63

Wu Y, Cosgrove DJ (2000) Adaptation of roots to low water potentials by changes in cell wall extensibility and cell wall proteins. J Exp Bot 51:1543–1553

Xu Y, Skinner DJ, Wu H et al (2009) Advances in maize genomics and their value for enhancing genetic gains from breeding. Int J Plant Genomics. https://doi.org/10.1155/2009/957602

Xu J, Yuan Y, Xu Y et al (2014) Identification of candidate genes for drought tolerance by whole-genome resequencing in maize. BMC Plant Biol 14:83

Yin DW, Jun M, Zheng GP et al (2012) Effects of biochar on acid black soil nutrient, soybean root and yield. Nat Resour Sustain Dev Ii 1–4(524–527):2278–2289

Ying S, Zhang D-F, Fu J et al (2012) Cloning and characterization of a maize bZIP transcription factor, ZmbZIP72, confers drought and salt tolerance in transgenic *Arabidopsis*. Planta 235:253–266

Zhang G, Guo G, Hu X et al (2010) Deep RNA sequencing at single base-pair resolution reveals high complexity of the rice transcriptome. Genome Res 20:646–654

Zhang M, Pan J, Kong X et al (2012) ZmMKK3, a novel maize group B mitogen-activated protein kinase kinase gene, mediates osmotic stress and ABA signal responses. J Plant Physiol 169:1501–1510

Zheng J, Zhao J, Tao Y et al (2004) Isolation and analysis of water stress induced genes in maize seedlings by subtractive PCR and cDNA macroarray. Plant Mol Biol 55:807–823

Zheng J, Fu J, Gou M et al (2010) Genome-wide transcriptome analysis of two maize inbred lines under drought stress. Plant Mol Biol 72:407–421

Zhu J-K (2002) Salt and drought stress signal transduction in plants. Annu Rev Plant Biol 53:247–273

Zhuang Y, Ren G, Yue G et al (2007) Effects of water-deficit stress on the transcriptomes of developing immature ear and tassel in maize. Plant Cell Rep 26:2137–2147

Zlatev Z, Lidon FC (2012) An overview on drought induced changes in plant growth, water relations and photosynthesis. Emir J Food Agric 24:57–72

Chapter 5
Physiological and Molecular Basis of Abiotic Stress Tolerance in Wheat

H. M. Mamrutha, Rajender Singh, Davinder Sharma, Karnam Venkatesh, Girish Chandra Pandey, Rakesh Kumar, Ratan Tiwari and Indu Sharma

Abstract Wheat is the second most important cereal crop of the world occupying about 220 million hectares area (mha) with a production of 716 million tons and accounts for 20% dietary protein of the people across world. Due to its significant contribution to global food security, it is very much essential to maintain the steady state in its production. Most wheat growing regions are affected by multiple abiotic stresses like drought, heat, salinity and water logging, either alone or in combination. It has been predicted that the yield of wheat will be lowered by 22% in the coming years due to significant changes in temperature and rainfall. An estimated 65 mha of wheat area worldwide would be affected by drought. Hence, having the comprehensive knowledge on existing information on major abiotic stresses of wheat and their physiological and molecular basis would help us to address in future research gaps. A compilation of different abiotic stress adaptive mechanisms, associated physiological traits, molecular markers, a cascade of gene networks involved in the development of transgenic wheat for drought and heat stress tolerance has been presented. This inclusive information will be useful to wheat researchers for further wheat improvement.

Keywords Wheat · Heat stress · Drought stress · Phenotyping
Molecular markers · Transgenics

5.1 Introduction

Wheat is the most widely grown crop in the world, with over more than 220 million hectares of cropland producing 715 million tons of food grain with a productivity of 3.2 tons per hectare (FAO 2015). Recent projected global demand indicates that

H. M. Mamrutha (✉) · R. Singh · D. Sharma · K. Venkatesh · R. Kumar · R. Tiwari · I. Sharma
ICAR-Indian Institute of Wheat and Barley Research, Karnal, India
e-mail: mamruthamadhu@gmail.com

G. C. Pandey
Department of Bioscience and Biotechnology, Banasthali University, Banasthali, India

V. R. Rajpal et al. (eds.), *Genetic Enhancement of Crops for Tolerance to Abiotic Stress: Mechanisms and Approaches, Vol. I*, Sustainable Development and Biodiversity 20, https://doi.org/10.1007/978-3-319-91956-0_5

world will need around 1090 million tons wheat by 2050 from its current production level. This production target has to be achieved with an annual production growth of 1.6% in the changing climate scenario (Paroda et al. 2013). According to the fifth assessment report of the Intergovernmental Panel on Climate Change (IPCC 2014), the globally averaged combined land and ocean surface temperature data shows a warming of 0.85°C (0.65–1.06) over the period from 1880 to 2012. For the three main staple crops i.e. wheat, rice and maize in tropical and temperate regions, climate change will negatively impact production and with increasing food demand, would pose large risks to food security globally. Half of the wheat-growing areas of the Indo-Gangetic Plains in India would likely become heat stressed by 2050 (Ortiz et al. 2008).

Heat and drought adversely affect various physiological processes like plant water status, electrolyte conductance, dissipation of excess heat, assimilate partitioning, accelerated senescence and finally reduced yield (Wahid et al. 2007; Hasanuzzaman et al. 2013). High temperature alters the integrity and functions of biological membranes by deforming the tertiary and quaternary structures of membrane proteins (Hemantaranjan et al. 2014). Tolerance to heat and drought stress is a complex phenomenon and wheat exhibits different tolerance mechanisms to adapt to maintain its yield. For example, it can minimize the exposure to heat stress by shortening its grain filling duration with an accelerated rate of dry weight accumulation or can exhibit the stay green behavior to prolong its grain filling to sustain its yield. Although there is considerable variation among genotypes in response to heat and drought stress, relatively little is known about the critical genes or quantitative trait loci (QTL) controlling heat and drought tolerance due to its quantitative nature. Often, consistent high-temperature and perfect drought conditions cannot be guaranteed in field and lack of various natural factors in controlled experiments makes it impossible to provide an effective solution. Further, tolerance at one growth stage may not be correlated with tolerance at other growth stages, which further complicates the situation. Hence, direct selection for these tolerances become extremely difficult by traditional breeding because of large genotype × environment interaction, lack of effective tolerance genes in different genetic backgrounds and different expression of tolerance depending on trials. In addition, there is no reliable selection criterion for thermo- and/or drought-tolerance that may be used over the years and generations. Consequently, the genetic dissection of the quantitative nature of complex traits is a prerequisite to allow cost-effective applications of genomics-based approaches (Collins et al. 2008). A detailed understanding of the wheat crops' responses to heat and drought stress, especially the physiological and molecular basis of yield loss and adaption mechanisms would assist not only in identifying future areas of research gaps but also will help in prioritizing the future roadmap to mitigate the effect of these stresses. Here, we have compiled the information on different abiotic stress adaptive mechanisms, associated physiological traits and cascade of gene networks operating in wheat under abiotic stresses. An update is also provided on molecular markers reported to be associated with different traits associated with abiotic stresses. This inclusive information will be useful to wheat researchers in developing climate-smart-improved wheat cultivars.

5.2 Heat and Drought Adaptive Mechanisms in Plants

Plants adapt to abiotic stresses mainly by three important mechanisms; stress escape, stress tolerance and stress avoidance. These mechanisms are in turn governed by many associated traits (Fig. 5.1). In escape mechanism, the plant senses the future occurrence of stress and adjusts its phenology in such a way that the critical stages (such as grain filling) of plant do not experience stress. In tolerance mechanism, the plant experiences the stress and withstands the stress condition by activation of different stress responsive genes, osmolytes and other pathways. Stay green behaviour involves tolerance to low water potential, cooler canopies, active photosynthetic state to sustain supply of current assimilates, better radiation use efficiency, and long grain filling period to maintain grain filling in elevated temperatures. Avoidance is the maintenance of an optimum plant water status by reducing water loss (by stomatal closure, development of trichomes or wax on stems and leaves, leaf rolling, better leaf angle to avoid direct sun exposure, senescence of older leaves, etc.) or maximizing water uptake (by better root architecture and growth).

5.3 Heat and Drought Stress Phenotyping Under Field Conditions

The phenotyping for heat stress under field condition is routinely done by comparing different traits under timely sowing (mid-November) and late sowing conditions (mid-December), whereas for drought stress, the phenotyping is done mainly by comparing the traits under irrigated and drought/rain fed conditions in the field. The different category of traits for comparison includes phenological, agronomical and physiological traits.

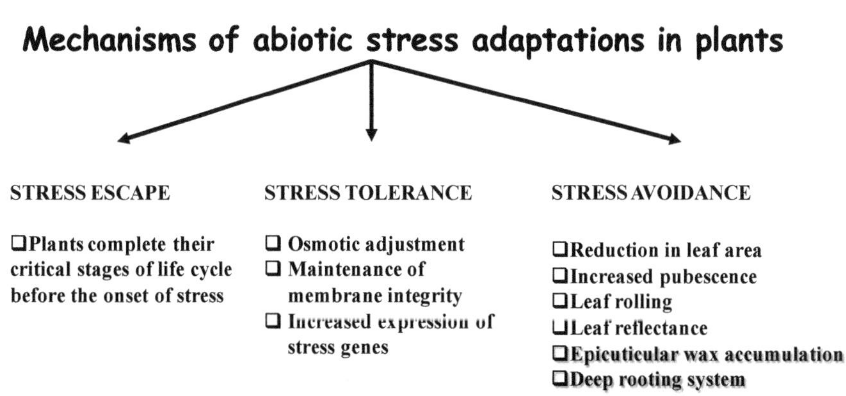

Fig. 5.1 Abiotic stress adaptive mechanisms and their associated traits in plants

5.3.1 Phenological Traits

In wheat, days to heading (DH), days to anthesis (DA), days to maturity (DM) and grain filling duration (GFD) are recorded under control and stress conditions for comparison. DH is calculated as days taken from sowing to the emergence of 75% of ears (spikes) in a plot. DA is days taken from sowing to the emergence of anthers in 75% of plants in a plot. DM is the total days taken from sowing to maturity when all plants in a plot show natural senescence and the grains become hard and fit for harvesting. GFD is the days between the date of anthesis and physiological maturity. It has been observed that under drought/heat stress conditions, the heading/anthesis/maturity occurs early compared to control conditions.

5.3.2 Agronomic Traits

Agronomic traits like plant height (PH), productive tillers (PT), biomass (BM), spike length (SL) and thousand grain weight (TGW) are known to be affected under drought/heat stress conditions. PH is measured at the time of maturity from the ground level up to the terminal spikelet, excluding the awns. Before harvesting, productive tillers per plant or per plot are counted. Biomass per plant is the weight of the whole plant with spikes and foliage after harvesting. Spike length is measured from the base to the tip of the spike excluding awns from uniform plants in a plot.

5.3.3 Physiological Traits

Various physiological traits are measured to quantify the extent of stress effects on wheat plants. The physiological traits and available methods for their measurements are listed in Table 5.1 (Tiwari and Mamrutha 2014).

5.4 Heat/Drought Susceptibility Index (HSI/DSI)

The HSI and DSI are routinely used for identifying heat/drought tolerant wheat genotypes. The HSI/DSI is calculated by the method suggested by Fischer and Maurer (1978) with the following formula: HSI/DSI = $(1 - Xh/X)/(1 - Yh/Y)$, Where, Xh and X are the phenotypic means for each genotype under stress and control conditions, respectively, and Yh and Y are the phenotypic means for all genotypes under stress and control conditions, respectively. The genotypes with

Table 5.1 Physiological traits and their measurements under normal and stress conditions in wheat

Sl. no.	Trait	Instrument used	Unit of measurement	Measurements under normal condition	Measurements under heat stress condition	References
Non destructive traits						
1	Canopy temperature	Infra red thermometer (IRT)	Degree centigrade (°C)/degree fahrenheit (°F)	Less than ambient temperature	Equal to or more than ambient temperature	Sharma et al. (2015)
2	Leaf chlorophyll content	Chlorophyll content meter	Chlorophyll content index (CCI) (0–99.9)	40–60 CCI	<40 CCI	Dwyer et al. (1991)
3	Photosynthesis using portable photosynthesis meter	Infra red gas analyser (IRGA)	µmol m^{-2} s^{-1}	15–30 µmol m^{-2} s^{-1}	5–20 µmol m^{-2} s^{-1}	Fracheboud (2004)
4	Stomatal conductance	Porometer	mmol m^{-2} s^{-1}	300–700 mmol m^{-2} s^{-1}	80–300 mmol m^{-2} s^{-1}	Rebetzke et al. (2001)
5	Chlorophyll fluorescence (CFL)	Chlorophyll fluorescence meter	F_v/F_m (0–0.84)	0.79–0.84	<0.75	Pandey et al. (2015)
6	Normalized difference vegetation index (NDVI)	NDVI meters	Index value from 0 to 1	0.4–0.6	<0.4	Mullan and Reynolds (2010)
7	Leaf area index (LAI)	LAI meter	Index value	>6	<4	Breda (2003)
8	Phenotyping for crop growth stages	Zadok's scale	Index	Varies with the plant developmental stage	Under stress the growth stages will be attained early compared to normal condition	Zadoks et al. (1974)
9	Soil moisture content	Soil moisture meter	Percentage	Varies with soil bulk density (BD) and soil depth	<non stress condition	Pearcy et al. (1989)

(continued)

Table 5.1 (continued)

Sl. no.	Trait	Instrument used	Unit of measurement	Measurements under normal condition	Measurements under heat stress condition	References
Destructive traits						
1	Leaf water potential (LWP)	The scholander pressure chamber/ pressure bomb	Bars	−5 to −10 bars	−20 to −40 bars	Turner (1988)
2	Leaf relative water content (RWC)	Gravimetric method	Percentage	95–98%	<60%	Barrs and Weatherley (1962)
3	Cell membrane stability (CMS)	Electrical conductivity meter (EC-meter)	Percentage	>60%	<60%	Sharma et al. (2015)
4	Osmotic potential	Vapour Pressure Osmometer	MPa	<−0.372	>−0.743	Morgan (1983)
5	Carbon isotope discrimination	Isotope ratio mass spectrophotometer (IRMS)	Per mil (‰)	−35 to −20‰	>−35‰	Farquhar et al. (1989)

HSI/DSI score of 1 and below 1 fall under the tolerant category and those with values greater than 1 fall under the susceptible category.

5.4.1 Novel and Precision Field Phenotyping for Heat Stress Tolerance Studies

Several efforts have been made to decipher traits/genes responsible for imparting high temperature tolerance in wheat. Both controlled as well as field-based studies have been undertaken in this regard. Lack of sufficient precision in simulating the ambient temperature dynamics and micro-environments prevailing in the field or repeatability of results in the field have been the severe bottlenecks. Hence, at ICAR-IIWBR, Karnal, India, a phenotyping method for screening wheat genotypes under high temperature using state-of- the-art Temperature Controlled Phenotyping Facility (TCPF) was developed, which ensures uniform crop stand. This allows screening of several wheat genotypes in a large plot size (simulating the fields) at a desired temperature at any stage of crop growth, while allowing plants to grow in the natural environment during rest of the period. To maintain the diurnal cycle during temperature stress treatment, temperature regulation in TCPF is manipulated based on the ambient temperature so that the desired difference between the temperature inside and outside the structure is maintained. A boiler-based heating system is utilized for increasing temperature in which the warm water runs through a network of pipelines hanging from the roof with several inlets and outlets that avoids formation of temperature gradient from one end to another in the structure. Integrated and automatically governed split air conditioners run through a control panel for cooling purposes. To maintain required humidity levels, a mist system provides fine-water droplets and the drip system provides irrigation (Fig. 5.2). Once the required temperature for stress treatment is over, the structure gets open and the crop again gets natural environment.

Lack of uniformity of plant stand while conducting a field experiment can substantially contribute to errors in the prediction of association between plant

Fig. 5.2 Temperature controlled phenotyping facility at ICAR-IIWBR, Karnal, India

phenotype and genotype. Among the several factors that contribute to experimental errors, inconsistent seeding depth and plant spacing are the important ones. This often occurs when seeds are sown by hand or seed drills. Hence, an improved planting method was devised for field experiments. The method involves a tool designed for dibbling seeds and a protocol to place seeds uniformly in the soil. The advantage of the new methods over conventional methods of sowing is that the new method improves the consistency in plant spacing and depth of seeding substantially. The reduction in error and the low coefficient of variation (CV) for the plant traits measured with the new method indicates enhanced precision in measured phenotypes under field condition relative to other methods. There is a clear advantage of this integrated method in differentiating high temperature response of a large number of genotypes of wheat with greater precision. The novel tool developed for ensuring uniform crop establishment and TCPF together can enhance the precision in field phenotyping for various abiotic stresses, in addition to heat stress (Sharma et al. 2016, 2018).

5.5 Generation of Genomic Resources and Their Utilization Under Heat and Drought Stress Tolerance

Breeding efforts to improve drought and heat tolerance have been hindered due to quantitative genetic basis of these traits. Marker-assisted selection (MAS) could be a great asset for plant breeders to attain this goal. Therefore, genomics-assisted improvement of abiotic stress tolerance of crops will increasingly rely on the quantitative trait locus (QTL) mapping approach (Collins et al. 2008). There are basic requirements for the genetic mapping of QTL: mapping population showing segregation for trait values, and genotypic data for the population.

Different types of mapping populations may be used for mapping depending on the genetics of the trait of interest. F_2 populations and back cross populations are the simplest type of mapping populations for self-pollinated species like wheat. The main advantages of these populations are that they are easy to develop and can be obtained in a short time. These populations are used for mapping traits having Mendelian inheritance, but are unsuitable for quantitative traits. Recombinant inbred lines (RILs) and doubled haploid (DH) lines represent permanent and immortal mapping populations as their genotypes are stable over generations. Selfing of individual F_2 plants derived from F_1 hybrids allow the development of RILs, which are near homozygous lines. Haploid production followed by chromosome doubling results in creation of genetically pure DH lines within a relatively short period of time. The major advantage of RILs and DH population are that they are true breeding lines. This allows replicated trials across different locations and years, which is a major requirement for phenotyping a quantitative trait influenced by environment.

The availability of molecular markers in the 1980s opened a new realm for quantitative genetics and breeding. In the last two decades, DNA-based molecular markers have become the dominant marker system for genetic analysis. These molecular markers are especially useful to breeders for selecting QTL and genetic linkage map construction. A genetic linkage map of a species or experimental population shows the position of its genetic markers relative to each other in terms of recombination frequency along a chromosome (Collard et al. 2005). The most commonly used recent molecular markers for construction of the physical and genetic linkage maps are Simple Sequence Repeat (SSR), Single Nucleotide Ploymorphism (SNP), Diversity Array Technology (DART) (Litt and Luty 1989; Tautz et al. 1986) and others include Restriction Fragment Length Polymorphisms (RFLPs) (Botstein et al. 1980), Amplified Fragment Length Polymorphisms (AFLPs) (Vos et al. 1995) and Random Amplified Polymorphic DNA (RAPD) (Williams et al. 1990). Detailed genetic linkage maps (Van Deynze et al. 1995; Nelson et al. 1995a, b; Marino et al. 1996) and physical maps (Delaney et al.1995; Mickelson-Young et al. 1995) using RFLP markers have been published for all seven homeologous chromosomes in wheat.

5.5.1 Chromosomal Regions Associated with Heat and Drought Stress Tolerance

Significant genetic variation for traits associated with drought and heat tolerance exist in wheat germplasm. Therefore, several QTL mapping studies related to heat and drought stress tolerance have been conducted in wheat (Quarrie et al. 1994; Morgan and Tan 1996; Byrne et al. 2002; Yang et al. 2002; Kuchel et al. 2007; Mathews et al. 2008; Rebetzke et al. 2008; Peleg et al. 2009; McIntyre et al. 2010; Pinto et al. 2010; Vijayalakshmi et al. 2010; Golabadi et al. 2010; Maccaferri et al. 2011; Alexander et al. 2012; Bennett et al. 2012; Kumar et al. 2012; Nezhad et al. 2012; Paliwal et al. 2012; Christopher et al. 2013; Tiwari et al. 2013; Mason et al. 2010, 2011, 2013; Acuna et al. 2014; Talukder et al. 2014; Mondal et al. 2015; Sharma et al. 2016). Kumar et al. (2012) identified three major and consistent QTLs for chlorophyll fluorescence, chlorophyll content and leaf temperature associated with drought tolerance on chromosome 2B, 3B and 4D. These QTLs explained up to 35% of the mean phenotypic variation with a LOD value of 6.3. QTLs for chlorophyll fluorescence were also reported on chromosomes 1B, 2A and 4A (Czyczyło-Mysza et al. 2011), whereas QTLs for chlorophyll content were identified on chromosomes 1A, 4A, 5A, 6A, 7A, 1B, 2B and 5B in Durum and wild Emmer wheat by (Peleg et al. 2009). Nezhad et al. (2012) detected six QTLs for thousand grain weight on chromosomes 7A and 7D under drought condition, which explained phenotypic variation ranged from 8.9 to 21.0%.

Morgan and Tan (1996) constructed RFLP linkage groups using an F_2 bread wheat mapping population and reported a single QTL for osmotic adjustment on the short arm of chromosome 7A. Meta-QTL analysis by Acuna et al. (2014) identified 736 QTLs associated with heat and drought tolerance in wheat. Yang et al. (2002) found two marker loci (*Xgwm11-1B* and *Xgwm293-5A*) for grain-filling duration under heat stress, which contributed 23% of the phenotypic variation. Kuchel et al. (2007) also found the association of *gwm11* on chromosome 1B with yield under high temperature stress. Mason et al. (2010) investigated the effect of heat shock on plant yield and yield components during early grain-filling and reported five stable QTL for HSI of single grain weight (1A and 2A), grain weight (3B), and grain number (2B and 3B) explaining up to 22% of phenotypic variation under short term heat stress in controlled conditions. In an analysis of spring wheat populations for heat tolerance, loci on chromosomes 2B and 5B were most important (Byrne et al. 2002). Few QTLs for the stay-green trait were found under heat stress (Kumar et al. 2010; Vijayalakshmi et al. 2010). Pinto et al. (2010) identified a QTL on chromosome 4A for canopy temperature under heat stress. Seven loci were found to co-localize for both HSI of main spike yield components and temperature depression indicating a strong genetic link between cooler organ temperature and heat tolerance (Mason et al. 2011). Paliwal et al. (2012) reported significant genomic regions on 2B, 7B and 7D to be associated with HSI of grain weight and grain filling duration and with the expression of canopy temperature depression under late sown condition in field experiment explaining more than 15% of phenotypic variation for these traits. Four QTLs were identified, located on linkage groups 1B, 1D, 4A, and 7A, associated with grain quality and quality stability (Beecher et al. 2012). Seven stable QTL were identified related to HSI of grain filling duration, thousand grain weight, grain yield and canopy temperature, mapping to chromosomes 1D, 6B, 2D and 7A (Tiwari et al. 2013). Talukder et al. (2014) identified five QTL regions significantly associated with plasma membrane damage, thylakoid membrane damage and chlorophyll content in a greenhouse experiment on chromosomes 1B, 1D, 2B, 6A and 7A explained up to 33.5% of the phenotypic variation. Mondal et al. (2015) reported stable QTL for leaf wax content and leaf temperature depression in controlled conditions on chromosomes 1B and 5A, explaining 8–12% of the phenotypic variation. Sharma et al. (2016) identified stable QTLs associated with grain filling duration, grain number and productive tillers on chromosomes 1B, 2B, 3B, 5A and 6B in field experiment explaining up to 22% phenotypic variation (Table 5.2).

5.6 Transcriptomics and Proteomics Studies for Heat and Drought Stress Tolerance

To cope with abiotic stresses, plants execute a number of physiological and metabolic responses which are regulated mostly at the gene expression level. In recent years, transcriptomics and proteomics have been applied to identify stress-responsive genes and proteins that are regulated by elevated temperatures and drought stress in several crops including wheat.

Table 5.2 A summary of major and stable QTLs for heat and drought tolerance reported in wheat

Chromosome	Marker	R^2	Trait	Condition	Germplasm	References
1A	*cfa2129*	27.4, 22.6	HSIGN, HSISGW	H	64 RILS, Halberd/Cutter	Mason et al. (2010)
1A	*wPt-9757*	13.5	DSICID	D	152 RILs, Langdon/G18-16	Peleg et al. (2009)
1A	*wmc469*	38	TGW	D	189 durum wheat elite accessions	Maccaferri et al. (2011)
1B	*gwm190*	13.1	SSIGW	H	144 RILs Kauz/MTRWA116	Mohammadi et al. (2008)
1B	*acg/cta-2*	14.1	PT	D	194 RILs, SeriM82/Babax	McIntyre et al. (2010)
1B	*wmc419*	36.6	PH	D	189 durum wheat elite accessions	Maccaferri et al. (2011)
1B	*2249474|F|0*	15	HSIGFD	H	92 RILs, K 7903/RAJ 4014	Sharma et al. (2016)
1B	*agg/cat-4*	24.2	NDVI	H	167 RILs, SeriM82/Babax	Pinto et al. (2010)
1D	*wmc216*	16.84	HSIYD	H	DH 138, Berkut/cv. Krichauff	Tiwari et al. (2013)
1D	*wPt9664*	12.43	HSICT	H	DH 138, Berkut/cv. Krichauff	Tiwari et al. (2013)
2A	*gwm294*	17.8	HSISGW	H	64 RILs, Halberd/Cutter	Mason et al. (2010)
2A	*gwm356*	21	HSISGW	H	64 RILs, Halberd/Cutter	Mason et al. (2010)
2A	*wmc407*	15	HSIGFD	H	64 RILs, Halberd/Cutter	Mason et al. (2010)
2A	*wPt-4855*	13	DSIYD	D	152 RILs, Langdon/G18-16	Peleg et al. (2009)
2A	*cgt.tgcg-349*	26	SEN	H	101 RILs Ventnor/Karl 92	Vijayalakshmi et al. (2010)

(continued)

Table 5.2 (continued)

Chromosome	Marker	R^2	Trait	Condition	Germplasm	References
2A	*gwm356*	17	SEN	H	101 RILs Ventnor/Karl 92	Vijayalakshmi et al. (2010)
2B	*barc200*	21.6, 25.9	HSIGN, FLW	H	64 RILs, Halberd/ Cutter	Mason et al. (2010)
2B	*gwm111*	24.8	HSIGW, HSIGN	H	64 RILs, Halberd/ Cutter	Mason et al. (2010)
2B	*gwm374*	13.1	YD	D	152 RILs, Langdon/ G18-16	Peleg et al. (2009)
2B	*gwm410*	17.4	CL	D	152 RILs, Langdon/ G18-16	Peleg et al. (2009)
2B	*wPt-0694*	23.6, 13.1	DM, YD	D	152 RILs, Langdon/ G18-16	Peleg et al. (2009)
2B	*Xcfa175*	46.5	Fv/Fm	D	104 RILs C306/ HUW206	Kumar et al. (2012)
2B	*Xgwm356*	50.1	CHL	D	104 RILs C306/ HUW206	Kumar et al. (2012)
2B	*1161184|F|0*	15	HSIGN	H	92 RILs, K 7903/RAJ 4014	Sharma et al. (2016)
2D	*cfd56*	23.5	DH	H	121 RILs, Halberd/ Karl92	Mason et al. (2011)
2D	*gwm261*	19.3	HSISGW	H	121 RILs, Halberd/ Karl92	Mason et al. (2011)
2D	*gwm484*	15.2, 32.8	DM,FLL	H	121 RILs, Halberd/ Karl92	Mason et al. (2011)
2D	*cfd233*	20.53	HSIGFD	H	DH 138, Berkut/cv. Krichauff	Tiwari et al. (2013)
3B	*wmc326*	21.2	HSIGW	H	64 RILs, Halberd/ Cutter	Mason et al. (2010)
3B	*wmc527*	19	HSIGW	H	64 RILs, Halberd/ Cutter	Mason et al. (2010)

(continued)

Table 5.2 (continued)

Chromosome	Marker	R^2	Trait	Condition	Germplasm	References
3B	*barc101*	45.19	TGW	D	151 RILs, Oste-Gata/ Massara	Golabadi et al. (2010)
3B	*gtg.agct-205*	18	SEN	H	101 RILs Ventnor/Karl 92	Vijayalakshmi et al. (2010)
3B	*P3622-400*	17.8	Fv/Fm	D	150 DHs, Hanxuan10/ Lumai 14	Yang et al. (2007)
3B	*gwm284*	13.75	Fv/Fm	D	150 DHs, Hanxuan10/ Lumai 14	Yang et al. (2007)
3B	*barc68*	59.1	CHL	D	104 RILs C306/ HUW206	Kumar et al. (2012)
3B	*1145590|F|0*	18	PT	H	92 RILs, K 7903/RAJ 4014	Sharma et al. (2016)
3B	*wPt-1804*	15.1	YD	D	167 RILs, SeriM82/ Babax	Pinto et al. (2010)
4A	*wmc89*	13.5	HSISGW	H	64 RILs, Halberd/ Cutter	Mason et al. (2010)
4A	*barc170*	15.5	HSIGN	H	64 RILs, Halberd/ Cutter	Mason et al. (2010)
4A	*wPt-11573*	19.1	SDM	D	152 RILs, Langdon/ G18-16	Peleg et al. (2009)
4A	*gwm0160*	21	TGW	D	100 Bread wheat elite accessions	Nezhad et al. (2012)
4A	*act/cag-5*	23.9	YD	D	167 RILs, SeriM82/ Babax	Pinto et al. (2010)
4B	*gwm368*	17	LA	H	101 RILs Ventnor/Karl 92	Vijayalakshmi et al. (2010)
4D	*cfa285*	46.4	CHL	D	104 RILs C306/ HUW206	Kumar et al. (2012)
5A	*barc197*	13.8	HSIGN	H	64 RILs, Halberd/ Cutter	Mason et al. (2010)

(continued)

Table 5.2 (continued)

Chromosome	Marker	R^2	Trait	Condition	Germplasm	References
5A	*gwm126*	32.1	HSIGN	H	64 RILs, Halberd/ Cutter	Mason et al. (2010)
5A	*gwm291*	21.9	HSIGW	H	64 RILs, Halberd/ Cutter	Mason et al. (2010)
5A	*wmc150*	16.4	VLW	H	64 RILs, Halberd/ Cutter	Mason et al. (2010)
5A	*gwm126*	14.6	CTD	H	121 RILs, Halberd/ Karl92	Mason et al. (2011)
5A	*gwm293*	13	TDM	D	152 RILs, Langdon/ G18-16	Peleg et al. (2009)
5A	*ksum024*	17.2	CID	D	152 RILs, Langdon/ G18-16	Peleg et al. (2009)
5A	*gwm156*	30	LA	H	101 RILs Ventnor/Karl 92	Vijayalakshmi et al. (2010)
5A	*1079678|F|0*	22	HSIGFD	H	92 RILs, K 7903/RAJ 4014	Sharma et al. (2016)
5B	*gwm213*	24.6	HSIGN	H	64 RILs, Halberd/ Cutter	Mason et al. (2010)
5B	*wmc160*	13	FLL, DH	H	64 RILs, Halberd/ Cutter	Mason et al. (2010)
5B	*gwm408*	13.4	HSIGN	H	121 RILs, Halberd/ Karl92	Mason et al. (2011)
5B	*gwm133*	29.5	SSIGW	H	144 RILs Kauz/ MTRWA116	Mohammadi et al. (2008)
5B	*wPt-11579*	27.4	CID	D	152 RILs, Langdon/ G18-16	Peleg et al. (2009)
5B	*wPt-6910*	27.4	CID	D	152 RILs, Langdon/ G18-16	Peleg et al. (2009)
5D	*fbb238b*	37.54	SRM	D	114 RILs, W7984/ Opata85	Salem et al. (2007)

(continued)

Table 5.2 (continued)

Chromosome	Marker	R^2	Trait	Condition	Germplasm	References
6A	*CAG. AGC-101*	26	SEN	H	101 RILs Ventnor/Karl 92	Vijayalakshmi et al. (2010)
6A	*CGT. GTG-343*	30	SEN	H	101 RILs Ventnor/Karl 92	Vijayalakshmi et al. (2010)
6A	*wmc417*	13.28	Fv/Fm	D	150 DHs, Hanxuan10/ Lumai 14	Yang et al. (2007)
6B	*1109194|F|0*	20	PT	H	92 RILs, K 7903/RAJ 4014	Sharma et al. (2016)
6B	*2280984|F|0*	12	HSIYD	H	92 RILs, K 7903/RAJ 4014	Sharma et al. (2016)
6D	*gwm325*	38.6, 13.1	HSIGW, HSIGFD	H	64 RILs, Halberd/ Cutter	Mason et al. (2010)
6D	*cfd42*	32.1	CTD	H	121 RILs, Halberd/ Karl92	Mason et al. (2011)
6D	*cfd49*	14.7	HSISGW	H	121 RILs, Halberd/ Karl92	Mason et al. (2011)
7A	*gwm282*	31.6	HSISGW	H	64 RILs, Halberd/ Cutter	Mason et al. (2010)
7A	*gwm60*	19	FLW	H	64 RILs, Halberd/ Cutter	Mason et al. (2010)
7A	*wmc422*	26.58	Fm	D	150 DHs, Hanxuan10/ Lumai 14	Yang et al. (2007)
7B	*gwm263*	15.6	DSIDH	D	152 RILs, Langdon/ G18-16	Peleg et al. (2009)
7B	*gwm263*	29.3, 42.4, 30.1, 22.4	DSIDM, DH, DM, HI	D	152 RILs, Langdon/ G18-16	Peleg et al. (2009)
7B	*gwm577*	21	LA	H	101 RILs Ventnor/Karl 92	Vijayalakshmi et al. (2010)

(continued)

Table 5.2 (continued)

Chromosome	Marker	R^2	Trait	Condition	Germplasm	References
7D	*gdm88*	17.39	Fv/Fm	D	150 DHs, Hanxuan10/ Lumai 14	Yang et al. (2007)
7D	*fbb189b*	21.01	SRM	D	114 RILs, W7984/ Opata85	Salem et al. (2007)

BM—biomass, CHL—chlorophyll content, CID—carbon isotope discrimination, CL—culm length, CTD—canopy temperature depression, DH—days to heading, DM—days to maturity, DSICID—drought susceptibility index for carbon isotope discrimination, DSIDH—drought susceptibility index for days to heading, DSIYD—drought susceptibility index for yield, FLL—flag leaf length, FLW—flag leaf width, Fm—maximal chlorophyll fluorescence, F_O—open chlorophyll fluorescence, Fv/Fm—chlorophyll fluorescence, GFR—grain filling rate, GW—grain weight, HSICT—heat susceptibility index for canopy temperature, HSIGFD—heat susceptibility index for grain filling duration, HSIGN—heat susceptibility index for grain number, HSIGW—heat susceptibility index for grain weight, HSISGW—heat susceptibility index for single grain weight, HSIYD—heat susceptibility index for yield, HT—plant height, LA—leaf area, NDVI—normalized difference vegetation index, PH—plant height, PT—productive tillers, SDM—spike dry matter, SEN—senescence, SRM—stem reserve mobilization, SSIGW—stress susceptibility index for grain weight, TDM—total dry matter, TGW—thousand grain weight, VLW—visual leaf wax, YD—yield

5.6.1 Heat Stress

Microarray analysis of gene expression has been used to investigate transcriptome changes in response to heat stress as well as combined stresses in wheat. A total of 6,560 probe sets displayed 2-fold or more changes in expression in the transcriptome of heat susceptible (Chinese Spring) and tolerant wheat (TAM107) genotypes by using a wheat genome array (Qin et al. 2008). The putative heat stress-responsive genes included those encoding heat shock proteins (HSPs), heat shock factors (HSFs), transcription factors and proteins involved in phytohormones biosynthesis/ signaling, calcium and sugar signal pathways, RNA metabolism, ribosomal proteins and primary and secondary metabolisms. A set of 313 probes were differentially expressed between the two genotypes, which could be responsible for the difference in heat tolerance of the two genotypes. Chauhan et al. (2011) identified heat-responsive genes through PCR-select subtraction technology. A total of 3,516 high quality expressed sequence tags (ESTs) were generated from three different developmental stages. Transcripts of many genes such as HSPs, transporters, protein modifiers, lipid transfer protein, L-myo-inositol-1-phosphate synthase, calcium binding proteins, membrane binding proteins, signaling molecules, helicase-like protein, alanine amino transferase, stress-induced protein Sti-1, activator of HSP90, peptidyl prolyl isomerase, heat shock factor and unknown functions were highly inducible by high temperature and remained stable at both temperature regimes.

A total of 148 transcripts of developing seed reverse subtracted library were checked for down-regulation by heat stress. Down-regulated genes were involved in carbohydrate metabolism, encoding components like sucrose synthase, amylase inhibitor, triose phosphate isomerase and soluble starch synthase. A number of genes encoding seed storage proteins (gliadins and glutenins) were also affected by high temperature. The proteome analysis in response to heat stress during grain filling in contrasting wheat cultivars revealed that proteins related to signal transduction (BRI1-KD interacting protein 114), heat shock (Hsp70), photosynthesis (Rubisco activase, sedoheptulose bisphosphatase and fructose-bisphosphate aldolase involved in RuBP generation, OEE1 involved in regulation of PSII, Peptidyl-prolyl-cis-trans isomerase required for the assembly and stabilization of PSII) and antioxidants (2-Cys peroxiredoxin BAS1) increased, while those related to nitrogen metabolism (Glutamine synthetase) decreased in the tolerant cultivar under heat stress (Wang et al. 2015). In heat tolerant genotype C306, expression level of *HSP101* showed up regulation during long term heat stress compared to heat susceptible genotype PBW343 which showed considerable reduction in *HSP101* transcripts (Almeselmani et al. 2012). Several discrete isoforms of the low molecular weight HSPs were observed as differentially expressed between the two cultivars; heat-susceptible (cv. Wyuna) and heat-tolerant (cv. Fang) of wheat (Skylas et al. 2002). Majoul et al. (2004) while working on proteomics of wheat seed development under heat stress found the constitutive accumulation of small HSPs, belonging to the family of 20 kDa small HSPs, but increased expression under heat stress treatment. In addition, three heat upregulated proteins showed similarities to elongation factors (EF) or eukaryotic translation initiation factors (eIF) indicating that translational activity was involved in the stress response. The wheat chloroplast HSP (HSP26) was highly inducible by heat stress in almost all the vegetative and generative tissues in wheat (Chauhan et al. 2012). *TaHsfA6f*, a member of the A6 subclass of heat shock transcription factors, expressed constitutively in green organs but markedly up-regulated during heat stress. Overexpression of *TaHsfA6f* in transgenic wheat resulted in up-regulation of HSPs and a number of other heat stress protection genes such as Golgi Anti-Apoptotic Protein (GAAP) and the large isoform of Rubisco activase. *TaHsfA6f* acts as transcriptional activator that directly regulates *TaHSP*, *TaGAAP*, and *TaRof1* genes in wheat and its gene regulatory network has a positive impact on thermotolerance (Xue et al. 2015).

MicroRNAs (miRNAs) are a class of small non-coding regulatory RNAs with large-scale regulatory effects on development and stress response in plants. Solexa high-throughput sequencing of wheat small RNAs revealed 9 putatively heat-responsive miRNA. The expression of miR172 was significantly decreased and 8 miRNAs, including miR156, miR159, miR160, miR166, miR168, miR169, miR827 and miR2005, were up-regulated after heat treatment (Xin et al. 2010).

5.6.2 Drought Stress

High-throughput transcriptome sequencing of wheat seedlings under normal condition and subjected to drought stress (DS), heat stress (HS) and their combination (HD) revealed that 1,328 transcription factors were responsive to stress treatments. The regulatory network analysis of HSFs and DREBs implicated that both are involved in the regulation of DS, HS and HD response and indicated a cross-talk between heat and drought stress. A large proportion (68.4%) of homeologous genes were found to exhibit expression partitioning in response to DS, HS or HD (Liu et al. 2015). The genome-wide comparison of transcript changes upon dehydration in the tolerant and sensitive wild emmer wheat (*T. turgidum* ssp. *diccocoides* (Korn.) Thell.) genotypes using the Affymetrix GeneChip® Wheat Genome Array revealed several unique genes or expression patterns such as *phospholipase C* gene, involved in 1,4,5-triphosphate (IP3)-dependent signal transduction pathways, ethylene- and abscisic acid (ABA)-dependent signaling. The preferential or faster induction of ABA-dependent transcription factors by the tolerant genotype as compared to the sensitive genotype indicated distinctive stress response pathways (Ergen et al. 2009). Two hundred and twenty-one uniquely expressed or highly abundant transcripts in the drought resistant wild emmer wheat revealed that 26% of them are involved in multilevel regulation such as transcriptional regulation, RNA binding, kinase activity and calcium and abscisic acid signaling implicated in stomatal closure (Krugman et al. 2010). Differential expression patterns were also identified in genes known to be involved in drought adaptation pathways, such as cell wall adjustment, cuticular wax deposition, lignification, osmoregulation, redox homeostasis, dehydration protection and drought-induced senescence, which demonstrated the potential of wild emmer wheat as a source for candidate genes for improving drought resistance (Krugman et al. 2010). The WRKY proteins belong to a superfamily of plant TFs involved in regulation of plant growth processes as well as biotic and abiotic stress responses. Thirty-five transcripts were detected having an identity to ten known *TaWRKY* genes through *in silico* approach using RNA-Seq data. The relative expression of *TaWRKY16/TaWRKY16-A, TaWRKY17, TaWRKY19-C, TaWRKY24, TaWRKY59, TaWRKY61*, and *TaWRKY82* were found to be up-regulated in root tissue of drought-tolerant cultivar Sivas 111/33 compared to susceptible cultivar Atay 85 (Okay et al. 2014). Genes involved in ABA, proline, glycine-betaine and sorbitol pathways were found to be up-regulated by drought stress in both bread and durum wheat. The expression levels of four 9-cis-epoxycarotenoid-dioxygenase (NCED)-related probes, the key enzyme of ABA biosynthesis, were strongly up-regulated by water stress. Several probe sets encoding enzymes involved in β-xanthophyll biosynthesis were also up-regulated by drought (Aprile et al. 2009). In plants, β-xanthophylls, violaxanthin and neoxanthin are biosynthetic precursors of ABA (Nambara and Marion-Poll 2005). The probe set encoding aldose reductase increased to seven times in bread wheat Chinese Spring under severe drought stress. This sequence expressed and regulated only in *T. aestivum*, is a typical example of a gene likely located on the D genome or regulated by genomic elements of the D genome (Aprile et al. 2009). Transgenic wheat lines overexpressing betaine aldehyde dehydrogenase (*BADH*) gene exhibited increased heat and drought tolerance. The over accumulation of glycine betaine resulted in stress tolerance through protecting the thylakoid membrane and promoting antioxidant activity, indirectly increasing photosynthesis and stabilizing water status when exposed to the combination of heat and drought (Wang et al. 2010a, b).

Drought stress-responsive miRNAs in the root and leaf of bread wheat (*T. aestivum* cv. Sivas 111/33) by miRNA microarray analysis showed that 285 miRNAs (207 upregulated and 78 downregulated) and 244 miRNAs (115 upregulated and 129 downregulated) were differentially expressed in leaf and root tissues, respectively. Among the differentially expressed miRNAs, 23 miRNAs were only expressed in the leaf and 26 miRNAs were only expressed in the root of wheat growth under drought stress. The regulatory network analysis showed that miR395 family was significantly up-regulated and has connections with a number of target transcripts, and miR159 and miR319 share a number of target genes (Akdogan et al. 2015).

5.7 Transgenic Wheat for Abiotic Stress Tolerance

It is a difficult task to genetically manipulate multigene-controlled traits through conventional breeding. Thus, introduction of an alien gene into wheat cultivars provides an alternative approach to facilitate the development of wheat varieties with improved stress tolerance for sustainable agriculture. Stress tolerance in transgenic plants has largely been achieved by over-expressing the stress-responsive gene transcription factors (Pellegrineschi et al. 2004; Kumar et al. 2017a, b; Rong et al. 2014; Saad et al. 2013), and heat shock transcription factor. Besides these, transgenic plants with enhanced expression of proteins involved in osmotic adjustment, reactive oxygen species removal, saturation of membrane-associated lipids, photosynthetic reactions, production of polyamines and protein biosynthesis process have improved stress tolerance (Grover et al. 2013) (Table 5.3).

5.8 Future Prospects

Drought and heat stresses are the most important environmental stresses affecting productivity of wheat crop worldwide. Since the degree of stresses vary in fields (degree, timing i.e. growth stage, and period of stress), the effect of genes may differ depending on environmental conditions. Elucidation of the complex mechanisms of these stresses regulated by large number of genes, requires a comprehensive and integrative approach of physiology, genetics, genomics and genetic engineering.

Table 5.3 Transgenic wheat for abiotic stress tolerance

Gene	Source	Promoter	Genotype	Trait	References
HVA1	*H. vulgare*	Maize ubiquitin (ubi)	Hi-Line	Drought stress	Sivamani et al. (2000)
DREB1A	*A. thaliana*	rd29A stress inducible	Bobwhite	Drought stress	Pellegrineschi et al. (2004)
AtNHX1	*A. thaliana*	CaMV35S	Hesheng 3, Yan 103	Salt stress	Xue et al. (2004)
GmDREB	*Glycine max*	ubi and rd29A stress inducible	Jimai 19	Drought stress, high salt, and freezing stress	Gao et al. (2005)
DREB	*A. thaliana*	rd29A stress inducible	8901, 5-98, Baofeng 104	Drought stress	Wang et al. (2006)
P5CS	*Vigna aconitifolia*	AIPC stress inducible	CD200126	Drought stress	Vendruscolo et al. (2007)
EF-Tu	*Z. mays*	ubi	Bobwhite and Xin Chun 9	Heat stress	Fu et al. (2008)
GhDREB	*G. hirsutum*	ubi and rd29A stress inducible	Yangmai, Lumai	Drought, high salt, and freezing stresses	Gao et al. (2009)
betA	*E. coli*	ubi	Jinan 17	Drought stress	He et al. (2011)
TaDREB2, TaDREB3	*T. aestivum*	Rab17 maize salt-inducible	Bobwhite	Drought stress	Morran et al. (2011)
TaNAC69	*T. aestivum*	HvDhn4 s Drought inducible	Bobwhite	Salt and drought stress	Xue et al. (2011)
AlSAP	*Aeluropus littoralis*		Karim	Salt and drought stress	Ben-Saad et al. (2012)
OsNAC1	*O. sativa*	ubi	Yangmai12	Salt and drought stress	Saad et al. (2013)
TaERF3	*T. aestivum*	ubi	Yangmai 12	Salt and drought stress	Rong et al. (2014)
TaHsfA6f	*T. aestivum*	HVA1 s Drought stress-inducible promoter	Bobwhite	Heat stress	Xue et al. (2015)
TaPIE1	*T. aestivum*	CaMV35S	Yangmai 12	Cold stress	Zhu et al. (2014)
mtlD	*E. coli*	ubi	Giza 163	Salt stress	El-Yazal et al. (2016)
TaCIPK25	*T. aestivum*	ubi	Chinese spring	Salt stress	Jin et al. (2016)
AtHDG11	*A. thaliana*	Actin1	Chinese spring	Drought stress	Li et al. (2016)

The state-of-the-art of unmanned aerial vehicle technology with high-throughput imaging systems may be integrated for efficient filed phenotyping for these stresses. Genome-wide association studies can facilitate dissection of complex genes controlling these abiotic stresses. The wheat genome sequence information can be explored to identify the candidate genes responsible for complex traits of agronomic importance to expedite the wheat improvement programs. Pyramiding and combination of different QTLs and transgenes through marker-assisted breeding approaches may lead to drought and heat tolerant wheat varieties.

References

Acuna TB, Rebetzke GJ, He X, Maynol E, Wade LJ (2014) Mapping quantitative trait loci associated with root penetration ability of wheat in contrasting environments. Mol Breed 1–12

Akdogan G, Tufekci ED, Uranbey S, Unver T (2015) Mirna-based drought regulation in wheat. Funct Integr Genomics 1–13. https://doi.org/10.1007/s10142-015-0452-1

Alexander LM, Kirigwi FM, Fritz AK, Fellers JP (2012) Mapping and quantitative trait loci analysis of drought tolerance in a spring wheat population using amplified fragment length polymorphism and diversity array technology markers. Crop Sci 52:253–261

Almeselmani M, Deshmukh PS, Chinnusamy V (2012) Effects of prolonged high temperature stress on respiration, photosynthesis and gene expression in wheat (*Triticum aestivum* L.) varieties differing in their thermotolerance. Plant Stress 6:25–32

Aprile A, Mastrangelo AM, De Leonardis AM, Galiba G, Roncaglia E, Ferrari F, De Bellis L, Turchi L, Giuliano G, Cattivelli L (2009) Transcriptional profiling in response to terminal drought stress reveals differential responses along the wheat genome. BMC Genom 10:279

Barrs HD, Weatherley PE (1962) A re-examination of the relative turgidity technique for estimating water deficits in leaves. Aust J Boil Sci 15:413–428

Beecher FW, Mason E, Mondal S, Awika J, Hays D, Ibrahim A (2012) Identification of quantitative trait loci (QTLs) associated with maintenance of wheat (*Triticum aestivum* Desf.) quality characteristics under heat stress conditions. Euphytica 188(3):361

Ben-Saad R, Ben-Ramdhan W, Zouari N, Azaza J, Mieulet D, Guiderdoni E, Ellouz R, Hassairi A (2012) Marker-free transgenic durum wheat cv. Karim expressing the *AlSAP* gene exhibits a high level of tolerance to salinity and dehydration stresses. Mol Breed Jun 130(1):521–33

Bennett D, Reynolds M, Mullan D, Izanloo A (2012) Detection of two major grain yield QTL in bread wheat (*Triticum aestivum* L.) under heat, drought and high yield potential environments. Theor Appl Genet 125:1473–1485

Botstein D, White R, Skolnick M, Davis R (1980) Construction of a genetic linkage map in man using RFLP. Am J Hum Genet 32:314–331

Breda NJ (2003) Ground-based measurements of leaf area index: a review of methods, instruments and current controversies. J Exp Bot 54:2403–2417

Byrne PF, Butler JD, Anderson GR, Haley SD (2002) QTLs for agronomic and morphological traits in spring wheat population derived from a cross of heat tolerant and heat sensitive lines. In: Proceedings of X plant, animal and microbe genomes conference San Diego, CA

Chauhan H, Khurana N, Nijhavan A, Khurana JP, Khurana P (2012) The wheat chloroplastic small heat shock protein (sHSP26) is involved in seed maturation and germination and imparts tolerance to heat stress. Plant Cell Environ 35:1912–1931

Chauhan H, Khurana N, Tyagi AK, Khurana JP, Khurana P (2011) Identification and characterization of high temperature stress responsive genes in bread wheat (*Triticum aestivum* L.) and their regulation at various stages of development. Plant Mol Biol 75:35–51

Christopher J, Christopher M, Jennings R, Jones S, Fletcher S, Borrell A, Manschadi AM, Jordan D, Mace E, Hammer G (2013) QTL for root angle and number in a population developed from bread wheats (*Triticum aestivum* L.) with contrasting adaptation to water-limited environments. Theor Appl Genet 126:1563–1574

Collard BCY, Jahufer MZZ, Brouwer JB, Pang ECK (2005) An introduction to markers quantitative trait loci (QTL) mapping and marker-assisted selection for crop improvement: the basic concepts. Euphytica 142:169–196

Collins NC, Tardieu F, Tuberosa R (2008) Quantitative trait loci and crop performance under abiotic stress: where do we stand? Plant Physiol 147:469–486

Czyczyło-Mysza I, Marcińska I, Skrzypek E, Chrupek M, Grzesiak S, Hura T, Stojałowski S, Myśków B, Milczarski P, Quarrie S (2011) Mapping QTLs for yield components and chlorophyll a fluorescence parameter in wheat under three levels of water availability. Plant Genet Resour 9:291–295

Delaney DE, Nasuda S, Endo TR, Gill BS, Hulbert SH (1995) Cytogenetically based physical maps of the group-2 chromosomes of wheat. Theor Appl Genet 91:568–573

Van Deynze AE, Dubcovsky J, Gill KS, Nelson JC, Sorrells ME, Dvorak J, Gill BS, Lagudah ES, McCouch SR, Appels R (1995) Molecular genetic maps for group 1 chromosomes of *Triticeae* species and their relation to chromosomes in rice and oat. Genome 38:45–59

Dwyer LM, Tollenaar M, Houwing L (1991) A nondestructive method to monitor leaf greenness in corn. Can J Plant Sci 71:505–509

El-Yazal MAS, Eissa HF, Ahmed SMAE, Howladar SM, Zaki SS, Rady MM (2016) The *mtlD* gene-overexpressed transgenic wheat tolerates salt stress through accumulation of mannitol and sugars. Plant 4(6):78–90

Ergen NZ, Thimmapuram J, Bohnert HJ, Budak H (2009) Transcriptome pathways unique to dehydration tolerant relatives of modern wheat. Funct Integr Genomics 9:377–396

FAO (2015) FAO statistical pocketbook. Food and Agriculture Organization, Rome

Farquhar GD, Ehleringer JR, Hubick KT (1989) Carbon isotope discrimination and photosynthesis. Annu Rev Plant Physiol 40:503–537

Fischer RA, Maurer R (1978) Drought resistance in spring wheat cultivars. I. Grain yield response. Aust J Agric Res 29:897–907

Fracheboud Y (2004) Using chlorophyll fluorescence to study photosynthesis. Institute of Plant Sciences, Zurich. Available at http://jaguar.fcav.unesp.br/download/deptos/biologia/durvalina/TEXTO-71.pdf. Accessed 23 Jan 2017

Fu J, Momčilović I, Clemente TE, Nersesian N, Trick HN, Ristic Z (2008) Heterologous expression of a plastid EF-Tu reduces protein thermal aggregation and enhances CO_2 fixation in wheat (*Triticum aestivum*) following exposure to heat stress. Plant Mol Biol 68:277–288

Gao SQ, Chen M, Xia LQ, Xiu HJ, Xu ZS, Li LC, Zhao CP, Cheng XG, Ma YZ (2009) A cotton (*Gossypium hirsutum*) DRE-binding transcription factor gene, *GhDREB*, confers enhanced tolerance to drought, high salt, and freezing stresses in transgenic wheat. Plant Cell Rep 28 (2):301–311

Gao SQ, Xu HJ, Cheng XG, Chen M, Xu ZS, Li LC, Ye XG, Du LP, Hao XY, Ma YZ (2005) Improvement of wheat drought and salt tolerance by expression of a stress-inducible transcription factor *GmDREB* of soybean (*Glycine max*). Chin Sci Bull 50:2714–2723

Golabadi M, Arzani A, Mirmohammadi Maibody SAM, Sayed Tabatabaei BE, Mohammadi SA (2010) Identification of microsatellite markers linked with yield components under drought stress at terminal growth stages in durum wheat. Euphytica 177:207–221

Grover A, Mittal D, Negi M, Lavania D (2013) Generating high temperature tolerance transgenic plants: achievements and challenges. Plant Sci 205–206:38–47

Hasanuzzaman M, Nahar K, Alam MM, Roy chowdhury R, Fujita M (2013) Physiological, biochemical and molecular mechanisms of heat stress tolerance in plants. Int J Mol Sci 14:9643–9684

He C, Zhang CW, Gao Q, Yang A, Hu X, Zhang J (2011) Enhancement of drought resistance and biomass by increasing the amount of glycine betaine in wheat seedlings. Euphytica 177:16–151

Hemantaranjan A, Nishant BA, Singh MN, Yadav DK, Patel PK, Singh R, Katiyar D (2014) Heat stress responses and thermotolerance. Adv Plants Agric Res 1:00012

IPCC (2014) Climate change 2014: impacts adaptation and vulnerability. Part B: regional aspects. Contribution of working group II to the fifth assessment report of Inter-governmental Panel on Climate Change [Barros VR, Field CB, Dokken DJ, Mastrandrea MD, Mach KJ, Bilir TE, Chatterjee M, Ebi KL, Estrada YO, Genova RC, Girma B, Kissel ES, Levy AN, MacCracken S, Mastrandrea PR, White LL (eds)]. Cambridge University Press, Cambridge, p 688

Jin X, Sun T, Wang X, Su P, Ma J, He G, Yang G (2016) Wheat CBL-interacting protein kinase 25 negatively regulates salt tolerance in transgenic wheat. Sci Rep 6:28884

Krugman T, Chagué V, Peleg Z, Balzergue S, Just J, Korol AB, Nevo E, Saranga Y, Chalhoub B, Fahima T (2010) Multilevel regulation and signalling processes associated with adaptation to terminal drought in wild emmer wheat. Funct Integr Genomics 10:167–186

Kuchel H, Fox R, Reinheimer J, Mosionek L, Willey N, Bariana H, Jefferies S (2007) The successful application of a marker-assisted wheat breeding strategy. Mol Breed 20:295–308

Kumar U, Joshi AK, Kumari M, Paliwal R, Kumar S, Roder MS (2010) Identification of QTLs for stay green trait in wheat (*Triticum aestivum* L.) in the ‘Chirya 3’ X ‘Sonalika’ population. Euphytica 174:437–445

Kumar S, Sehgal SK, Kumar U, Vara Prasad PV, Joshi AK, Gill BS (2012) Genomic characterization of drought tolerance-related traits in spring wheat. Euphytica 186(1):265–276

Kumar R, Mamrutha HM, Kaur A, Venkatesh K, Grewal A, Kumar R, Tiwari V (2017a) Development of an efficient and reproducible regeneration system in wheat (*Triticum aestivum* L.). Physiol Mol Biol Plants 23:945–954

Kumar R, Mamrutha HM, Kaur A, Grewal A (2017b) Synergistic effect of cefotaxime and timentin to suppress the *Agrobacterium* overgrowth in wheat (*Triticum aestivum L.*) transformation. Asian J Microbiol Biotechnol Environ Sci 19(4):961–967

Li L, Zheng M, Deng G, Liang J, Zhang H, Pan Z, Long H, Yu M (2016) Overexpression of *AtHDG11* enhanced drought tolerance in wheat (*Triticum aestivum* L.). Mol Breed 36:23

Litt M, Luty JA (1989) A hypervariable microsatellite revealed by *in vitro* amplification of a dinucleotide repeat within the cardiac muscle actin gene. Am J Hum Gen 44:397–401

Liu Z, Xin M, Qin J, Peng H, Zhongfu N, Yao Y, Sun Q (2015) Temporal transcriptome profiling reveals expression partitioning of homeologous genes contributing to heat and drought acclimation in wheat (*Triticum aestivum* L.). BMC Plant Biol 15:152. https://doi.org/10.1186/s12870-015-0511-8

Maccaferri M, Sanguineti MC, Demontis A, El-Ahmed A, Moral LG, Maalouf F, Nachit M, Nserallah N, Ouabbou H, Rhouma S, Royo C, Villegas D, Tuberosa R (2011) Association mapping in durum wheat grown across a broad range of water regimes. J Exp Bot 62:409–438

Majoul T, Bancel E, Tribol E, Hamida JB, Branlard G (2004) Proteomic analysis of the effect of heat stress on hexaploid wheat grain: characterization of heat-responsive proteins from non-prolamins fraction. Proteomics 4:505–513

Marino CL, Nelson JC, Lu YH, Sorrells ME, Lopes CR, Hart GE (1996) Molecular genetic linkage maps of the group 6 chromosomes of hexaploid wheat (*Triticum aestivum* L. em. Thell). Genome 39:359–366

Mason RE, Hays DB, Mondal S, Ibrahim AMH, Basnet BR (2013) QTL for yield, yield components and canopy temperature depression in wheat under late sown field conditions. Euphytica 194:243–259

Mason RE, Mondal S, Beecher FW, Hays DB (2011) Genetic loci linking improved heat tolerance in wheat (*Triticum aestivum* L.) to lower leaf and spike temperatures under controlled conditions. Euphytica 180:181–194

Mason RE, Mondal S, Beecher FW, Pacheco A, Jampala B, Ibrahim AMH, Hays DB (2010) QTL associated with heat susceptibility index in wheat (*Triticum aestivum* L.) under short-term reproductive stage heat stress. Euphytica 174:423–436

Mathews K, Malosetti M, Chapman S, McIntyre L, Reynolds M, Shorter R, van Eeuwijk F (2008) Multi-environment QTL mixed models for drought stress adaptation in wheat. Theor Appl Genet 117(7):1077–1091

McIntyre CL, Mathews KL, Rattey A, Drenth J, Ghaderi M, Reynolds M, Chapman SC, Shorter R (2010) Molecular detection of genomic regions associated with grain yield and yield components in an elite bread wheat cross evaluated under irrigated and rainfed conditions. Theor Appl Genet 120(3):527–541

Mickelson-Young L, Endo TR, Gill BS (1995) A cytogenetic ladder map of the wheat homoeologous group-4 chromosomes. Theor Appl Genet 90:1007–1011

Mohammadi V, Zali AA, Bihamta (2008) Mapping QTL for heat tolerance in wheat. J Agric Sci Technol 10:261–267

Mondal S, Mason RE, Huggins T, Hays DB (2015) QTL on wheat (*Triticum aestivum* L.) chromosomes 1B, 3D and 5A are associated with constitutive production of leaf cuticular wax and may contribute to lower leaf temperatures under heat stress. Euphytica 201(1):123–130

Morgan JM (1983) Osmoregulation as a selection criterion for drought tolerance in wheat. Aust J Agric Res 34:607–614

Morgan JM, Tan MK (1996) Chromosomal location of a wheat osmoregulation gene using RFLP analysis. Aust J Plant Physiol 23:803–806

Morran S, Eini O, Pyvovarenko T, Parent B, Singh R, Ismagul A, Eliby S, Shirley N, Langridge P, Lopato S (2011) Improvement of stress tolerance of wheat and barley by modulation of expression of DREB/CBF factors. Plant Biotechnol J 9:230–249

Mullan DJ, Reynolds MP (2010) Quantifying genetic effects of ground cover on soil water evaporation using digital imaging. Funct Plant Biol 37:703–712

Nambara E, Marion-Poll A (2005) Abscisic acid biosynthesis and catabolism. Annu Rev Plant Biol 56:165–185

Nelson JC, Van Deynze AE, Autrique E, Sorrells ME, Lu YH, Negre S, Bernard M, Leroy P (1995b) Molecular mapping of wheat homoeologous group 3. Genome 38:525–533

Nelson JC, Deynze AE, Sorrells ME, Autrique E, Lu YH, Merlino M, Atkinson M, Leroy P (1995a) Molecular mapping of wheat. Homoeologous group 2. Genome 38:516–524

Nezhad K, Weber WE, Roder MS, Sharma S, Lohwasser U, Meyer RC, Saal B, Borner A (2012) QTL analysis for thousand-grain weight under terminal drought stress in bread wheat (*Triticum aestivum* L.). Euphytica 186:127–138

Okay S, Derelli E, Unver T (2014) Transcriptome-wide identification of bread wheat WRKY transcription factors in response to drought stress. Mol Genet Genomics. https://doi.org/10.1007/s00438-014-0849-x

Ortiz R, Sayre KD, Govaerts B (2008) Climate change: can wheat beat the heat. Agric Ecosyst Environ 126:46–58

Paliwal R, Roder MS, Kumar U, Srivastava JP, Joshi AK (2012) QTL mapping of terminal heat tolerance in hexaploid wheat (*T. aestivum* L.). Theor Appl Genet 125(3):561–575

Pandey GC, Mamrutha HM, Tiwari R, Sareen S, Bhatia S, Siwach P, Tiwari V, Sharma I (2015) Physiological traits associated with heat tolerance in bread wheat (*Triticum aestivum L*). Physiol Mol Biol Plants 21(1):93–99

Paroda RS (2013) Indian seed sector: the way forward. Special lecture delivered at Indian seed congress, Friday, 8 Feb 2013. Published by NSAI, New Delhi

Pearcy RW, Ehleringer J, Mooney HA, Rundel PW (eds) (1989) Plant physiological ecology. Field methods and instrumentation. Chapman & Hall, London

Peleg Z, Fahima T, Krugman T, Abbo S, Yakir D, Korol AB, Saranga Y (2009) Genomic dissection of drought resistance in durum wheat × wild emmer wheat recombinant inbreed line population. Plant Cell Environ 32:758–779

Pellegrineschi A, Reynolds M, Pacheco M, Brito RM, Almeraya R, Yamaguchi-Shinozaki K, Hoisington D (2004) Stress-induced expression in wheat of the *Arabidopsis thaliana DREB1A* gene delays water stress symptoms under greenhouse conditions. Genome 47:493–500

Pinto RS, Reynolds MP, Mathews KL, McIntyre CL, Olivares-Villegas JJ, Chapman SC (2010) Heat and drought adaptive QTL in a wheat population designed to minimize confounding agronomic effects. Theor Appl Genet 121:1001–1021

Qin D, Wu H, Peng H, Yao Y, Ni Z, Li Z, Zhou C, Sun Q (2008) Heat stress-responsive transcriptome analysis in heat susceptible and tolerant wheat (*Triticum aestivum* L.) by using Wheat Genome Array. BMC Genom 9:432

Quarrie SA, Gulli M, Calestani C, Steed A, Marmiroli N (1994) Location of a gene regulating drought-induced abscisic acid production on the long arm of chromosome 5A of wheat. Theor Appl Genet 89:794–800

Rebetzke GJ, Condon AG, Farquhar GD, Appels R, Richards RA (2008) Quantitative trait loci for carbon isotope discrimination are repeatable across environments and wheat mapping populations. Theor Appl Genet 118:123–137

Rebetzke GJ, Condon AG, Richards RA, Read JJ (2001) Phenotypic variation and sampling for leaf conductance in wheat (*Triticum aestivum* L.) breeding populations. Euphytica 121:335–341

Rong W, Qi L, Wang A, Ye X, Du L, Liang H, Xin Z, Zhang Z (2014) The ERF transcription factor TaERF3 promotes tolerance to salt and drought stresses in wheat. Plant Biotechnol J 12:468–479

Saad AS, Li X, Li HP, Huang T, Gao CS, Guo MW, Cheng W, Zhao GY, Liao YC (2013) A rice stress-responsive NAC gene enhances tolerance of transgenic wheat to drought and salt stresses. Plant Sci 203–204:33–40

Salem KFM, Röder MS, Börner A (2007) Identification and mapping quantitative trait loci for stem reserve mobilisation in wheat (*Triticum aestivum* L.). Cereal Res Commun 35(3):1367–1374

Sharma D, Mamrutha HM, Gupta VK, Tiwari R, Singh R (2015) Association of SSCP variants of HSP genes with physiological and yield traits under heat stress in wheat. Res Crops 16(1):139–146

Sharma D, Tiwari R, Gupta VK, Rane J, Singh R (2018) Genotype and ambient temperature during growth can determine the quality of starch from wheat. J Cer Sci 79:240–246

Sharma D, Singh R, Rane J, Gupta VK, Mamrutha HM, Tiwari R (2016) Mapping quantitative trait loci associated with grain filling duration and grain number under terminal heat stress in bread wheat (*Triticum aestivum* L.). Plant Breed 135(5):538–545

Sivamani E, Bahieldinl A, Wraith JM, Al-Niemi T, Dyer WE, David Ho TH, Qu RD (2000) Improved biomass productivity and water use efficiency under water deficit conditions in transgenic wheat constitutively expressing the barley *HVA1* gene. Plant Sci 155:1–9

Skylas DJ, Cordwell SJ, Hains PG, Larsen MR, Basseal DJ, Walsh BJ, Blumenthal C, Rathmell W, Copeland L, Wrigley CW (2002) Heat shock of wheat during grain filling: proteins associated with heat-tolerance. J Cereal Sci 35:175–188

Talukder SK, Babar MA, Vijayalakshmi K, Poland J, Prasad PV, Bowden R, Fritz A (2014) Mapping QTL for the traits associated with heat tolerance in wheat (*Triticum aestivum* L.). BMC Genet 15:97

Tautz D, Trick M, Dover GA (1986) Cryptic simplicity in DNA is a major source of genetic variation. Nature 322:652–656

Tiwari C, Wallwork H, Kumar U, Dhari R, Arun B, Mishra VK, Reynolds MP, Joshi AK (2013) Molecular mapping of high temperature tolerance in bread wheat adapted to the eastern Gangetic Plain of India. Field Crop Res 154:201–210

Tiwari R, Mamrutha HM (2014) Precision phenotyping for mapping of traits for abiotic stress tolerance in crops. In: Salar RK (ed) Biotechnology: prospects and applications. Springer India, New Delhi, p 79

Turner NC (1988) Measurement of plant water status by the pressure chamber technique. Irrig Sci 9:289–308

Vendruscolo EC, Schuster I, Pileggi M, Scapim CA, Molinari HB, Marur CJ, Vieira LG (2007) Stress-induced synthesis of proline confers tolerance to water deficit in transgenic wheat. J Plant Physiol 164(10):1367–1376

Vijayalakshmi K, Fritz AK, Paulsen GM, Bai G, Pandravada S, Gill BS (2010) Modeling and mapping QTL for senescence-related traits in winter wheat under high temperature. Mol Breed 26:163–175

Vos P, Hogers R, Bleeker M, Reijans M, Lee T, Hornes M, Frijters A, Pot J, Peleman J, Kuiper M, Zabeau M (1995) AFLP: a new technique for DNA fingerprinting. Nucleic Acids Res 23:4407–4414

Wahid A, Gelani S, Ashraf M, Foolad MR (2007) Heat tolerance in plants: an overview. Environ Exp Bot 61:199–223

Wang X, Dinler BS, Vignjevic M, Jacobsen S, Wollenweber B (2015) Physiological and proteome studies of responses to heat stress during grain filling in contrasting wheat cultivars. Plant Sci 230:33–50

Wang GP, Li F, Zhang J, Zhao MR, Hui Z, Wang W (2010b) Over accumulation of glycine betaine enhances tolerance of the photosynthetic apparatus to drought and heat stress in wheat. Photosynthetica 48:30–41

Wang JW, Yang FP, Chen XQ, Liang RQ, Zhang LQ, Geng DM, Zhang XD, Song YZ, Zhang GS (2006) Induced expression of DREB transcriptional factor and study on its physiological effects of drought tolerance in transgenic wheat. Acta Genet Sin 33(5):468–476

Wang GP, Zhang XY, Li F, Luo Y, Wang W (2010a) Over accumulation of glycine betaine enhances tolerance to drought and heat stress in wheat leaves in the protection of photosynthesis. Photosynthetica 48:117–126

Williams JGK, Kubelik AR, Livak KJ, Rafalski JA, Tingey SV (1990) DNA polymorphisms amplified by arbitrary primers are useful as genetic markers. Nucleic Acids Res 18:6531–6535

Xin M, Wang Y, Yao Y, Xie C, Peng H, Ni Z, Sun Q (2010) Diverse set of microRNAs are responsive to powdery mildew infection and heat stress in wheat (*Triticum aestivum* L.). BMC Plant Biol 10:123

Xue GP, Drenth J, McIntyre CL (2015) *TaHsfA6f* is a transcriptional activator that regulates a suite of heat stress protection genes in wheat (*Triticum aestivum* L.) including previously unknown Hsf targets. J Exp Bot 66:1025–1039

Xue GP, Way HM, Richardson T, Drenth J, Joyce PA, McIntyre CL (2011) Overexpression of *TaNAC69* leads to enhanced transcript levels of stress up-regulated genes and dehydration tolerance in bread wheat. Mol Plant 4(4):697–712

Xue ZY, Zhi DY, Xue GP, Zhang H, Zhao YX, Xia GM (2004) Enhanced salt tolerance of transgenic wheat (*Triticum aestivum* L.) expressing a vacuolar Na^+/H^+ antiporter gene with improved grain yields in saline soils in the field and a reduced level of leaf Na^+. Plant Sci 167 (4):849–859

Yang DL, Jing RL, Chang XP, Li W (2007) Quantitative trait loci mapping for chlorophyll fluorescence and associated traits in wheat (*Triticum aestivum*). J Integr Plant Biol 49:646–654

Yang J, Sears RG, Gill BS, Paulsen GM (2002) Quantitative and molecular characterization of heat tolerance in hexaploid wheat. Euphytica 126:275–282

Zadoks JC, Chang TT, Konzak CF (1974) A decimal code for the growth stages of cereals. Weed Res 14:415–421

Zhu X, Qi L, Liu X, Cai S, Xu H, Huang R, Li J, Wei X, Zhang Z (2014) The wheat ethylene response factor transcription factor pathogen-induced ERF1 mediates host responses to both the necrotrophic pathogen *Rhizoctonia cerealis* and freezing stresses. Plant Physiol 164:1499–1514. https://doi.org/10.1104/pp.113.229575

Chapter 6
Molecular Chaperones: Key Players of Abiotic Stress Response in Plants

Suchismita Roy, Manjari Mishra, Om Prakash Dhankher, Sneh L. Singla-Pareek and Ashwani Pareek

Abstract Plants counter an array of stresses by generation of a group of stress-related proteins, often referred to as the chaperones. Expression of these chaperones is induced in response to almost all kinds of stress. However, there are numerous evidences showing that these chaperones are vital for survival even under normal physiological conditions. They act as key modulators in physiological stress response and acquired tolerance. Research carried out over the past several years has clearly established that these chaperones are involved in diverse cellular functions such as folding, accumulation, translocation and degradation of proteins. Thus, these evolutionarily conserved proteins affect a broad array of cellular processes. Gaining knowledge about this cellular chaperone machinery is of immense significance to understand the mechanism of interdependent stress-related cross talk in plants and ultimately, for the crop improvement programs.

Keywords Calnexin · Calreticulin · Chaperones · Cyclophilins
Heat shock proteins

Suchismita Roy and Manjari Mishra have equally contributed to this work.

S. Roy · M. Mishra · A. Pareek (✉)
School of Life Sciences, Jawaharlal Nehru University, New Delhi, India
e-mail: ashwanip@mail.jnu.ac.in

O. P. Dhankher
Stockbridge School of Agriculture, University of Massachusetts,
Amherst, MA 01003, USA

S. L. Singla-Pareek
Plant Stress Biology, International Centre for Genetic Engineering and Biotechnology,
New Delhi, India

V. R. Rajpal et al. (eds.), *Genetic Enhancement of Crops for Tolerance to Abiotic Stress: Mechanisms and Approaches, Vol. I*, Sustainable Development and Biodiversity 20, https://doi.org/10.1007/978-3-319-91956-0_6

6.1 Introduction

Plants are sessile and thus are exposed to several harsh environmental cues. Abiotic stresses are one of the main causes for yield loss throughout the world (Suzuki et al. 2014). Unavoidable forms of these stresses as drought, salinity, extremes of temperature and air pollutants act concurrently on plants causing disturbances in their overall cellular homeostasis. In order to cope with such conditions, plants have devised a wide range of mechanisms for their survival, sustained growth and development. In plant system, stress tolerance is a myriad of various stress responsive mechanisms that act in an orchestrated manner (Sewelam et al. 2016). Chaperones and foldases, such as heat shock proteins (Hsps), immunophilins, calnexins and calreticulins are the primary requisites for survival in unsuitable environments. Recent studies have shown that these chaperone families are also effective in neutralizing phenotypic variation (Rutherford 2003). Research shows that the heat shock response is highly elastic in its alteration under different environments, even in similar species. This chapter largely focuses on critically presenting the evidence suggesting the role of chaperones in plants in relation to varied abiotic stresses, normal physiological situation and their comprehensive network of actions.

The cells perceive sudden fluctuation in the cellular repertoire as "shock". This shock hinders plant's normal physiological, metabolic, morphological and anatomical features, and causes a drastic reduction in plant's performance and yield. The sudden and abrupt changes cause transcriptional activation of several shock-related genes, which cause rapid synthesis and accumulation of several chaperone-like proteins (Verghese et al. 2012). The reaction to heat-shock response is well adapted in different organisms of diverse ecological niche. The algae experience heat-shock response at 5°C, whereas the thermophylic archaea experiences such kind of stress at temperatures above 100°C (Rutherford 2003). Chaperone system existing in the plant, forms a complex network involving different molecules at different cellular and subcellular level (Verghese et al. 2012). Recently, it has become quite evident that they interact with other stress responsive components and thus administer stress tolerance to the plant. Amongst these chaperones, Hsps are the most abundant ones, expressed in response to all other forms of stresses (Fulda et al. 2010). These diverse class of proteins are known to have allied functions, but with a distinctive mechanism of action. Hsps rearrange themselves along with nascent or stress-related proteins, help in rescuing misfoldings and aggregation of proteins via binding to various intermediate complexes. Due to their stress-related behavioural pattern, these are often called as the "stress proteins" (Benjamin and McMillan 1998). The best explained chaperone systems till date are the Hsp70/DnaK and Hsp60/Gro. Hsps, initially, were identified as a class of proteins that show enhancement in their expression when the cells are exposed to high temperature. Ritossa (1962) observed these heat shock proteins while observing the gene expression of the puffing polytene chromosomes of *Drosophila* subjected to high temperature stress. Subsequently, he observed this phenomenon of increase in protein synthesis by added stressors such as azide, 2,

4-dinitrophenol and salicylate. Later, these proteins were acknowledged as heat shock proteins (Tissieres et al. 1974). These classes of proteins are present in all organisms (Lindquist 1986). Based on their intra and extracellular localization, these proteins have different roles. Intracellular Hsps pose a defensive mechanism, allowing cells to endure lethal conditions. Several Hsps have been identified to act together with other components of the coordinated cell death or apoptosis. Extracellularly located Hsps or other membrane-bound Hsps are known to trigger several immunological functions as well. They elicit an immune reaction nurtured by the adaptive or innate immunity of the organism. This system is not well studied in plants though.

Hsps have a mass ranging from 10 to 200 kDa and they take part in the perception and induction of the signaling cascade during stress (Schoffl et al. 1999). Numerous forms of Hsps have been analysed in almost all organisms. Schlesinger (1990) concluded that the principle Hsps of *Homo sapiens* do not differ from those of bacteria, except for the presence of Hsp33. Most of the Hsps have a characteristic C terminal domain called heat-shock domain (Helm et al. 1993).

Other chaperones include, peptidyl-prolyl cis/trans isomerases (PPIases) and protein disulfide isomerases (PDIs), which aid in catalyzing the arrangement and reorganization of *cis-trans* isomerization of peptide linkages presiding proline residues and disulfide bonds, respectively. These chaperones are collectively known as foldases as they act in similar manner, in folding or unfolding of proteins, assembly of multi-protein units, trafficking proteins to and from the subcellular compartmnets, controling cell cycle, signaling and safeguarding of cells against stress or cell death (Kumari et al. 2013). The common property amongst them is their ability to bind to substrate proteins in their unsteady structural forms and reconstitute them back to their native form. Other endoplasmic reticulum based organellar chaperones are the calreticulin and calnexins. These are Ca^{2+} binding chaperones of eukaryotic systems, acting in glycoprotein folding mechanics, protein quality control and Ca^{2+} concentration (Boston 1996). Table 6.1 enlists chaperones in plants and their characteristics including: (a) nomenclature (b) subcellular localizations (c) functions (d) homologues.

6.2 Functional Characterization of Molecular Chaperones in Plants

Molecular chaperones generally identify the hydrophobic regions on unfolded polypeptides facilitating proper protein folding and preventing their aggregation. Free energy of the residing amino acids of a protein decides the pattern in which protein folds and this is the main determining factor for the function of any protein (Levitt et al. 1997). However, factors like temperature, acidity or alkalanity, salt concentration, and total protein concentration affect the proper folding of proteins. The molecular chaperones bind to and stabilize unfolded and partially folded polypeptides by preventing unnecessary interaction via minimizing the protein aggregation or promoting dissociation of aggregates and direct them to proper

Table 6.1 Chaperones in plants, suggested prokaryotic homologues location and functions

Chaperones in plants	Function	Homologues in prokaryotes	Location	References
Hsp100	Show ATPse activity; dissociates aggregates, facilitates proteolysis, resolubilization of non-functional protein aggregates, protein remodelling	Clp B	Cytosol	Parsell et al. (1994)
Hsp90	Stabilizes proteins prior to complete folding or activation; forms stable complexes with inactive receptor and other transcription factors	HtpG	Endoplasmic reticulum (ER)	Taipale et al. (2010)
Hsp70	Show ATPase activity, protein folding, transport across membranes and proteolysis; and prevents aggregation of misfolded and unassembled proteins, signaling	DnaK/Ssa	Cytosol, ER, mitochondria, chloroplasts, peroxisomes	Miernyk (1997)
Hsp60	Show ATPase activity, promotes efficient folding, forms stable inactive aggregates	Chaperonin/ Cpn60/ GroEL	Plastids, mitochondria, and cytoplasm	Hemmingsen et al. (1988)
Hsp40	Gene expression and translational initiation, folding and unfolding as well as translocation and degradation of proteins	DnaJ	Cytosol, ER	Kampinga and Craig (2010)
sHsp	Blocks aggregation of protein; stabilizes misfolded protein involved in regulation of actin assembly/ disassembly	–	Nucleus, membrane, cytoskeleton	Mogk et al. (2003)
Protein di sulphidase	Breakage and alternative isomerization of incorrectly formed peptide linkages	–	Lumen of ER, mitochondria, nucleus, and cytosol	Hatahet and Ruddock (2007)
Peptidyl prolyl *cis trans* isomerase	Protein stabilization and protein folding, hormonal signaling	–	Cytosol, chloroplast, nucleus, mitochondria	Kumari et al. (2009)
Calnexins and calreticulins	Maintaining protein folding machinery that enhances correct folding of proteins	–	ER	Michalak et al. (2009)

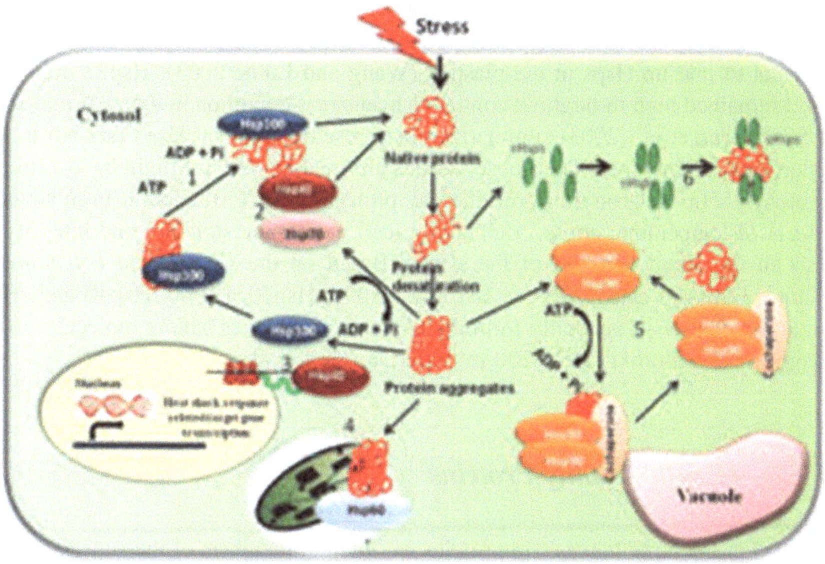

Fig. 6.1 Role of chaperones in plants under stress condition.(1) Protein remodelling via Hsp100. (2) Protein refolding via Hsp70 and Hsp40. (3) Preventing of misfolding in newly synthesized protein via Hsp40. (4) Protein stabilization in the chloroplast via Hsp60. (5) Protein refolding *via* ATPase activity of Hsp90 and co-chaperonin complex. (6) Protein stabilization via dimeric sHsps

protein folding, transportation and degradation pathways (Miernyk 1997; Sigler et al. 1998). Chaperones bind to the nascent polypeptide chain, while its translation is going on, to prevent misfolding and aggregation of amino terminal of peptidal chain, until it folds in a proper manner. Chaperones are also known to bind to and stabilize protein in unfolded confirmation during their subcellular repositioning (Young et al. 2004). For example, during the transfer of protein from cytosol to the mitochondria, cytosolic chaperones bind to and stabilize the protein in cytosol in a partially unfolded conformation; mitochondrial chaperones facilitate their transfer across mitochondrial membrane and assist in protein folding within the same. They assist in assembly of poypeptide chains, assembly of macromolecular structures and regulation of protein degradation (Young et al. 2004). Figure 6.1 enlists the known chaperones, their related functions under stress and summarizing their role in plants.

6.2.1 *Heat Shock Proteins (Hsps)*

Hsp gene family is well characterized in plants and they are required for normal build up and regulation of the plant as well as to tide over unfavorable environmental conditions. Hsps are primarily concerned in heat stress-related response in association with the response of downstream proteins. Mutants, not capable of producing these Hsps, are prone to severe heat-induced damage (Burke 2001). Heat sensitivity towards heat stress was allied with lower efficiency of the bentgrass variants to pile up Hsps in the plastids (Wang and Luthe 2003). Hsp22 transcript level remained high throughout continous heat stress condition in maize (Lund et al. 1998). Barua et al. (2003) summarized the presence of small Hsps (sHsps) in the chloroplast membranes. Their observations affirmed that sHsps might have a role in photosynthetic electron transport, thereby protecting the PSII subunit from adverse effects of temperature stress. Maintaining cell membrane structure and integrity is also an important function of the sHsps. Based on the *Escherichia coli* nomenclature, Hsps are categorized as Hsp100, Hsp90, Hsp70, Hsp60, Hsp40 and sHsp proteins. The sHsps generally form multisubunit complexes having molecular mass ranging from approximately 200 to 800 kDa (Kim et al. 1998).

6.2.2 Hsp100 Family Proteins

Hsp100 or the homologues Clp proteins are integral members of the superfamily of ATPase with an extensive variety of different active properties such as protein disaggregation and degradation (Parsell et al. 1994). Hsp100 proteins perform binary function as a chaperone and also as a regulatory protein. These proteins act via altering the portion of fully degraded or distorted protein substrates in the cell (Liberek et al. 2008). The removal of polypeptides that are not functional and are potentially harmful as a result of misfolding, accumulation or denaturation is crucial for the maintenance of homeostasis in the cells. Another unique task of this class is the recovery of protein aggregates by resolubilization of protein aggregates, which are in their unfolded state. This also serves to degrade totally damaged polypeptides (Kim and Schoffl 2002). The mission of rescuing aggregated proteins involves the support of Hsp70 protein. Hsp70 is a different ATP-dependent chaperone system. Hsp100 helps in solubilizing the protein aggregate and delivers it in a state that Hsp70 complex can revert back (Zolkiewski et al. 2012). Hsp100 proteins are typically hexameric rings, with a middle domain (M domain), an N-terminal domain (N domain) followed by a nucleotide-binding domain (NBD-1). In addition to these domains, they also contain a second nucleotide-binding domain (NBD-2), next to the C terminal domain. N and M domains are indistinctly related (Lipinska et al. 2013). The NBDs are competent enough to bind and cause hydrolysis of ATP. Once ATP binding takes place, it mutually stabilizes the oligomeric status and its connections with the substrate protein. Protein remodeling in this case also occurs by the hydrolysis of ATP. The M domain is distinctive for Hsp104, which is also found to be vital for protein remodeling (Lipinska et al. 2013).

Hsp100 has been surveyed in many plant species including *Arabidopsis*, soybean (*Glycine max*), tobacco (*Nicotiana tabaccum*), rice (*Oryza sativa*), maize (*Zea mays*) and wheat (*Triticum* spp.) (Pareek et al 1995). Hsp100 like molecular chaperones are frequently expressed throughout plant life. Environmental stresses such as extreme temperatures, dessication, saline or dark-induced etiolation enhances their expressions. Hsp101 in association with the 5-leader sequence of tobacco mosaic virus (TMV) has also been recognized as a *trans*-acting feature accountable for mediating enhancement of translation (Verchot 2012). In the same report, Hsp101 has been suggested to function as a RNA binding protein, which gets associated to a poly CAA region inside the leader sequence.

6.2.3 Hsp90 Family Proteins

Hsp90 relatively comprises of 1–2% of total protein present in the eukaryotic cytosol (Taipale et al. 2010). Hsp90 adorns the much important central regulatory protein of many of the biological pathways regulating growth and development. It also serves an important role in evolutionary processes. Hsp90s expression profiles in *Arabidopsis* have been shown to keep pace both with the plant's developmental stages and with different stressers such as salt stress, severe temperature conditions, heavy metals etc. Their expression profile also gets affected by the phytohormones and circadian rhythm (Li et al. 2013). Hsp90s serve as molecular chaperones in dimerised forms often in harmony with their ATPase activity. Hsp90s are constitutively expressed and show their induced expression level in response to stress across diverse genera of organism (Taipale et al. 2010). Hsp90 family of proteins form a conserved set of proteins, containing a conserved ATP-binding domain towards the N-terminus, M domain and a domain at the C-terminus. The C domain is responsible for the proteins dimerization. These are implicated in the maturation of a variety of meta-stable protein substrates. Hsp90s exert their chaperone like activity on a plethora of target proteins including several receptors, cell cycle kinases, signaling cascade-related components, microtubule dynamics, proteolytic machinery, and related proteins (Taipale et al. 2010). Hsp90s have shown elevated expression level under stress condition in the endoplasmic reticulum (ER), chloroplasts and mitochondria, suggesting their contribution to stress response. Cytosolic, ER and plastidial Hsp90 genes have been characterized from different plant species and share nearly 63–71% similarity with *Hsp90* of *Saccharomyces* and Hsps of animal source (Krishna and Gloor 2001). Genome-wide analysis of *Hsp90* genes in *Arabidopsis* revealed the presence of 7 members of Hsp90 family from which *At*Hsp90-1-*At*Hsp90-4 are cytosolic, whereas *At*Hsp90-5, *At*Hsp90-6-*At*Hsp90-7 are predicted to be plastidial-, mitochondrial- and ER-localized, respectively (Krishna and Gloor 2001). Hsp90s are known to act in co-ordination with other co-chaperones, such as Hsp70s, which includes Hip i.e. Hsp70 interacting protein and Hop i.e. Hsp70/ Hsp90 organizing protein (Carrigan et al. 2006). Mammalian co-chaperone homologues isolated from *Glycine max*, *Gm*Hop-1, Hop protein, (Zhang et al. 2003) showed indued level of transcripts under stress condition. It was reported that under normal condition, Hsp90.2 negatively regulates the transcription of heat-induced genes by elimination of heat shock factors (HSFs) (Zhang et al. 2003). Heat shock reaction includes inactivation of Hsp90.2 and activation of subsequent genes harbouring HSF elements. Subsequently, it was ascertained that Hsp90.2 overexpression downregulates *HsfA2* transcription, leading to *HsfA2* induction in a way to acclimatize to reactive oxygen species (ROS) genarated stress (Nishizawa-Yokoi et al. 2010). Similar results were noticed in rhizosphere fungus, where an inhibitor of Hsp90, affected the overall growth and development, but boosted up the level of resistance in *Arabidopsis* (McLellan et al. 2007). Hsp90 is also identified to interact with proteasomal 26S and aids in packaging and preservation of this complex (Imai et al. 2003).

Sangster et al. (2008) stated the role of Hsp90 as a 'buffer' of morphological evolution as well. Geldanamycin treatment on Hsp90 of *Arabidopsis* generated several morphological phenotypes, presumably the genetic variations that are usually neutralized by Hsp90. The buffering activity was allied to varied development and morphogenesis relating the signaling cascades. Pertaining to physiological environment, the genetic variations that are masked by the Hsp90 buffering activity or neutralizing outcome is subdued (Sangster et al. 2008). Authors thus reported a potent evolutionary mechanism in which Hsp90 undertakes a crucial position in maintaining genetic stability at usual physiological conditions, while allowing mutational changes that would be apparent under the prevailing stress conditions (Sangster et al. 2008).

6.2.4 Hsp70 Family Proteins

Till date, Hsp70s are the best characterized Hsps in plants. Hsp70s are, at hand, in the cytosol, mitochondria, endoplasmic reticulum and as well as in the plant-specific organelles including chloroplasts, peroxisomes, glyoxysomes and P-bodies (Wimmer et al. 1997). The identity between bacterial and eukaryotic Hsp70s pertains to around 50%, indicating their immense importance and functions in various life forms. Numerous cellular activities are envisaged by Hsp70s. These proteins prefer to undertake this task via binding to hydrophobic amino acid chains, which get exposed when a protein is in its unfolded state during or before folding. It helps to prevent aggregation of misfolded proteins, necessiates renaturation of protein aggregates and keeps proteins in a competent mode, which can easily be translocated to various other compartments as condition prevails (Boston et al. 1996). Hsp70s partake in molecular folding all the way through the cycle that involve steps of binding ATP, ATP hydrolysis and further peptide discharge (Miernyk 1997). Wheat germ extract with exhausted cytosolic Hsp70 showed inefficient co-translocational machinery and reduced processing power of its precursor proteins. The functions got restored on addition to the same Hsp (Miernyk et al. 1992). Some Hsp70s participate in framing the activity of additional regulatory proteins acting as negative regulators of HSF-mediated transcription. It was observed by Sandqvist et al. (2009) that the Hsp70 and its associated transcription factors prohibit trimerization and binding of heat shock elements to their respective transcription factors. Hsp70s contains two major domains i.e., ATPase domain at its N terminal that binds to ATP and causes hydrolysis of ADP and the substrate-binding domain (SBD) that helps in folding of intermediate complexes (Mayer and Bukau 2005). The C sub-domain acts as a 'lid' for the SBD (Mayer and Bukau 2005). HSP70-ATP together in a bound state causes the protein lid to open, following which peptides associates and is released swiftly. The reverse happens in the ADP bound state. The lid is clogged and peptides are firmly adhered to the SBD (Mayer and Bukau 2005).

In a study, 11 out of the 14 *Hsp70* genes present in *Arabidopsis* showed significant increase in their expression level under heat stress. Rest of the Hsp70s did not show enhancement of expression by heat stress (Guy and Li 1998). In spinach, there are 12 genes encoding Hsp70s (Sung et al. 2001). Some Hsp70 homologues are constitutively expressed in the cytosol and are referred to as heat shock cognate 70 (HSC70) proteins. They also help in stabilizing newly developed proteins before being freed from the ribosomal complex, which thus prevents the possible misfolds and aggregation of the newly formed proteins before its complete synthesis. Hsp70 homologue has also been isolated from tomato (*Lycopersicon esculentum*), which is much similar to the BiP/GRP proteins present in other eukaryotic organism. Cooper and Ho (1987) reported the presence of a 72-kDa Hsp, enriched in endoplasmic reticulum membranes from heat-stressed corn root system, and a heat shock cognate protein (Hsc70) characterized from microsomes of growing caryopsis of wheat.

6.2.5 Hsp60 Family Proteins

Hsp60s, also known as the chaperonins, are the most conserved and form a ubiquitous class existing in the plastids, mitochondria, and in the cytoplasm of plants. The term 'chaperonin' was coined to portray a class of chaperones, having evolutionary homology to *E. coli* GroEL (Yamada et al. 2002). Chaperonins form a part of elaborate co-operative network of chaperones. Plant chaperonins in general are the stromal chaperones. Hsp60 and Hsp70 are mainly concerned in attaining serviceable conformation of recently synthesized chloroplast target proteins (Jackson-Constan et al. 2001). Functional characterization of chaperonins in plants is very confined. Hsp60s are critical in maintaining the integrity of plastidial proteins like Rubisco (Young et al. 2004). Chaperonins are also present in chloroplasts and mitochondria. Chaperonins aid in forming high molecular weight oligomeric complexes of approximately 800 kDa (Hemmingsen et al 1988). Chaperonins falls into two groups; GroEL chaperonins mostly present in prokaryotes and endosymbionts and the chaperonins containing TCP-1 (CCT), which exist in archaebacterial and eukaryotic cytosol. The CCTs isolated from *Bruguiera*, a halophyte bearing Group II chaperonin, improves tolerance toward osmotic and salinity stress in *E. coli* (Yamada et al. 2002).

Hsp60s are characterized by their capability to identify proteins with abnormally exposed hydrophobic residues and form stable inactive aggregates. Hsp60s show an inherent ATPase activity implying an ATP-dependent augment in the energy of misfolded or aggregated substrate molecules. Sequence similarity of *groEL* gene with its complementarity with the cDNA of the chloroplast Cpn60 protein, proved them as chaperonins and that these proteins might be evolutionarily homologues (Bukau et al 2006). These chaperones come together as stacked seven ringed structure in combination with Cpn10, another co-chaperone (Hsp10/GroES homologue). They possibly form a heptameric caps at both ends of the tetra decamer Cpn60. Other proteins then enter the newly formed Cpn60 core, which apparently provids a secluded section where protein folding can happen without much obstacle. The entire process is supervised by the dynamics of folding and binding phases of the ATP-dependent cycling of Cpn60 subunits. Boston et al. (1996) documented the character of the Cpn60/Cpn10 chaperonin in the folding of larger and smaller subunits of Rubisco. Maize mitochondrial Hsp60 protein levels increased up to three times in expression during a prolonged heat treatment in maize seedlings (Lund et al. 1998). Hsp60s also constitute a larger portion of total soluble protein during imbibition and early stages of seedling growth, as compared to mature seedlings (Lund et al. 1998). Hsp60s are required during germination and other phases of active mitochondrial biogenesis. The end product of the *E. coli groEL* is indispensable for viability of the cell and is also essential for the clustering of capsid proteins of bacteriophages.

6.2.6 Hsp40 Family Proteins

Accurate assembly of cellular proteins is essential for cell's overall functions. Hsp70/DNAK chaperones play an important role in such cases. It is hypothesized that Hsp40 recognizes the denatured proteins and handovers the "client" protein to Hsp70 for further folding via stimulating the Hsp70 ATPases. Consequently, certain nucleotide exchange factors (NEFs) operate to let the bound client proteins released and help them to transform to their indigenous form. Diverse Hsp40 proteins identify different set of client proteins and its related substrates. Plasticity of the SBD domain of Hsp70 allows it to lodge a large array of client proteins (Schlecht et al. 2011). Hsp40 has a conserved J domain sequence of 70 amino acid residues long at the N-terminus, which is responsible for its association with Hsp70. On the other hand, the C-terminus is variable amongst different Hsp40s and is responsible for providing specificity to substrate "client" proteins. Hsp40 proteins are broadly divided into three groups on the basis of their protein sequences, followed by the presence of a J domain. Type I has a Glycine- and phenylalanine-enriched region and four cysteine residue repeats in zinc finger domains. Glycine and phenylalanine residues are present in type II only, whereas type III comprises of a J domain only (Walsh et al. 2004; Kampinga and Craig 2010). Transgenic *Arabidopsis* plants that constitutively overexpress Hsp40 exhibit increased tolerance towards salinity in comparision to wild type plants. DnaJ was identified as a vital force in providing tolerance towards salinity in *Arabidopsis*. Only a few genetic studies in the green algae *Chlamydomonas* have indicated roles for the Hsp70 and Hsp40 chaperone systems in its microtubule dynamics (Silflow et al. 2011). *BIL2* gene also codes for another mitochondrial Hsp40. *BIL2*-over-expressing plants showed cell elongation on treatment with a brassinosteroid (BR) biosynthesis inhibitor Brz. It increases the development of plant inflorescence and its root system. It also regulates BR-responsive gene expression. These plants showed tolerance to oligomycin (ATPase inhibitor of the mitochondria) and showed increased vigour of exogenous ATP in the treated plants than wild type plants. *BIL2* participates in increasing the endurance against abiotic stress tolerance like salinity and photoperiodic treatment (Bekh Ochir et al. 2013).

6.2.7 sHsps Family Proteins

sHsps are low molecular-weight Hsps of approximately 12-40 kDa. As other chaperones prevailing in plant system, sHsps are also produced ubiquitously in prokaryotes and eukaryotes cells under the condition of heat and other forms of stress. sHsps have the ability to form active oligomers, which gets disassembled for effective chaperoning activity. sHsps monomers have highly conserved alpha-crystallin domain (ACD) at its C-terminal and are enriched in beta-strands, which is accountable for its dimeric structure. On the other terminal, resides the less conserved N terminal domain (Kriehuber et al. 2010). *Triticum* Hsp16.9ACDs are arranged as trimers of dimers, forming a dodecamer 2-ringed structure, as studied by its diffraction pattern (van Montfort et al. 2001). sHsps have a tenacity to bind to non-native proteins employing hydrophobic linkages as alone it cannot reframe non-native proteins structures. They cause stabilization and prevention of aggregates of proteins, which are non-native in their structure. This, thereby, facilitates their subsequent refolding by ATP-dependent chaperones such as DnaK/ClpB/Dna system. Currently, sHsps from *Pisum sativum* and *Synechocystis,* Hsp18.1 and Hsp16.6 respectively, under *in vitro* conditions, binds to unfolded proteins and aids in dynamic refolding by Hsp70/Hsp100 protein complex and its formation (Mogk et al. 2003).

Six multigene families of sHsps have been characterized in plants. Individual gene family has proteins residing in vivid cellular compartments. Scharf et al. (2001) grouped six classes of 13 different *Arabidopsis* sHsps relying on their intracellular localization and their similarities to protein residues (Scharf et al. 2001). sHsps probably reflects molecular adaptation to stress as increasing evidence suggests a positive link between sHsps accumulation and plant tolerance under stress conditions. Maize mitochondrial sHsps (*Zm*Hsp) showed better mitochondrial electron transport under saline condition, primarily by protection of the Complex I of the ETC (Electron Transport Chain). Chloroplastic sHsp26.2 was reported in the *Agrostis stolonifera.* But point mutated sHsp26.2 m, identical to sHsp26.2 created a stop codon isolated from the heat-sensitive variant. It was unable to show stress responsiveness (Wang et al. 2003). Plants synthesize significant amount of sHsps when subjected to high temperatures, drought stress, oxidative stress, cold acclimation, salts, and ABA treatment. It was noticed that there was a positive qualitative relation between the accumulation of sHsps in the plastids and thermotolerance of heat shock from temperature ranging from 28 to 40°C in different *Anthophyta* species, including C-3, C-4, CAM, monocotyledonous and dicotyledonous species (Downs and Heckathorn 1998). Similar results were obtained for other *Anthophyta* species. Downs and Heckathorn (1998) suggested the role of mitochondrial sHsps in protecting ubiquinone oxidoreductase (complex I) under heat treatment in *Pyrus pumila*. A recent report suggested sHsps playing a significant role in the quality membrane control and hence has the potential contribution in the membrane integrity maintenance especially under the conditions of stress. Liming et al. (2008) transformed plants with Hsp24 isolated from *Trichoderma*, which conferred higher tolerance to heat stress when expressed constitutively in *S. cerevisiae*.

6.2.8 Co-chaperones

Co-chaperones are the proteins, which participate in the function of other chaperone. They have similar functions as to prevent polypeptide aggregation and thus have chaperone activity. These co-chaperones have regulatory function in chaperone action. Co-chaperones act as a mediator of chaperone specificity by assisting the selection of the client protein. For example, in case of Hsp70 or Hsp90, co-chaperones bind and release by Hsp70 or Hsp90 in a manner that facilitates protein folding and disassembly. Co-chaperones have been categorized on the basis of the domain architecture. First is the J domain found in Hsp40 co-chaperone of Hsp70 and another one is tetra tricopeptide repeats (TFR) domain found in co-chaperones that act together with Hsp70 and Hsp90. Substrate binding and release cycles of Hsp70 require Hsp70 co-chaperones (DnaJ/Hsp40 and GrpE). Aha1 is the activator of Hsp90 ATPase and is stress-regulated co-chaperone, i.e. required for *in vivo* Hsp90-mediated activation of its patron protein (Obermann et al. 1998). *In vitro* Aha1 and its homolog Hch1 (suppressor of Hsp90) stimulates inherent Hsp90 ATPase activity in eukaryotes (Panaretou et al. 1998). Unc45 is a co-chaperone for Hsp90, which has TPR domain and helps in myosin assembly. P^{23} is a co-chaperone for Hsp90, which helps in maturation of Hsp90 client proteins at a later stage. Cdc37 is co-chaperone of Hsp-90 and helps in protein kinase folding (Joo et al. 2011). PP5 is a Hsp90 co-chaperone, which has a TPR domain and is involved in protein phosphatase activity (de la Fuente van Bentem et al. 2005). Cytosolic Hip50 has a TPR domain and can act with the ATPase domain of Hsc70, further enhancing its interaction with substrate by establishing its ADP-associated form (Hohfeld et al. 1995). Link between Hsp70 and Hsp90 in various systems is also provided by HOP protein, which is also another TPR domain associated co-chaperone (Dittmar et al. 1996; Chen and Smith 1998; Zhang et al. 2003). Hsp40 is J domain containing co-chaperone, which triggers Hsp70 ATPase activity (Wall et al. 1994). DnaJ and GrpE (nucleotide exchange factor) stimulate DnaK ATPase activity (Liberek et al.1991). Bag1 is GrpE homologue and regulates nucleotide exchange and ATPase activity of Hsp70 (Hohfeld and Jentsch 1997;

Sondermann et al. 2001). Fes1, another Hsp70 nucleotide exchange factor, is involved in ubiquitin-dependent degradation of misfolded proteins present in the cytosolic fraction (Gowda et al. 2012). Chip (a TPR domain containing Hsp70/Hsp90 co-chaperonre) is also involved in proteasomal degradation. Tom70 is TPR domain containing Hsp70/Hsp90 co-chaperonre and is involved in mitochondrial preprotein transport (Fan and Young 2011). FKB52/51 and Cyp40 are TPR domain containing Hsp90 co-chaperonre and are implicated in peptidyl propyl *cis-trans* isomerase activity. Cns1 also have TPR domain containing Hsp90 co-chaperone activity.

6.3 Other Chaperones

6.3.1 Disulphide Isomerases

Disulphide isomerases (PDIs) are generally involved in dealing with nascent polypeptides in the ER lumen to catalyze the formation of new disulphide bonds. During folding and maturation processes of proteins in eukaryotes, these PDIs aid in the breakage and alternative isomerization of incorrectly formed peptide linkages. Even though most of them are endoplasmic reticulum dependent, PDIs have predominant existence at other intracellular locations as in nucleus, cytosol and also in mitochondria. PDI was the first protein-folding catalyst reported (Hatahet and Ruddock 2007). It contains a typical thioredoxin (TRX) domain, which catalyses disulfide bond configuration in endoplasmic reticulum under the condition of oxidative stress. It helps in stabilizing the structure of protein during its folding state. Homologs of PDIs, as present in plants have a conserved amino acid active site Ala-Pro-Trp-Cys-Gly-His-Cys-Lys and endoplasmic reticulum nascent peptide (Gruber et al. 2007), which is absent in vertebrates. Plant PDIs form quite a large family and have more diversity as compared to the animal kingdom. *Arabidopsis*, soybean, rice and maize have nearly 10 to 20 members of the PDIs with their TRX domains (Yuen et al. 2016). These proteins have some unique roles in storage protein folding. PDI having a mass of 55-kDa proteins is characterized by TRX domain sequence in the order of "a-b-b′-a′" domains. The catalytic motifs involved with redox regulation are found in the 2"a" domains and TRX fold structure lies in the "b" domains. The "b" regions are rich with hydrophobic residues and are concerned with substrate detection and binding activity (Selles et al. 2011). A short inter domain region called x-linker comprising of 19 residues connects the b′ and the a′ domains, and a highly acidic extension at the C-terminus (c domain) is involved in calcium sensing and binding. It contains the typical ER retrieval motif 'KDEL' (D' Aloisio et al. 2010). The 4 TRX domains are arranged in the form of a "U" like shape having the active sites opposite to each other across its arms. The inside surface of the "U" is rich in hydrophobic residues, thereby promoting interactions with misfolded proteins (D'Aloisio et al. 2010). *AtPDI* gene isolated from *Arabidopsis* has been studied with altered expression in different tissues, in response to chemically-treated unfolded protein resonse (UPR), and in null mutants of UPR signaling intermediaries (AtIRE1-2 and AtbZIP60). These experiments provide the evidence of higher expression of 6 *AtPDI* genes by UPR and its expression was found to be perturbed by actinomycin D treatment (transcription inhibitor), indicating UPR-induced *AtPDI* gene transcription. PDIs response was also studied in the grass family, alfalfa and *Nicotiana* cells on treatment with the antibiotic and tunicamycin. Genome-wide analysis of PDIs sequences have revealed the presence of 10 of its types in plants. In the primitive algae *Chlamydomonas*, PDI RB60 served as a redox sensor of an mRNA-binding complex concerned with photoregulation. It affects translation of psbA, which is the RNA, encoding for the protein D1 of the photosystem II complex (Levitan et al. 2005). Thus PDIs also play a role in the regulation of dynamic disulfide bonds in chloroplasts.

6.3.2 Peptidyl Prolyl **cis-trans** *Isomerase*

Cyclophilins were first isolated by Handschumacher et al. (1984) from the bovine thymocytes. They are conserved across genera, giving prominence to their role in overall cellular processes. Majority of the existing cyclophilins display peptidyl prolyl *cis-trans* isomerase (PPIase) activity. Their diversity, assorted cellular locations and dynamics in folding of protein highlight them to be integrated in the group of molecular chaperones. Cyclophilins are the members of group of proteins quite oftenly called "immunophillins". Cyclophilins are highly conserved proteins and are ubiqutously present, having diverse functions. Several plant cyclophilins have been reported from *Lycopersicon*, *Vicia faba*, maize, sorghum, pigeon pea, *Arabidopsis*, tomato, wheat, rice and even the algae. Cyclophilins assumed to exercise cellular protection, helping the plant to adapt to specific unfavorable changes in the environment. The universal presence and diverse roles of cyclophilins in plant system serves as a notion for several studies related to plant cyclophilins at a global level (Kumari et al. 2013). An affirmative workflow between cyclophilin protein family and stress protection has been deciphered. Cyclophilin's exact mode of action to bring about stress protection is yet to be identified. All cyclophillins have conserved domain i.e. cyclophillin like domain (CLD). Some are single domain cyclophillins having only CLD domain, whereas others are multiple domain cyclophillins having CLD domains like TPR, Zinc finger etc. (Taylor et al. 2001; Kumari et al. 2013). Cyclophilin A (CypA) is present in cytosol, whereas other cyclophilins having single domain or multiple domains are found in mitochondria, endoplasmic reticulum and nucleus. Cyclophillins having nuclease activity possess two active sites, one having PPIase activity and other for catalytic degeneration of DNA in a Ca^{2+} and Mg^{2+}-dependent manner. There are 16 cyclophilins proteins in humans, 29 cyclophilins proteins in rice and 8 cyclophilins proteins in *Saccharomyces* (Kumari et al. 2015). Immunophilins like chaperones have enzymatic properties and have vivid localization and roles in protein folding. They have peptidyl-propyl isomerase activity (Fisher et al. 1984), which is required for protein folding. They also act as scaffolding proteins due to their involvement in assembly of supramolecular complexes (Goel et al. 2001). Due to their *cis-trans* isomerisation catalytic activity, they act as acceleration factor for protein folding. In proteins, *trans* state is mainly favoured in unfolded state and in such state process of *cis-trans* isomerisation is very slow i.e. rate limiting step for final folding and *cis*-confirmation of proteins (Herzberg and Moult 1991). Abiotic stresses such as thermal stress, UV treatment, changes in acidic or alkaline nature of cell environment, treatment with oxidants etc. enhance the expression of cyclophilins (Kumari et al. 2013). These stresses generally lead to unfolding or misfolding of intracellular proteins and cause the induction of transcriptional response due to heat shock and hence induction of Hsps. TLP46 is the best characterized cyclophillin protein from spinach, which is involved in the regulation of D1 photosystem II protein (Fulgosi et al. 1998). *Arabidopsis* cyclophillin AtCyp38 plays a major role in the assembly of photosystem II supercomplexes (Fu et al. 2007). *Arabidopsis* cyclophillin cyp20-3, also known as ROC4, is highly sensitive to oxidative stress and enables thiol biosynthesis pathway to overcome stress effects (Dominguez-Solis et al. 2008). AtCyp59 is involved in activities that connects transcription and splicing (Leverson and Ness 1998). *Rc*Cyp1, a castor bean cyclophillin, plays role in refolding of protein entered through phloem translocation stream (Chou and Gasser 1997).

6.3.3 Calnexins and Calreticulins

Calnexins (Cnxs) and calreticulins (Crts) are calcium sensing molecular chaperones present in endoplasmic reticulum (Michalak et al. 2009). Calnexins along with calreticulins are responsible for checking the protein folding machinary that enhance appropriate folding of proteins that go into the secretory channel and promotes degradation of misfolded proteins. Both of them are reported to be present in all the plants across genera. They are also known to diverge from a common ancestral source (Del Bem 2011). Duplication event of Cnxs and Crts happens to occur in the early tracheophytas. Calnexin, as a founder gene member, was inherited from the early chlorophytas in a low copy number (Del Bem 2011). Crts, as present in plants, help in development, growth as well as resistance to various environmental stresses. Crt from *Arabidopsis* (*AtCRT*), has been established as an alleviator of unfolded protein response as it gets induced by the effector drug tunicamycin. Two distinguished isoformic sets of Crts have been identidied in higher plants (Jia et al. 2009). Ca^{2+}-dependent processes, endoplasmic reticulum mediated chaperone response, programmed cell death and necrosis are many of the cellular responses, which gets altered by Crt gene expression. High mRNA levels of Crts were markedly observed in *Brassica* seedlings under saline and high temperatures (Georges et al. 1999). *TaCrt* overexpressing lines showed superior drought tolerance in wheat (Jia et al. 2008). These proteins help in combating dessication stress in *Arabidopsis*. Recently, calreticulins Crt1, Crt2 and Crt3 from *Arabidopsis* have been suggested to mediate defensive responses against both viral and microbial inoculums. Crt1, Crt2 and Crt3 have amino acid sequence identity as well. Co-expression of a maize Crt1 helps in mitigating the related side effects via enhancing the Ca^{2+} content (Wu et al. 2012). Perhaps, expression of Crts could mitigate the hypersensitivity for ion inequity in *Nicotiana*. Furthermore, enhanced Crts expressions mitigated blossom end rot disease (BER) in *Crt1* expressing lines of *Lycopersicon* (Wu et al. 2012).

Calnexins (Cnxs) in plant system were first characterized in *Arabidopsis*, following which they were characterized from *Pisum*, spinach, barley, tobacco and many other plant species (Schrag et al. 2001). Calnexins are 90 kDa proteins harbouring a large calcium-binding luminal domain at N terminal region, a distinct transmembrane helix and an additional short acidic cytosolic domain. Calreticulins are well conserved Ca^{2+}-binding proteins having 3 different domains: N-terminus domain which is globular in form, a middle domain which is proline rich, and a C terminus domain which is acidic, following which is the ER related signal peptide, (K/H)DEL (Michalak et al. 2009). In osmotic or other kinds of stresses, Cnxs show reduced buildup in developing soybean root system. The spinach Crt's glycan structure when resolved (Navazio et al. 1996), revealed it to be explicitly phosphorylated by the casein kinase (CK2) (Baldan et al. 1996). Crts play role in Ca^{2+} sensing and binding (Michalak et al. 1992). It has role in Ca^{2+} signaling as second messenger (Mery et al.1996), cell adherence (Coppolino et al. 1997) and subsequent gene regulation (Michalak et al. 1992).

Chaperoning activity for both the Crts and Cnxs is considered from their significant sequence homology (Rajagopalan et al. 1994; Tatu and Helenius 1997). *Nicotiana* Crt is present in stress condition and works in a nucleotide-dependent method (Denecke et al. 1995). The major disparity between Crts of plant and animal origin is the prominence of N-glycosylation sites, which are primarily taken by glycan-like chains, as reported in several plant species. Several possible phosphorylation consensus sequences for the protein CK2 are phosphorylated competently in plant Crts (Mariani et al. 2013).

Both Crts and Cnxs have properties similar to lectin and can attach to the monoglucosylated glycans, which are Asn-linked of the glycoproteins (Hammond et al. 1994). The current understanding for substrate binding and liberation of these two proteins is dependent on the turnover of the enzymes that are accountable for trimming the glucose moiety and reglucosylation of high-mannose Asn-linked glycans (Rodan et al. 1996). UDP-glucose, a glycoprotein glucosyl transferase reglucosylate nonglucosylated glycans to monoglucosylated forms, reverting them into ligands. Reglucosylation occurs only on glycopolypeptides that have not reached their destined forms, suggesting that the calnexin-calreticulin/glucosyltransferase system regulates the conformational maturation stages and renders them to the ER (Sousa and Parodi 1995).

6.4 Role of Chaperone in Protein Translocation in Plants

6.4.1 Chaperone Mediated Translocation of Protein in Mitochondria

Translocation of protein to and from the double membrane of mitochondria is carried out by the pre-sequence translocation coupled motor complex (PAM) associated with the chaperone Hsp70 (mtHsp70), J domain containing proteins (PAM16/TIM16, PAM18/TIM14, Mdj2) (Mokranjac et al. 2003; Kozany et al. 2004; Mayer 2004; Vasiljev et al. 2004) and Mge1 (nucleotide exchange factor) that are responsible for this kind of interaction. The ATP-bound mitochondrial Hsp70 (mtHsp70) remain associated with the translocating protein substrate and also to the mitochondrial import canal by its affiliation to the Translocase of the Inner membrane of Mitochondria (Tim). mtHsp70 remains tightly bound to the incoming polypeptide in the ADP bound state (Matouschek et al. 2000; Liu et al. 2003; Mayer 2004). Brownian ratchet model stated that the transitory binding of mtHsp70 to the incoming translocating chain prevents its backward movement, thus favouring its forward movement (Simon et al. 1992; Neupert and Brunner 2002). The energy requirement is fulfilled by the ATP hydrolysis. Power stroke model proposed that ATP hydrolysis cause conformational change within the substrate via ATP hydrolysis and results in high affinity state with the substrate. It produces mechanical force i.e. sufficient to pull polypeptide into matrix causing it to unfold into the cytoplasmic side (Glick 1995; Matouschek et al. 1997, 2000). Figure 6.2a represents chaperone-mediated translocation across mitochondrial membrane.

6.4.2 Chaperone-Mediated Translocation Across Chloroplast Membrane

Chloroplast contains sub-organellar compartments such as the outer and inner membrane, intermembranal space, stroma, membrane and lumen of the thylakoid. Nuclear-targeted chloroplast proteins are generated in the cytosol with a nascent peptide, which in alliance with cytosolic Hsp70 and 14-3-3 proteins target the precursor protein to the plastid (Jackson-Constan et al. 2001). Precursor protein gets linked with the component of external membrane translocon, COM70 (Wu et al. 1994) via GTP hydrolysis (Olsen and Keegstra 1992; Young et al. 1999). Interaction of precursor protein to the component of inner membrane translocon is mediated by ATP hydrolysis and is assisted by heat shock protein Hsp70 residing in the intermembrane space, which helps in the precursor proteins' relocation to the Translocon at the inner membrane of the chloroplast (Tic) complex. Final transfer of precursor protein into the chloroplastic interior happens, where the signal peptide is detached in an energy-dependent manner by Hsp93 (Nielsen et al. 1997). Finally,

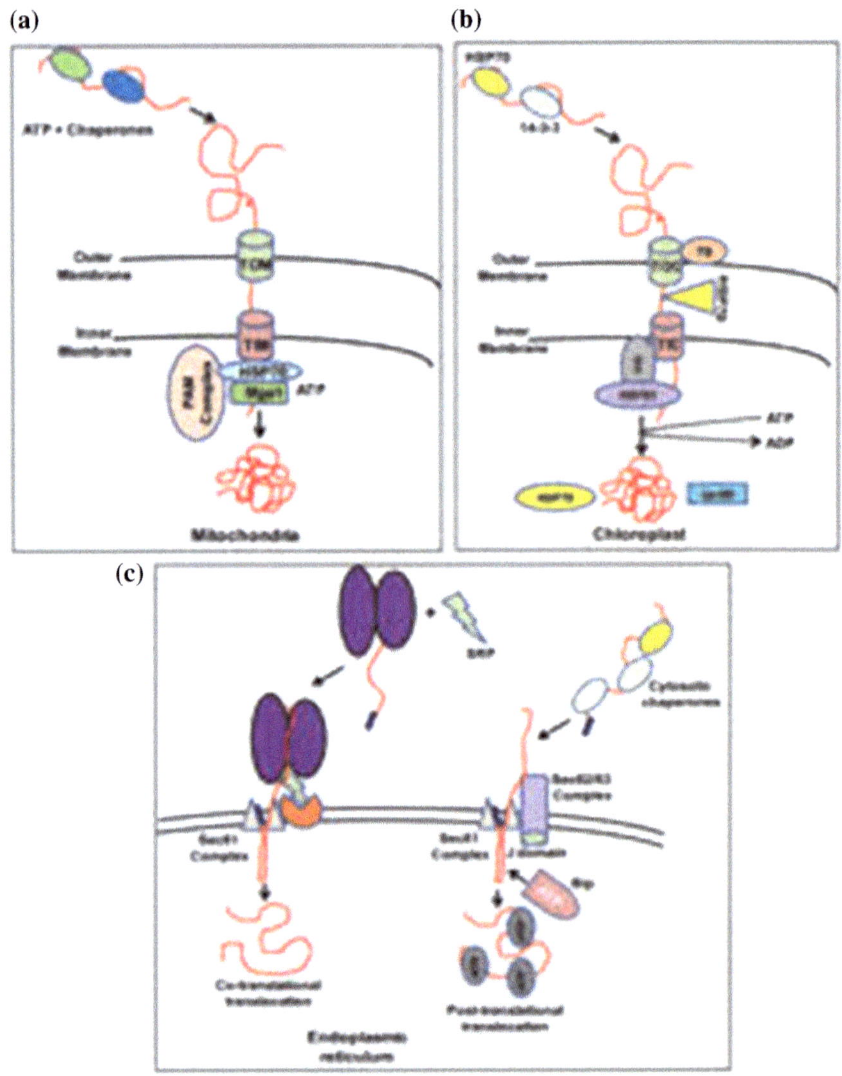
(a)
(b)
(c)
ATP + Chaperones
Outer Membrane
Inner Membrane
ATP
Mitochondria
HSP70
14-3-3
Outer Membrane
Inner Membrane
ATP
ADP
Chloroplast
SRP
Cytosolic chaperones
Sec62/63 Complex
Sec61 Complex
Sec61 Complex
J domain
Bip
Co-translational translocation
Post-translational translocation
Endoplasmic reticulum

◂**Fig. 6.2** Chaperone mediated protein translocation: **a** schematic presentation of the role of Hsp70 in protein translocation through mitochondrial membrane. MtHsp70ATP anchors to theTim44 protein. The nearby J domain of membrane-anchored PAM complex triggers ATP hydrolysis. ATP hydrolysis cause conformational change within the chaperone which result in high affinity state of HSP70 with the substrate and produce mechanical force i.e. sufficient to pull polypeptide into matrix and its unfolding into the cytoplasmic side. **b** Schematic presentation of Hsp70's role in protein translocation through chloroplast membrane. HSP70 and 14-3-3 remain associated with incoming peptide and keep it in unfolded state. GTP hydrolysis stimulates association of component of outer membrane (com70) with precursor protein. ATP hydrolysis in intermembrane space promotes association of precursor protein with inner membrane components i.e. facilitated by HSP70 chaperone. Stromal ATP hydrolysis promotes translocation of protein into chloroplast interior where HSP70 and cpn60 promotes its proper folding. **c** Schematic view of SRP and Sec61 mediated co-translational tanslocation across ER membrane. Post translational translocation involves channel association with Sec62/Sec63 tetrameric complex and Bip i.e. a member of HSP70 family. J domain of Sec63 triggers ATP hydrolysis which favours polypeptide translocation across endoplasmic reticulum membrane

chaperones in the chloroplastic stromal region such as Hsp70 and CPN60 aid in the folding of imported protein into its native confirmation. After that, protein may enter into the thylakoid lumen by a variety of pathways such as general secretory (SEC) pathway, TAT (Twin Arginine Translocation) pathway and SRP (Signal Recognition Particle) pathway (Jackson-Constan et al. 2001). Chaperone-mediated translocation across chloroplast membrane has been shown in Fig. 6.2b.

6.4.3 *Chaperone-Mediated Translocation Across Endoplasmic Reticulum*

Co-translational translocation is used for secretory and membrane proteins translocation. In this process, the signal sequence present on the elongating polypeptide chain from translating ribosomes is captured by the signal recognition particle (SRP). Subsequently, the ribosome and SRP complex adheres to the membrane via SRP and membrane receptor complexes and the association happens by the interaction amongst the ribosome and the translocation channel (Halic and Beckmann 2005). The elongating polypeptide enters the ER membrane from the ribosomal tunnel. This reaction happens by GTP hydrolysis but hydrolysis is not required for polypeptide movement through channel (Connolly and Gilmore 1986). In case of membrane proteins, certain peptide segments emerge from the ribosome channel junction but do not enter into the channel and make cytosolic domains (Mothes et al. 1997). In most cells, protein transportation occurs post-translationally. This pathway is generally adapted by the lower organisms such as bacteria where translocation is not in parity with translational mechanisms. Soluble secretory proteins having moderately hydrophobic signal sequences favour this path (Ng et al. 1996; Huber et al. 2005). These proteins should be in unfolded state after their release from ribosome (Huber et al. 2005). In eukaryotes, channel proteins remain associated with Sec62/Sec63, which is a tetrameric complex along with the luminal chaperone Bip i.e. Hsp70 family member (Panzner et al. 1995). During translocation of polypeptide into the channel, all cytosolic chaperones are detached from the peptide chain (Plath and Rapoport 2000). Once the polypeptide chain enters the channel, binding immunoglobulin protein (Bip) prevents its backward movement inside the cytosol, favouring translocation in forward direction. ATP bound Bip interacts with the Sec63 residue of the J domain resulting in ATP hydrolysis and further closing of peptide binding pocket surrounding the translocating polypeptide. After significant movement of polypeptide in the forward direction, Bip molecule can also bind next to it. The process keeps going on until the complete movement of polypeptide chain occurs (Rapoport 2007). Figure 6.2c represents involvement of chaperone and other component in tanslocation across ER membrane.

6.5 Protein Degradation via Chaperone in Plants

6.5.1 Protein Quality Control in Chloroplast and Mitochondria

Chaperones are mandatory to sustain the misfolded protein in soluble state by preventing their aggregation i.e. essential for the degradation by protease (Wagner et al. 1994). Different types of proteases have been identified in *Arabidopsis*, which are analogues of bacterial proteases and help in degradation of misfolded proteins.

Clp chaperone protease: Clp proteases play role in chloroplast and mitochondria (Nakabayashi et al. 1999). These belong to the serine type proteases having two different polypeptides (Proteolytic subunit and regulatory ATPase subunit) assigned to different functions. Only one isomer is encoded by plastidic genome, while others are nuclear-encoded. ClpP4, 5, 6 are targeted to stroma (Sokolenko et al. 1998) and ClpP2 is directed to mitochondria (Halperin et al. 2001). Four other sequences having sequence similarity with ClpP proteins are present in *Arabidopsis* genome but they don't have catalytic triad.

Lon Protease: Lon protease possesses both chaperone and protease function (Goto-Yamada et al. 2015). *Arabidopsis* genome contains three sequences, which encode Lon proteases. Lon proteases are characterized by the presence of single polypeptide having both functions (catalytic as well as regulatory). Lon1 is targeted to mitochondria (Sarria et al. 1998), Lon2 and Lon3 are targeted to chloroplast.

FtsH Protease: *Arabidopsis* genome contains nine sequences, which encode Ftsh protease. Like Lon proteases, Ftsh proteases are also characterized by the presence of single polypeptide having both functions (catalytic as well as regulatory). FtsH1 and FtsH2 are present in thylakoid membrane (Lindahl et al.1996, Chen et al. 2000). FtsH3 is predicted to be targeted in mitochondria. Mitochondrial homologue of FtsH4 has been reported in yeast (Leonhard et al. 1996). FtsH5, 6, 7, 8 are present in chloroplast.

DegP Protease: DegP protease possesses both protease and chaperone activity. *Arabidopsis* genome contains thirteen sequences, which encode DegP protease, all having catalytic triad His-Asp-Ser. DegP1 and DegP2 are targeted to thylakoid membrane (Peltier et al. 2000). The localization of other isoforms has been predicted to be varying from chloroplast to mitochondria, nuclei and cytosol.

6.5.2 Chaperone and Ubiquitin Proteasome-Mediated Degradation

In archea, eukaryotes and in some bacteria, chaperones are required for degradation of abnormal proteins through ubiquitin-proteasome pathway. This comprises of active participation of enzymes that link polypeptide co-factor ubiquitin on to the protein and direct their degradation. These tagged proteins are accepted by 26S proteasome complex, which consists of multicatalytic proteases that cause degradation of ubiquitinated proteins to small peptides (Burger and Seth 2004). Attachments of ubiquitin to the target protein require the active participation of enzymes. E1 is the UB-activating enzyme, E2 is the UB-conjugating enzyme and E3 is the UB ligase enzyme. E1 hydrolyze ATP, adenylates UB at the C-terminus end and promotes a thioester bond accompanied by C-terminus of UB and active site of cysteine of E1 (Pickart 2001; Schulman and Harper 2009). UB is linked by another thioester bond to the active site of cysteine of E2. E3 helps in the transport of UB from E2 to the lysine residue of E3, which results into isopeptide bond formation between lysine and UB at the C-terminus. Further, UB molecules can be linked to first one to form a poly UB chain via its activity of UB elongation enzyme i.e. E4 (Koegl et al. 1999). Poly UB chains linked at different arrangement alters the fate of target protein. Lys (11), Lys (29) and Lys (48) linked polyubiquitin chain targets the protein for proteasome. Lys (6) and Lys (63)-linked polyubiquitin chain alters protein activity, location or trafficking. The 26S proteasome is a big complex having 20S barrel shaped proteolytic center consisting of alternating α and β subunits and 19S regulatory caps present at both ends. The 19S caps recognize de-ubiquitinylate and unfold the target protein before it is pulled through the hollow core of the catalytic centre, where it gets dissociated into the native amino acids residues that can be used again.

6.5.3 Endoplasmic Reticulum Associated Degradation

In endoplasmic reticulum, oligosaccharyl transferase complex transfers oligosaccharide (Glc3Man9GlcNAc2) from the lipid-associated precursor proteins to the residue asparagine of the target polypeptides. The α-glucosidase is responsible for removing 2 terminal glucose moieties. The resulting polypeptides with monoglucosylated N-glycan are then subjected to calnexin/calreticulin cycle (Huttner and Stasser 2012). Both Cnxs and Crts influence folding along with other folding catalysts, similar to the PDI family members. Accurately folded glycoproteins are detached from the Cnx/Crt cycle and further processing is done by α-mannosidases in the Golgi complex. The misfolded proteins are subjected to degradation (Huttner and Stasser 2012). In ER, degradation of misfolded protein occurs with the help of HRD1-SEL1L/HRD3-OS9 pathway where misfolded glycoproteins are identified and mannose trimmed by MNS (Mannosidase) proteins (Huttner and Stasser 2012). Mannosidase proteins remove mannose residues and produce a glycan chain. OS9 recognizes this glycan chain and then removal of misfolded protein needs HRD1-SEL1L/HRD3-OS9 complex (Su et al. 2011; Huttner et al. 2012), which results in ubiquitylation and degradation (Huttner and Stasser 2012). Further investigation is required for endoplasmic reticulum associated protein degradation (ERAD) pathway in plants.

6.6 Role of Chaperones in Plants

6.6.1 Chaperones in Plant Growth and Development

Chaperones actively participate in normal plant metabolism. Lin et al. (1984) observed that the induction of Hsp synthesis in *Glycine max* var. wayne seedlings was accompanied by the reduction of other proteins synthesis after exposure to heat shock (from 28 to 45°C) for 10 min (longer periods killed the seedlings). Hsps production and accumulation occur differently in different parts of the plant. Constant high temperature during seed maturation is certainly detrimental in all instances. Pea pods experience high temperature during later stages of development under field conditions. Hernandez and Vierling (1993) provided a conclusive study on Hsps, which are expressed in standard conditions during pea pod development. Also, Hsp expression is reported to occur at different developmental stages of plant, in which cooling mechanisms forms the limiting factor, such as during the period of seedling emergence. Seed temperatures are sufficient to stimulate Hsps, which may occur under normal conditions for many plant species. Hsp transcripts and HSP protein synthesis take place in all proliferative parts, for example in seed development, in the aleurone layer of imbibed seeds and in the developing embryos as well (Hong and Vierling 2001). A pea *hsp70* (*LP19*) gene showed a complex pattern of expression, as it was shown to be expressed not only in response to heat shock, but also during normal developmental events in the plant, including pod lignification and seed maturation (Dhankher et al. 1997). Simarly, various other HSPs, such as HSP 104 and HSP 90, have been shown to be induced in response to multiple stresses and various stages of development in rice (Pareek et al. 1995, Singla and Grover 1993).

sHsp mRNAs are reported to be present during early embryogenesis and also in totally developed pea pods and wheat grains (DeRocher and Vierling 1994). sHsp mRNAs expression is followed by accumulation of the corresponding proteins during the later stages of seed development. These emphasizes that sHsps are also expressed in seeds as a safety guard against tentative stress conditions. sHsps mRNAs expression has been shown during some stages of microsporogenesis in *Lilium* and maize (Bouchard 1990). Cross-linking and immunoprecipitations methods have shown Hsp70 to play an active role in import of premature protein by interacting with chloroplast precursor transit peptides. It was recommended that precursor target peptides be translocated from cytoplasmic chloroplastic Hsp70s (Kourtz and Ko 1997). DnaK was reported to act with Hsp70s and can also interact with chloroplast signal peptides (Ivey and Bruce 2000; Rial et al. 2000). These Hsp70s interactions were quite selective, as this happened only with the precursor proteins and not with the mature complexes (Rial et al. 2000). Most probably, Hsp70 is implicated in the import of late and early protein-mediated functions (Schnell et al. 1994).

Singla et al. (1997) provided evidence regarding the developmental and stress-related regulation of Hsp100 in *Oryza sativa*. High transcript level was reported in developing and mature rice grains even under un-induced stress condition. High induction levels of Hsp100 protein has also been reported in maize and wheat grains. The HSP100 transcript abundance in rice grains decreased during seed germination (Singh et al. 2010). Expression of Hsp101 during maize growth has revealed its excess in the tassel development, ear formation, embryo and endosperm formation etc. and lower expression level in the roots and foliar appendages (Young et al. 2001). Organs which are implicated in growth of the germ cells, such as green tissues including the shoot and floral meristematic regions have higher Hsp101 transcript level, while other organs involved in the general growth of the plant like leaves and roots do not need Hsp101 expression under control conditions. Hsp101 expression level was reported to be low in the mature pollen grains and was not triggered by heat (Young et al. 2001). Constitutive Hsp100 expression in seeds and remaining developing organs is a pre-planned mode of adaptation that is not required for prolonged period under non-stress condition but actively required under unfavourable conditions. Hsp60 is necessary during mitochondrial biogenesis. The requirement of this protein lies in assisting the rapid assembly of the protein complexes. The presence of Hsp60 in pumpkin cotyledons has provided evidence regarding its role in earlier stages of seed germination (Tsugeki et al. 1992). The chloroplastic GroEL/Hsp60 homologue is the Rubisco subunit binding protein, often referred to as the chaperonin. It is hypothesized that Hsp60 is associated with the assembly of Rubisco holoenzyme. Mutation in *Arabidopsis* plastid chaperonin protein Cpn60 is associated with defect in plastid growth and, consequently reduction in embryo and seedling development (Suzuki et al. 2009). Cpn60b antisense lines of transgenic tobacco plants showed strong phenotypic changes like delayed growth pattern, deferred flowering, chlorosis and stunting of leaves (Apuya et al. 2001).

PDIs gets upper hand in endoplasmic reticulum related stress conditions. They are expressed ubiquitously in cotyledonary tissues. Soybean PDI sequence showed major conservation with *Arabidopsis* and *Oryza* genome (Wadahama et al. 2008). During seed development and germination, these PDIs accompany changes of the storage proteins at their reduced or oxidised level. A disulfide bond promotes packaging of these proteins into storage organelles. This process converts back the oxidized forms to reduced form, which in turn makes them more susceptible for proteolysis and increases the accessibility of macronutrients like nitrogen and sulfur, which are essential for germination. PDIs regulate several signaling cascades through their alliance with other transcriptional factors. In the algae, *Chlamydomonas* the RB60 (PDI like protein) is drawn in regulating the photosynthetic efficiency along with the redox regulatory protein complex which controls translation of chloroplastic protein (Levitan et al. 2005).

PDI expression level is abundant in the developing caryopses as compared to other tissues. *In vivo* and *in vitro* reports strongly suggest that members of the PDI family play different physiological roles in different types of plant cells. For example, *Arabidopsis* AtPDIL1, which restores the wild-type phenotype in the *E. coli* protein folding mutant dsbA, is highly expressed in root tips and developing seeds and interacts with BiP in the ER, thereby establishing its involvement in oxidative protein folding, whereas, the ER-localized AtPDIL2, which can restore the wild-type phenotype in a yeast PDI1 null mutant, is extremely expressed in the micropylar area of the ovule and assists in the development of embryonal sac (Onda 2013). Crts are the most abundant in the tissues actively involved with secretion, for example in vasculature development and germination of the seeds in floral organs. *Arabidopsis* Crt3, the founder of the plant Crts family, acts in keeping a defective brassinosteroid receptor in the ER in right form (Jin et al. 2009).

6.6.2 Chaperones in Abiotic and Biotic Stresses

It is known that heat and other stress treatments cause a transient spurt in cytosolic Ca^{2+} level and further enhancement of ROS production in plant cells, which in turn activates stress-related genes, i.e. the chaperones. Furthermore, plant's mitochondria participates in maintaining the homeostasis of intracellular calcium level and serves as a source of ROS generation. These results provide grounds to believe that mitochondria takes part in heat-activated expression of Hsp genes by modulating ROS production and the Ca^{2+} level in the cytoplasm. For example, the treatment of *Arabidopsis* cells with the bacterial elicitor harpin disrupted normal functioning of mitochondria and activated Hsp expression (Rhoads et al. 2006). Analogous effect in plants was also observed under anoxia and in mutants with damaged mitochondrial functions. In some cases, Ca^{2+} may cause cell death, whereas in other conditions it trigerrs the expression of stress-related genes, thus protecting the cell from being completely perished. The treatment of *A. thaliana* cell culture with the programmed cell death causing agent amiodarone (AMD) and the inhibitor of oxidative phosphorylation carbonyl cyanide m-chlorophenyl hydrazone (CCCP) at optimal temperature elevated the Ca^{2+} concentration in the cytoplasm and, at the same time, activated Hsp101 expression (Pyatrikas et al. 2014). The presence of CCCP during heat stress inhibited the heat induced activation of Hsp101 expression, whereas no such effect was observed with the drug AMD (Pyatrikas et al. 2014). The excessive rise in the cytosolic Ca^{2+} level was accompanied by the inhibition of Hsp101 expression. Hsp activation as well as expression under stress depends on particular spatio-temporal dynamics of cytosolic Ca^{2+} changes (Pyatrikas et al. 2014). Numerous pathways related to signaling are involved in the heat shock response, several of them direct Hsps while others regulate downstream effector molecules activation. Promising evidences have shown that any shock response is followed by some extent of oxidative stress. There is a likelihood of a cross-talk between temperature and oxidative stress-mediated signaling pathways (Desikan et al. 2004). In plants, elevated temperature favours H_2O_2 accumulation which resulted in increased NADPH oxidase enzyme activity (Desikan et al. 2004). This spurt was interconnected with heat shock responsive genes activation. This process has been hypothesized to be directed through the sensing of the released redox molecules as H_2O_2 by HSFs (Miller and Mittler 2006). Prior treatments with H_2O_2 or the compound menadione also caused increased tolerance to heat shock, whereas mutants such as *atrbohB* and *atrbohD* (NADPH oxidases delpeted) showed deformity in tolerance. Moreover, in maize mitochondrial mutants having defects in their respiratory chain, specific Hsps were upregulated. Mostly, the Hsp90s chaperones substrates are kinases and the factors that play role in signaling pathways via activation or inhibition of diverse defense-associated gene expression. They are also implicated in the regulation between different signaling pathways. Plants Hsp90 provides resistance not only against abiotic but also biotic stresses. Hsp90 interacts with SGT1 i.e. G2 allele kinetochore protein suppressor and is required for Mla12 (Mildew resistance locus a12) resistance (RAR1). It also confers permanence to resistance (R) protein providing identification of pathogen effectors. Interaction of SGT1 with Hsp90 is mediated through binding of its CS and TPR domains with Hsp90 ATPase and C-terminal sites, respectively. SGT1 interacts with the R protein through the leucine rich repeats (LRR) (Shirasu and Schulze-Lefert 2003), while RAR1 interacts with the Hsp90ATP binding region. In AtHsp90-2 mutant lines, another disease resistant signature RPM1 was weakened. Hsp90 provides tomato Pto gene-mediated resistance to *Pseudomonas* cloves. The interaction of Hsp90 with RAR1 and TIR-NB-LRR provides resistance to TMV. Virus-induced gene silencing has supported the notion that Hsp90, SGTl, and RARl are required for Mla13-directed resistance against powdery mildew in *Hordeum* and resistance against leaf rust disease in *Triticum* mediated through *Lr21*.

Current findings by Bhattarai et al. (2007) revealed that plant Hsp90s are involved in signaling resistance against pests and insects. Tomato Mi-1 protein contains NBS and LRR motifs that directs resistance against root-knot nematodes, aphids and *Bemecia*. SGTl and Hsp90 knockouts lowered the Mi-1-directed resistance against nematodes and aphids. *Ulva* Hsp90, positively regulated the diurnal and temperature fluctuations. It also affected the prolonged heavy metal stress in its sterile condition (Tominaga et al. 2012). Yeast Hsp90 is helpful in maintaining permeability of cell wall and stress effects by maintaining the glycerol content via HOG pathway and several mitogenesis associated protein kinases (MAPK) (Hawle et al. 2007). *Arabidopsis* ROF1, another FK506 binding protein family member, interacts with the co-chaperone Hsp90.1 through its TPR domain (Meiri and Breiman 2009).

Under physiological conditions, ROF1-Hsp90.1 resides in the cytosol. Upon stress stimulation, HsfA2 combines with the ROF1-Hsp90.1complex, which further translocates to the nucleus and regulates the expression of targeted stress tolerant genes (Meiri and Breiman 2009). ROF1 or ROF2 silenced lines showed increased stress tolerance (Xu et al. 2012). Some plant Hsp70s show enhanced expression on cold shock (Li et al. 1999), such as in tomato and spinach. Low temperature stress was perceived by cytosol and mitochondria was reported in *Lycopersicon*, spinach and *Arabidopsis* (Guy and Li 1998). The cause behind specific response of these Hsps at low temperature is yet to be deciphered.

Calreticulins play defensive role in plant inherent immunity. They are recruited by the immune receptors to provide complete defense against different pathogens (Caplan et al. 2009). In case of TMV, *Arabidopsis* Crt1, Crt2 and Crt3 mediated protection against pathogenesis has also been reported. Mi-Crt secreted by the nematode into the apoplast of diseased tissues, imparts an essential role in pathogenesis through suppressing basal defence of the plants. RNA interference of Mi-Crton, on the other hand, makes plants more susceptible to nematode infection compared to the wild type (WT) (Jaouannet et al. 2013). Matsukawa et al. (2013) have shown that *Nicotiana* Crt 3a (*NbCrt3a*) is employed for resistance of *Nicotiana* to *Phytophthora* infection. *NbCrt3a* codes for an ER quality-control (ERQC) chaperone, which helps in glycoproteins maturation. Shaterian et al. (2005) explored the expression of Crts and concluded its involvement in ABA-dependent tolerance mechanism, controlled mainly by its root system. The Crts also showed their role in providing drought tolerance in wheat crop (Jia et al. 2008). Co-expression of *ZmCrt1* mitigated the adverse effects via enhancing the Ca^{2+} concentration as reported in *Zea mays* (Nouri et al. 2012). *LEN1* (encoding Cpn60b) deletion causes cell death in *Arabidopsis*. This activates systemic acquired resistance (SAR), triggered by necrotic lesions induced from infections.

6.6.3 An Interaction Map of Chaperones and Other Stress Responsive Mechanisms

Environmental stresses are liable for photosynthetic ROS production. There is an existing notion of cross-talk between all the stress signaling components and complexes forming a complex interaction map. This ubiquitous system plays key functions in stress as well as non-stressed conditions. There are several evidences suggesting the active participation of chaperones with other stress-responsive mechanisms, creating an avenue for complex network in the plant system. In plants a burst of H_2O_2 occurs at short intervals as a result of NADPH oxidase activity gets enhanced at elevated temperature conditions (Desikan et al. 2004). This sudden excitation was directly linked with the expression profile of the heat responsive genes, which was reported to occur by sensing of H_2O_2 by HSFs. This report further shed light on the over expression of other chaperones as Hsp70 and Hsp17 under different stress conditions (Desikan et al. 2004). On the other hand, osmolytes which are low weight organic compounds also render a positive correlation. It accumulates in plants under any unfavourable conditions. The disaccharide trehalose prohibits the denatured protein segregation and maintains protein in a moderately folded state, which can be reactivated by other chaperones. In *E. coli*, other osmolytes such as as glycine betaine may act as 'chemical chaperones' by rising the steadiness of native proteins and supporting the refolding of added unfolded polypeptides (Singer and Lindquist 1998). Trehalose and a sHsp26 from *Artemia* (a crustacean), can act synergistically *in vitro* throughout heat stress treatment (Viner and Clegg 2001). It is thus well documented that stability of cells protein and chaperone directed disaggregation and protein refolding can be modulated by different osmolytes at different cellular level. These studies help us to relate the complex network among Hsps, chaperones and stress responsive mechanisms.

Hsp90 and Hsp70 chaperones families along with their co-chaperones bind with a plethora of signaling related molecules, which involves hormone receptors, kinases and regulators of cell cycle and cell-death, demonstrating interplay of interdependent networks. The oxidized and reduced state of thiol containing molecules is crucial in carrying out cellular functions. Even though most of these reports were in systems other than plants, related cross-talk interactive maps might exist in plant system. For example, HSFs enhanced the expression of antioxidants like ascorbate peroxidases (APX) in *Arabidopsis*. It is suggested that HSFs might be implicated in Hsp synthesis and also in the regulation of antioxidant gene expression. Further, ER chaperones and foldases function together *in vivo* to fold proteins but the mechanism is still unclear. Their association into multimember complexes has been proposed to focus their activities on nascent forms.

6.7 Engineering Plant Chaperones for Enhanced Abiotic Stress Tolerance

With the availaibility of whole genome sequences of crop species, for example rice, *Arabidopsis*, maize etc., recombinant DNA technology has become a suitable approach to generate transgenic varieties with improved tolerance to abiotic stresses. Hsp family members have been widely used for this purpose (Agarwal et al. 2001; Batra et al. 2007). Recently, there are several reports where researchers have manipulated different Heat Shock Response (HSR) components to raise the transgenic lines, which can withstand increased temperature and other allied stresses. A large number of heat shock genes have been cloned (Agarwal et al. 2003). Tolerance to elevated temperature can be achieved either by directly manipulating the Hsps or by manipulating regulatory factors that control Hsps, osmolytes and/or regulatory molecules affecting membrane fluidity and cell detoxification system (Grover et al. 2000). Lee et al. (1995) changed the Hsp expression profile by changing the level of HSFs in *A. thaliana* and observed that over expression of *A. thaliana* transcription factor *Athsf* successfully led to increase in thermotolerance. There are reports which clearly showed that at high temperature, Heat shock elements (HSEs) interact with positively acting heat shock transcription factors for enhanced expression of heat shock proteins (Wu 1995). It has been proposed that Hsp70 regulates HSF and thermotolerance acquisition. Hsp70 antisense *A. thaliana* lines were also found to have loss of thermotolerance (Lee and Schoffl 1996). Prandl et al. (1998) over expressed *Athsf3* in *Arabidopsis* under the CaMv35S constitutive promoter and observed increased thermotolerance in transgenic lines. Modified expression of carrot sHsp gene 17.7 altered the range of thermotolerance (Malik et al. 1999). Queitsch et al. (2000) showed that transgenic *A. thaliana* plants over expressing AtHsp100 protein had better survival rates at 45° C temperature. Yeh et al. (2002) noticed that overexpression of OsHsps 16.9 in *E. coli* provided thermal stress tolerance to bacteria. Pike et al. (2001) reported the role of sHsps in providing thermotolerance to *E. coli* as well as the blue green algae *Synechococcus*. They over-expressed *OsHsp* from rice cytoplasm, *tom111* from *Lycopersicon* chloroplasts and 6803 HSP from *Synechocytis* sp. PCC6803 in *E. coli.* They observed that all these proteins protect the enzyme malate dehydrogenase from aggregation *in vitro*. These proteins were also able to protect several other soluble proteins from thermal aggregation *in vitro* in *E.coli* extract as well as the extract of the pigment phycocyanin of *Synechococcus* sp. Down regulation of *Hsp100* in *A. thaliana* and *Z. mays* mutants generally lead to the loss of both basal and acquired thermotolerance (Hong and Vierling 2001; Nieto-Sotelo et al. 2002). Panchuk et al. (2002) overexpressed *hsf3* in *A. thaliana* and resulted in enhanced thermotolerance. Overexpression of *hsfA1* leads to enhanced thermotolerance in tomato (Mishra et al. 2002). Further, overexpressiopn of AtHsp100 protein in rice plants resulted in better recovery after temperature stress compared to the control one (Katiyar-Agarwal et al. 2003). Overexpression of rice *Os*Hsps resulted in enhanced thermotolerance and resistance to ultraviolet radiation (Murakami et al. 2004). Tobacco plants overexpressing tomato *LeHsp* gene resulted in inceased thermotolerance (Sanmiya et al. 2004). Yang et al. (2006) observed that silencing of Hsp100/ClpB protein in tomato plants lead to loss of thermotolerance. Hsa32 is a heat inducible protein and its silencing leads to loss of thermotolerance even after the preconditioning of *A. thaliana* plants with the sub-lethal temperature treatment (Charng et al. 2006). Charng et al. (2007) observed that *HsfA2* knockout mutants are more sensitive to heat stress than wild type suggesting a major role of *HsfA2* in acquired thermotolerance in *Arabidopsis*. Schramm et al. (2008) characterized DREB2A as a heat stress responsive regulator and emphasized its role in the induction of *HsfA3* under temperature stress. Malik et al. (1999) provided evidence that overexpression of *sHsps 17.7* in carrot cell lines and plants resulted in increased endurance of transgenic cell lines to high temperature. Overexpression of rice chloroplast *OsHsp26* in *E. coli* provided heat tolerance and tolerance to other oxidative stresses (Lee and Vierling 2000). Ono et al. (2001) reported that overexpression of DnaK from halotolerant cyanobacteriun *Aphanothece halophytica* in tobacco showed endurance towards thermal stress. Transgenic tobacco plants overexpressing class 1 sHsps also showed thermotolerance (Park and Hong 2002). Introduction of tomato mitochondrial sHsps in tobacco also provided tolerance to thermal stress (Sanmiya et al. 2004). Li et al. (2005) reported that *A. thaliana AthsfA2* overexpression increased themotolerance, whereas *AthsfA2* mutant lines showed reduced basal tolerance and oxidative stress tolerance in *Arabidopsis*. *Arabidopsis* transgenic lines overexpressing *OshsfA2e* under the control of maize ubiquitin promoter also showed thermotolerance (Yokotani et al. 2007). Molecular chaperones, which belong to heat shock cognate 70KDa family (hsc70), are highly conserved and their altered expression assists in biotic and abiotic stress tolerance (Cazale et al. 2009). Preconditioning of plants with mild stress conditioned them to overcome severe stress (Vasquez-Robinet et al. 2010).

Calreticulin's role in response to environmental stress has also been investigated. To elucidate the functions of the calreticulin in reaction to drought stress, Jia et al. (2008) overexpressed full-length wheat calreticulin *TaCrt* in tobacco plants. *TaCrt* over-expressing plants showed drought resistance compared to the control plants under water-deficit condition. Transgenic tobacco plants showed higher water uptake effectiveness, water retention capacity, relative water content and low level of membrane damage under water-deficit conditions compared to the wild type control plants (Jia et al. 2008). Wheat calreticulin is thus implicated in the drought stress tolerance. Calreticulins have the ability to increase the capacity to rapidly store calcium or release it from the ER as well. Overexpression of maize Crt in *Arabidopsis* lines led to increase in total calcium, chlorophyll level and yield compared to control wild type plant (Tsou et al. 2012). *Arabidopsis* transgenic lines overexpressing maize Calreticulin also showed increased root growth and better survival under intermittent drought stress through up-regulation of *CIPK6* (Tsou et al. 2012). These transgenics have proven that chaperones are suitable candidate genes for transgenic studies. Table 6.2 provides comprehensive detail on plant transgenics raised by genetic manipulation of chaperones for different abiotic stress tolerance.

6.8 Conclusion and Future Perspectives

Chaperones provide tolerance to plants as ellucidated in several plants systems and their role in plants overall sustenance have also been widely accepted. Their presence in all the cellular organelle as well their multifaceted functionality reaffirms their regulatory character. The natural phenomenon of stress adaptation is

Table 6.2 Representative examples of chaperones reported to be suitable for generating transgenic plants for various abiotic stresses

Molecular chaperones	Transformants	Role in abiotic stress tolerance in plants	References
AtHsp70	*A. thaliana*	Overexpression lines show thermotolerance and antisense lines show loss of thermotolerance	Lee and Schoffl (1996)
DcHsp17.7	*E. coli*	Transformants exhibit thermotolerance	Malik et al. (1999)
AtHsp100	*A. thaliana*, *Z. mays*	Transformants are tolerant to sudden exposure of extreme temperature	Hong and Vierling (2001), Nieto-Sotelo et al. (2002)
OsHsp26	*E. coli*	Overexpression provides thermotolerance and tolerance to other oxidative stresses	Lee et al. (2000)
OsHsp16.9	*E. coli*	Overexpression provide thermotolerance	Yeh et al. (2002)
AtHsp101	*O. sativa*	Transformants show tolerance to high temperature	Katiyar-Agarwal et al. (2003)
Lehsp	*N. tabacum*	Overexpression lines show thermotolerance, whereas, antisense lines are thermosensitive	Sanmiya et al. (2004)
Hsp17.7	*O. sativa*	Overexpression confers both heat tolerance and UV-resistance	Murakami et al. (2004)
Tomato MtsHsp	Tomato	Overexpression lead to thermotolerance	Nautiyal et al. (2005)
Hsp21	*N. tabacum*	Protect photosystem II from temperature induced oxidative stress	Neta-Sharir et al. (2005)
OaPDI (Oldenlandia affinis)	*O. sativa*	Tolerance towards oxidative stress.	Gruber et al. (2007)
TaCRT	*T. aestivum*	Overexpression shows drought stress tolerance under water deficit condition	Jia et al. (2008)
Hsc70	*A. thaliana*	Overexpression lead to tolerance to salt, cadmium and arsenic.	Cazale et al. (2009)
ZmCrt	*A. thaliana*	Overexpression lead to increase in calcium content, chlorophyll content and yield	Tsou et al. (2012)
GhCyp1	*N. tabacum*	Salt and biotic stress tolerance	Zhu et al. (2011)
MTH1745 (PDIL)	*O. sativa*	Heavy metal tolerance	Chen et al. (2012)
OsCyp2	*N. tabacum*	Overexpression shows tolerance towards salinity and oxidative stress	Kumari et al. (2015)
AtHsp90.5	*Arabidopsis*	Tolerance towards salt and drought	Oh et al. (2014)

complicated and requires active participation of many biological micro- and macromolecules, including proteins and lipids. Depiction of their impending sites of stress perception and understanding the way plants respond to any form of stress forming a complex network is yet to be studied. It would be remarkable to map these chaperones to quantitative trait loci linked to tolerance. Chaperones synergistically form alluring molecular networks that plants exploit to tolerate different stress forms. These proteins generally show pleiotropy, interact with multiple pathways in varied fashion. Cellular stress response has implications on, and is subjected by, mechanisms at all steps of organismal constitution. Some Hsps are quite exclusive to the plants, solely because of the localization of these chaperones in chloroplasts and other plant specific subcellular compartments. Many reports for the presence of Hsp70, Hsp60, sHsps, cyclophilins in the organelles do exits. These chaperones are quite related to their homologs in both eukaryotic and prokaryotic organisms. They show similar impact to the retention of chloroplast structure and organellar maintainance, but further investigations are underway.

Acknowledgements Authors would like to thank the financial support received from DST-PURSE and UPOE-II, Government of India. SR and MM would like to thank the support received from CSIR as JRF and SRF.

References

Agarwal M, Katiyar-Agarwal S, Sahi C, Gallie D, Grover A (2001) *Arabidopsis thaliana* Hsp100 proteins: Kith and kin. Cell Stress Chaperon 63:219–224

Agarwal M, Sarkar N, Grover A (2003) Low molecular weight heat shock proteins in plants. J Plant Biol 30:141–149

Apuya N, Yadegari R, Fischer RL, Harada JJ, Zimmerman JL et al (2001) The *Arabidopsis* embryo mutant schlepperless has a defect in the chaperonin-60α gene. Plant Physiol 126:717–730

Baldan B, Navazio L, Friso A, Mariani P, Meggio F (1996) Plant calreticulin is specifically and efficiently phosphorylated by protein kinase CK2. Biochem Biophys Res Commun 221 (498):502

Barua D, Downs CA, Heckathorn SA (2003) Variation in chloroplast small heat-shock protein function is a major determinant of variation in thermotolerance of photosynthetic electron transport among ecotypes of *Chenopodium album*. Funct Plant Biol 30:1071–1079

Batra G, Chauhan VS, Singh A, Sarkar NK, Grover A (2007) Complexity of rice Hsp100 gene family: lessons from rice genome sequence data. J Biosci 32:611–619

Bekh-Ochir D, Shimada S, Yamagami A, Kanda S, Ogawa K et al (2013) A novel mitochondrial DnaJ/Hsp40 family protein BIL2 promotes plant growth and resistance against environmental stress in brassinosteroid signaling. Planta 237(6):1509–1525

Benjamin IJ, McMillan DR (1998) Stress (heat shock) proteins: molecular chaperones in cardiovascular biology and disease. Circ Res 83:117–132

Bhattarai KK, Li Q, Liu Y, Dinesh-Kumar SP, Kaloshian I (2007) The *Mi-1*-mediated pest resistance requires *Hsp90* and *Sgt1*. Plant Physiol 144(1):312–323

Boston RS, Viitanen PV, Vierling E (1996) Molecular chaperones and protein folding in plants. Plant Mol Biol 32:191–222

Bouchard RA (1990) Characterization of expressed meiotic prophase repeat transcript clones of Liluim: meiosis-specific expression, relatedness, and affinities to small heat shock protein genes. Genome 33:68–79

Bukau B, Weissman J, Horwich A (2006) Molecular chaperones and protein quality control. Cell 125(3):443–451

Burger AM, Seth AK (2004) The ubiquitin mediated protein degradation pathway in cancer: therapeutic implication. Eur J Cancer 40(15):2217–2229

Burke JJ (2001) Identification of genetic diversity and mutations in higher plant acquired thermotolerance. Physiol Plant 112:167–170

Caplan JL, Zhu X, Mamillapalli P, Marathe R, Anandalakshmi R et al (2009) Induced ER chaperones regulate a receptor-like kinase to mediate antiviral innate immune response in plants. Cell Host Microbiol 6(5):457–469

Carrigan PE, Sikkink LA, Smith DF, Alvarado MR (2006) Domain: domain interactions within Hop, the Hsp70/Hsp90 organizing protein, are required for protein stability and structure. Protein Sci 15(3):522–532

Cazale AC, Ment MC, Chiarenza S, Roncato MA, Pochon N et al (2009) Altered expression of cytosolic/nuclear HSC70-1 molecular chaperone affects development and abiotic stress tolerance in *Arabidopsis thaliana.* J Exp Bot 60:2653–2664

Charng YY, Liu HC, Liu NY, Hsu FC, Ko SS (2006) Arabidopsis Hsa32, a novel heat shock protein, is essential for acquired thermotolerance during long recovery after acclimation. Plant Physiol 140:1297–1305

Charng YY, Liu HC, Liu NY, Chi WT, Wang CN et al (2007) A heat-inducible transcription factor, HsfA2, is required for extension of acquired thermotolerance in *Arabidopsis*. Plant Physiol 143:251–262

Chen M, Choi Y, Voytas DF, Rodermel S (2000) Mutations in the Arabidopsis VAR2 locus cause leaf variegation due to the loss of a chloroplast FtsH protease. Plant J 22:303–313

Chen S, Smith DF (1998) Hop as an adaptor in the heat shock protein 70 (Hsp70) and Hsp90 chaperone machinery. J Biol Chem 273:35194–35200

Chou IT, Gasser CS (1997) Characterization of the cyclophilin gene family of *Arabidopsis thaliana* and phylogenetic analysis of known cyclophilin proteins. Plant Mol Biol 35:873–892

Connolly T, Gilmore R (1986) Formation of a functional ribosome–membrane junction during translocation requires the participation of a GTP-binding protein. J Cell Biol 103:2253–2261

Cooper P, Ho TH (1987) Intracellular localization of heat shock proteins in maize. Plant Physiol 84:1197–1203

Coppolino MG, Woodside MJ, Demaurex N, Grinstein S, St-Arnaud R, Dedhar S (1997) Calreticulin is essential for integrin-mediated calcium signalling and cell adhesion. Nature 386:843–847

D'Aloisio E, Paolacci AR, Dhanapal AP, Tanzarella OA, Porceddu E, Ciaffi M (2010) The protein disulfide isomerase gene family in bread wheat (*T. aestivum* L.). BMC Plant Biol 10:101

de la Fuente van Bentem S, Vossen JH, de Vries KJ, van Wees S, Tameling WI, Dekker HL et al (2005) Heat shock protein 90 and its co-chaperone protein phosphatase 5 interact with distinct regions of the tomato I-2 disease resistance protein. Plant J 43:284–98

Del Bem LE (2011) The evolutionary history of calreticulin and calnexin genes in green plants. Genetica 139(2):255–259

Denecke J, Carlsson LE, Vidal S, Höglund AS, Ek B, Van Zeijl MJ et al (1995) The tobacco homolog of mammalian calreticulin is present in protein complexes *in vivo*. Plant Cell 7:391–406

DeRocher AE, Vierling E (1994) Developmental control of small heat shock protein expression during pea seed maturation. Plant J 51:93–102

Desikan R, Hancock JT, Neill SJ (2004) Oxidative stress signaling. In: Hirt H, Shinozaki K (eds) Plant responses to abiotic stress. Springer, Berlin, pp 73–93

Dhankher OP, Drew JE, Gatehouse JA (1997) Characterization of a pea hsp70 gene which is both developmentally and stress regulated. Plant Mol Biol 34:345–352

Dittmar KD, Hutchison KA, Owens-Grillo JK, Pratt WB (1996) Reconstitution of the steroid receptor–hsp90 heterocomplex assembly system of rabbit reticulocyte lysate. J Biol Chem 271:12833–12839

Dominguez-Solis JR, He Z, Lima A, Ting J, Buchanan BB, Luan S (2008) A cyclophilin links redox and light signals to cysteine biosynthesis and stress responses in chloroplasts. Proc Natl Acad Sci USA 105:16386–16391

Downs CA, Heckathorn SA (1998) The mitochondrial small heat-shock protein protects NADH: ubiquinone oxidoreductase of the electron transport chain during heat stress in plants. FEBS Lett 430:246–250

Fan AC, Young JC (2011) Function of cytosolic chaperones in Tom70-mediated mitochondrial import. Protein Pept Lett 18(2):122–131

Fisher G, Bang H, Mech C (1984) Determination of enzymatic catalysis for the cis-trans-isomerization of peptide binding in proline-containing peptides. Biomed Biochim Acta 43:1101–1111

Fu A, He Z, Lima A, Buchanan BB, Luan S (2007) A chloroplast cyclophilin functions in the assembly and maintenance of photosystem II in Arabidopsis thaliana. Proc Natl Acad Sci USA 104:15947–15952

Fulgosi H, Vener AV, Altschmied L, Herrmann RG, Andersson B (1998) A novel multi-functional chloroplast protein: identification of a 40 kDa immunophilinlike protein located in the thylakoid lumen. EMBO J 17:1577–1587

Fulda S, Gorman AM, Hori O, Samali A (2010) Cellular stress responses: cell survival and cell death. Intl J Cell Biol 214074:23

Georges F, Hussain A, Keller WA (1999) Transcription patterns of the Calreticulin gene in *Brassica napus* seedlings under different environmental stress conditions. In: Proceedings of the 1st international conference in Egypt on plant tissue culture and its applications, Cairo, Egypt, pp 26–40

Glick BS (1995) Can Hsp70 proteins act as force-generating motors? Cell 80(1):11–14

Goel M, Garcia R, Estacion M, Schilling WP (2001) Regulation of Drosophila TRPL channels by immunophilin FKBP59. J Biol Chem 276:38762–38773

Goto-Yamada S, Mano S, Yamada K, Oikawa K, Hosokawa Y et al (2015) Dynamics of the light-dependent transition of plant peroxisomes. Plant Cell Physiol 56(7):1264–1271

Gowda NKC, Kandasamy G, Froehlich MS, Dohmen RJ, Andreasson C (2012) Hsp70 nucleotide exchange factor Fes1 is essential for ubiquitin-dependent degradation of misfolded cytosolic proteins. Proc Natl Acad Sci USA 110(15):5975–5980

Grover A, Agarwal M, Katiyar-Agarwal S, Sahi C, Agarwal S (2000) Production of high temperature tolerant transgenic plants through manipulation of photosynthetic membrane lipids. Curr Sci 79:557–559

Gruber CW, Cemazar M, Clark RJ, Horibe T et al (2007) A novel plant protein-disulfide isomerase involved in the oxidative folding of cystine knot defense proteins. J Biol Chem 282:20435–20446

Guy CL, Li QB (1998) The organization and evolution of the spinach stress 70 molecular chaperone gene family. Plant Cell 10:539–556

Halic M, Beckmann R (2005) The signal recognition particle and its interactions during protein targeting. Curr Opin Struct Biol 15:116–125

Halperin T, Ostersetzer O, Adam Z (2001) ATP dependent association between subunits of Clp protease in pea chloroplasts. Planta 213(4):4–619

Hammond C, Braakman I, Helenius A (1994) Role of N-linked oligosaccharide recognition, glucose trimming, and calnexin in glycoprotein folding and quality control. Proc Natl Acad Sci USA 91:913–917

Handschumacher RE, Harding MW, Rice J, Drugge RJ, Speicher DW (1984) Cyclophilin: a specific cytosolic binding protein for cyclosporine A. Science 226:544–554

Hatahet F, Ruddock LW (2007) Substrate recognition by the protein disulfide isomerases. FEBS J 274:5223–5234

Hawle P, Horst D, Bebelman JP, Yang XX, Siderius M et al (2007) Cdc37p is required for stress-induced high-osmolarity glycerol and protein kinase C mitogen-activated protein kinase pathway functionality by interaction with Hog1p and Slt2p (Mpk1p). Eukaryot Cell 6:521–532

Helm KW, Lafayete PR, Nago RT, Key JL, Vierling E (1993) Localization of small heat shock proteins to the higher plant endomembrane system. Mol Cell Biol 13:238–247

Hemmingsen SM, Woolford C, Van der Vies SM, Tilly K, Dennis DT et al (1988) Homologous plant and bacterial proteins chaperone oligomeric protein assembly. Nature 333:330–334

Hernandez LD, Vierling E (1993) Expression of low molecular weight heat-shock proteins under field conditions. Plant Physiol 101:1209–1216

Herzberg O, Moult J (1991) Analysis of the steric strain in the polypeptide backbone of protein molecules. Proteins 11:223–229

Hohfeld J, Minami Y, Hartl FU (1995) Hip, a novel cochaperone involved in the eukaryotic Hsc70/Hsp40 reaction cycle. Cell 83:589–598

Hohfeld J, Jentsch S (1997) GrpE-like regulation of the hsc70 chaperone by the anti-apoptotic protein BAG-1. EMBO J 16:6209–6216

Hong SW, Vierling E (2001) Hsp101 is necessary for heat tolerance but dispensable for development and germination in the absence of stress. Plant J 27:25–35

Huber D, Boyd D, Xia Y, Olma MH, Gerstein M et al (2005) Use of thioredoxin as a reporter to identify a subset of *Escherichia coli* signal sequences that promote signal recognition particle-dependent translocation. J Bacteriol 187:2983–2991

Huttner S, Stasser R (2012) Endoplasmic reticulum-associated degradation of glycoproteins in plants. Front in Plant Sci 3:67. https://doi.org/10.3389/fpls.2012.00067

Huttner S, Veit C, Schoberer J, Grass J, Strasser R (2012) Unraveling the function of *Arabidopsis thaliana* OS9 in the endoplasmic reticulum-associated degradation of glycoproteins. Plant Mol Biol 79:21–33

Imai J, Maruya M, Yashiroda H, Yahara I, Tanaka K (2003) The molecular chaperone Hsp90 plays a role in the assembly and maintenance of the 26S proteasome. EMBO J 22(14):3557–3567

Ivey RA, Bruce BD (2000) *In vivo* and *in vitro* interaction of DnaK and a chloroplast transit peptide. Cell Stress Chaperones 5:62–71

Jackson-Constan D, Akita M, Keegstra K (2001) Molecular chaperones involved in chloroplast protein import. Biochim Biophys Acta 1541:102–113

Jaouannet M, Magliano M, Arguel MJ, Gourgues M, Evangelisti E et al (2013) The root-knot nematode calreticulin Mi-CRT is a key effector in plant defense suppression. Mol Plant Microbe Interact 26(1):97–105

Jia XY, Xu CY, Jing RL, Chang XP (2008) Molecular cloning and characterization of wheat calreticulin (CRT) gene involved in drought-stressed responses. J Expt Bot 59:739–751

Jia XY, He LH, Jing RL, Li RZ (2009) Calreticulin: conserved protein and diverse functions in plants. Physiol Plant 136(2):127–138

Jin H, Hong Z, Su W, Li J (2009) A plant-specific calreticulin is a key retention factor for a defective brassinosteroid receptor in the endoplasmic reticulum. Curr Issue 106(32):13612–13617

Joo JH, Dorsey FC, Joshi A, Hennessy-Walters KM, Rose KL et al (2011) Hsp90-Cdc37 chaperone complex regulates Ulk1- and Atg13-mediated mitophagy. Mol Cell 43:572–585

Kampinga HH, Craig EA (2010) The HSP70 chaperone machinery: J proteins as drivers of functional specificity. Nat Rev Mol Cell Biol 11:579–592

Katiyar-Agarwal S, Agarwal M, Grover A (2003) Heat tolerant basmati rice engineered by overexpression of hsp101 gene. Plant Mol Biol 51:677–686

Kim BH, Schoffl F (2002) Interaction between *Arabidopsis* heat shock transcription factor 1 and 70 KDa heat shock proteins. J Exp Bot 53:371–375

Kim KK, Kim R, Kim SH (1998) Crystal structure of a small heat-shock protein. Nature 394:595–599

Koegl M, Hoppe T, Schlenker S, Ulrich HD, Mayer TU et al (1999) A novel ubiquitination factor, E4, is involved in multiubiquitin chain assembly. Cell 96:635–644

Kourtz L, Ko K (1997) The early stage of chloroplast protein import involves Com70. J Biol Chem 272:2808–2813
Kozany C, Mokranjac D, Sichting M, Neupert W, Hell K (2004) The J domain-related cochaperone Tim16 is a constituent of the mitochondrial TIM23 preprotein translocase. Nat Struct Mol Biol 11:234–241
Kriehuber T, Rattei T, Weinmaier T, Bepperling A, Haslbeck M et al (2010) Independent evolution of the core domain and its flanking sequences in small heat shock proteins. FASEB J 24(10):3633–3642
Krishna P, Gloor G (2001) The Hsp90 family of proteins in *Arabidopsis thaliana*. Cell Stress Chaperones 6(3):238–246
Kumari S, Singh P, Singla-Pareek SL, Pareek A (2009) Heterologous expression of a salinity and developmentally regulated rice cyclophilin gene (OsCyp2) in E. coli and S. cerevisiae confers tolerance towards multiple abiotic stresses. Mol Biotechnol 42:195–204
Kumari S, Roy S, Singh P, Singla-Pareek SL, Pareek A (2013) Cyclophilins: Proteins in search of function. Plant Signal Behav 8:22734–22741
Kumari S, Joshi R, Singh K, Roy S, Tripathi AK et al (2015) Expression of a cyclophilin OsCyp2-P isolated from a salt-tolerant landrace of rice in tobacco alleviates stress via ion homeostasis and limiting ROS accumulation. Funct Integr Genomics. https://doi.org/10.1007/s10142-014-0429-5
Lee GJ, Vierling E (2000) A small heat shock protein cooperates with heat shock protein 70 systems to reactivate a heat-denatured protein. Plant Physiol 122:189–198
Lee JH, Hubel A, Schoffl F (1995) Derepression of the activity of genetically engineered heat shock factor causes constitutive synthesis of heat shock proteins and increased thermotolerance in transgenic *Arabidopsis*. Plant J 8(4):603–612
Lee JH, Schoffl F (1996) An Hsp70 antisense gene affects the expression of HSP70/HSC70, the regulation of HSF, and the acquisition of thermotolerance in transgenic *Arabidopsis thaliana*. Mol Gen Genet 252:11–19
Leonhard K, Herrmann JM, Stuart RA, Mannhaupt G, Neupert W et al (1996) AAA proteases with catalytic sites on opposite membrane surfaces comprise a proteolytic system for the ATP-dependent degradation of inner membrane proteins in mitochondria. EMBO J 15:4218–4229
Leverson JD, Ness SA (1998) Point mutations in v-Myb disrupt a cyclophilin-catalyzed negative regulatory mechanism. Mol Cell 1:203–211
Levitan A, Trebitsh T, Kiss V, Pereg Y, Dangoor I, Danon A (2005) Dual targeting of the protein disulfide isomerase RB60 to the chloroplast and the endoplasmic reticulum. Proc Natl Acad Sci USA 102(17):6225–6230
Levitt M, Gerstein E, Huang S, Tsai SJ (1997) Protein folding: the endgame. Annu Rev Biochem 66:549–579
Li QB, Haskell DW, Guy CL (1999) Coordinate and noncoordinate expression of the stress 70 family and other molecular chaperones at high and low temperature in spinach and tomato. Plant Mol Biol 39:21–34
Li W, Wei Z, Qiao Z, Wu Z, Cheng L et al (2013) Proteomics analysis of alfalfa response to heat stress. PLoS ONE. https://doi.org/10.1371/journal.pone.0082725
Li C, Chen Q, Gao X, Qi B, Chen N et al (2005) AtHsfA2 modulates expression of stress responsive genes and enhances tolerance to heat and oxidative stress in *Arabidopsis*. Sci China Ser C Life Sci 48:540–550
Liberek K, Marszalek J, Ang D, Georgopoulos C, Zylicz M (1991) *Escherichia coli* DnaJ and GrpE heat shock proteins jointly stimulate ATPase activity of DnaK. Proc Natl Acad Sci USA 88:2874–2878
Liberek KO, Lewandowska A, Ziętkiewicz S (2008) Chaperones in control of protein disaggregation. EMBO J 27(2):328–335
Liming Y, Qian Y, Pigang L, Sen L (2008) Expression of the HSP24 gene from *Trichoderma harzianum* in *Saccharomyces cerevisiae*. J Therm Biol 33:1–6

Lin CY, Roberts JK, Key JL (1984) Acquisition of thermotolerance in soybean seedlings: synthesis and accumulation of heat shock proteins and their cellular localization. Plant Physiol 74(1):152–160

Lindahl M, Tabak S, Cseke L, Pichersky E, Andersson B et al (1996) Identification, characterization, and molecular cloning of a homologue of the bacterial FtsH protease in chloroplasts of higher plants. J Biol Chem 271:29329–29334

Lindquist S (1986) The heat-shock response. Annu Rev Biochem 55:1151–1191

Lipinska N, Ziętkiewicz S, Sobczak A, Jurczyk A, Potocki W et al (2013) Disruption of ionic interactions between the nucleotide binding domain 1 (NBD1) and middle (M) domain in Hsp100 disaggregase unleashes toxic hyperactivity and partial independence from Hsp70. J Biol Chem 288:2857–2869

Liu QL, D Silva P, Walter W, Marszalek J, Craig EA (2003) Regulated cycling of mitochondrial Hsp70 at the protein import channel. Science 300:139–141

Lund AA, Blum PH, Bhattramakki D, Elthon TE (1998) Heat-Stress response of maize mitochondria. Plant Physiol 116(3):1097–1110

Malik MK, Slovin JP, Hwang CH, Zimmerman JL (1999) Modified expression of a carrot small heat shock protein gene, hsp17.7 results in increased or decreased thermotolerance. Plant J 20:89–99

Mariani P, Navazio L, Zuppini (2013) Calreticulin and the endoplasmic reticulum in plant cell biology. Landes Bioscience, Austin

Matouschek A, Azem A, Ratliff K, Glick BS, Schmid K et al (1997) Active unfolding of precursor proteins during mitochondrial protein import. EMBO J 16(22):6727–6736

Matouschek A, Pfanner N, Voos W (2000) Protein unfolding by mitochondriaThe Hsp70 import motor. EMBO Rep 1(5):404–410

Matsukawa M, Shibata Y, Ohtsu M, Mizutani A, Mori H et al (2013) *Nicotiana benthamiana* calreticulin 3a is required for the ethylene-mediated production of phytoalexins and disease resistance against oomycete pathogen *Phytophthora infestans*. Mol Plant Microbe Interact 26 (8):880–892

Mayer MP (2004) Timing the catch. Nat Struct Mol Biol 11:6–8

Mayer MP, Bukau B (2005) Hsp70 chaperones: cellular functions and molecular mechanism. Cell Mol Life Sci 62(6):670–684

McLellan CA, Turbyville TJ, Wijeratne EM, Kerschen A, Vierling E et al (2007) A rhizosphere fungus enhances *Arabidopsis* thermotolerance through production of an HSP90 inhibitor. Plant Physiol 145:174–182

Meiri D, Breiman A (2009) *Arabidopsis* ROF1 (FKBP62) modulates thermotolerance by interacting with HSP90.1 and affecting the accumulation of HsfA2-regulated sHSPs. Plant J 59:387–399

Mery L, Mesaeli N, Michalak M, Opas M, Lew DP, Krause KH (1996) Overexpression of calreticulin increases intracellular Ca2+-storage and decreases store-dependent Ca influx. J Biol Chem 267:2557–2562

Michalak M, Milner RE, Burns K, Opas M (1992) Calreticulin. Biochem J 285:681–692

Michalak M, Groenendyk J, Szabo E, Gold LL, Opas M (2009) Calreticulin, a multi-process calcium-buffering chaperone of the endoplasmic reticulum. Biochem J 417:651–666

Miernyk JA, Duck NB, Shatters RG Jr, Folk WR (1992) The 70-kilodalton heat shock cognate can act as a molecular chaperone during the membrane translocation of a plant secretory protein precursor. Plant Cell 4:821–829

Miernyk JA (1997) The 70 kDa stress-related proteins as molecular chaperones. Trends Plant Sci 2:80–87

Miller G, Mittler R (2006) Could heat shock transcription factors function as hydrogen peroxide sensors in plants? Ann Bot 98:279–288

Mishra SK, Tripp J, Winkelhaus S, Tschiersch B, Theres K et al (2002) In the complex family of heat stress transcription factors, HsfA1 has a unique role as master regulator of thermotolerance in tomato. Genes Dev 16:1555–1567

Mogk A, Schlieker C, Friedrich KL, Schönfeld H, Vierling E, Bukau B (2003) Refolding of substrates bound to small hsps relies on a disaggregation reaction mediated most efficiently by ClpB/DnaK. J Biol Chem 278:31033–31042

Mokranjac D, Sichting M, Neupert W, Hell K (2003) Tim14, a novel key component of the import motor of the TIM23 protein translocase of mitochondria. EMBO J 22:4945–4956

Mothes W, Heinrich SU, Graf R, Nilsson I, von Heijne G, Brunner J (1997) Molecular mechanism of membrane protein integration into the endoplasmic reticulum. Cell 89:523–533

Murakami T, Matsuba S, Funatsuki H, Kawaguchi K, Saruyama H, Tanida M, Sato Y (2004) Overexpression of a small heat shock protein, sHSP17.7, confers both heat tolerance and UV-B resistance to rice plants. Mol Breed 13:165–175

Nakabayashi K, Ito M, Kiosue T, Shinozaki K, Watanabe A (1999) Identification of clp genes expressed in senescing *Arabidopsis* leaves. Plant Cell Physiol 40:504–514

Nautiyal PC, Shono M, Egawa Y (2005) Enhanced thermotolerance of the vegetative part of MT-sHSP transgenic tomato line. Sci Horticult 105(3):393–409

Navazio L, Baldan B, Mariani P, Gerwig GJ, Vliegenthart JFG (1996) Primary structure of the N-linked carbohydrate chains of calreticulin from spinach leaves. Glycoconj J 13:977–983

Neupert W, Brunner M (2002) The protein import motor of mitochondria. Nat Rev Mol Cell Biol 3:555–565

Ng DT, Brown JD, Walter P (1996) Signal sequences specify the targeting route to the endoplasmic reticulum membrane. J Cell Biol 134:269–278

Nielsen E, Akita M, Davila-Aponte J, Keegstra K (1997) Stable association of chloroplastic precursors with protein translocation complexes that contain proteins from both envelope membranes and a stromal HSP100 molecular chaperone. EMBO J 16:935–946

Nieto-Sotelo J, Martinez LM, Ponce G, Cassab GI, Alagon A et al (2002) Maize HSP101 plays important roles in both induced and basal thermotolerance and primary root growth. Plant Cell 14:1621–1633

Nishizawa-Yokoi A, Tainaka H, Yoshida E, Tamoi M, Yabuta Y et al (2010) The 26S proteasome function and Hsp90 activity involved in the regulation of HsfA2 expression in response to oxidative stress. Plant Cell Physiol 51:486–496

Nouri MZ, Hiraga S, Komatsu S (2012) Characterization of calnexin in soybean roots and hypocotyls under osmotic stress. Phytochem 74:20–29

Obermann WMJ, Sondermann H, Russo AA, Pavletich NP, Hartl FU (1998) *In vivo* function of Hsp90 is dependent on ATP binding and ATP hydrolysis. J Cell Biol 443:901–910

Olsen LJ, Keegstra K (1992) The binding of precursor proteins to chloroplasts requires nucleoside triphosphates in the intermembrane space. J Biol Chem 267(1):433–439

Onda Y (2013) Oxidative protein-folding systems in plant cells. Int J Cell Biol 585431:15

Ono K, Hibino T, Kohinata T, Suzuki S, Tanaka Y, Nakamura T et al (2001) Overexpression of DnaK from a halotolerant cyanobacterium *Aphanothece halophytica* enhances the high temperature tolerance of tobacco during germination and early growth. Plant Sci 160:455–461

Panaretou B, Prodromou C, Roe SM, Brien RO, Ladbury JE et al (1998) ATP binding and hydrolysis are essential to the function of the Hsp90 molecular chaperone *in vivo*. EMBO J 17:4829–4836

Panchuk II, Volkov RA, Schoffl F (2002) Heat stress and heat shock transcription factor-dependent expression and activity of ascorbate peroxidase in *Arabidopsis*. Plant Physiol 129:838–853

Panzner S, Dreier L, Hartmann E, Kostka S, Rapoport TA (1995) Post translational protein transport in yeast reconstituted with a purified complex of Sec proteins and Kar2p. Cell 81:561–570

Pareek A, Singla SL, Grover A (1995) Immunological evidences for accumulation of two high molecular weight (104 and 90 kDa) Hsps I response to different stresses in rice and response to heat shock in diverse plant genera. Plant Mol Biol 29:293–301

Park SM, Hong CB (2002) Class I small heat shock protein gives thermotolerance in tobacco. J Plant Physiol 159:25–30

Parsell DA, Kowal AS, Singer MA, Lindquist S (1994) Protein disaggregation mediated by heat-shock protein Hsp104. Nature 373:475–478

Peltier JB, Friso G, Kalume DE, Roepstorff P, Nilsson F et al (2000) Proteomics of the chloroplast: systematic identification and targeting analysis. Plant Cell 12(3):319–341
Pickart CM (2001) Mechanisms underlying ubiquitination. Annu Rev Biochem 70:503–533
Pike CS, Grieve J, Badger MR, Price GD (2001) Thermoprotective properties of small heat shock proteins from rice, tomato and Synechocystis sp. PCC6803 overexpressed in, and isolated from, *Escherichia coli*. Aust J Plant Physiol 28:1219–1229
Plath K, Rapoport TA (2000) Spontaneous release of cytosolic proteins from post-translational substrates before their transport into the endoplasmic reticulum. J Cell Biol 151:167–178
Prandl R, Hinderhofer K, Eggers-Schumacher G, Schoffl F (1998) HSF3, a new heat shock factor from *Arabidopsis thaliana*, derepresses the heat shock response and confers thermotolerance when overexpressed in transgenic plants. Mol Gen Genet 258:269–278
Pyatrikas DV, Rikhvanov EG, Fedoseeva IV, Varakina NN, Rusaleva TM et al (2014) Mitochondrial retrograde regulation of HSP101 expression in *Arabidopsis thaliana* under heat stress and amiodarone action. Russian J Plant Physiol 61(1):80–89
Queitsch C, Hong SW, Vierling E, Lindquist S (2000) Hsp101 plays a crucial role in thermotolerance in *Arabidopsis*. Plant Cell 12:479–492
Rajagopalan S, Xu Y, Brenner MB (1994) Retention of unassembled components ofintegral membrane proteins by calnexin. Science 263:387–389
Rapoport TA (2007) Protein translocation across the eukaryotic endoplasmic reticulum and bacterial plasma membranes. Nature 450(7170):663–669
Rhoads DM, Umbach AL, Subbaiah CC, Siedow JN (2006) Mitochondrial reactive oxygen species. Contribution to oxidative stress and interorganellar signaling. Plant Physiol 141 (2):357–366
Rial DV, Arakaki AK, Ceccarelli EA (2000) Interaction of the targeting sequence of chloroplast precursors with Hsp70 molecular chaperones. Eur J Biochem 267:6239–6248
Ritossa F (1962) A new puffing pattern induced by temperature shock and dnp in drosophila. Experientia 18:571–573
Rodan AR, Simons JF, Trombetta ES, Helenius A (1996) N-linked oligosaccharides are necessary and sufficient for association of glycosylated forms of bovine RNase with calnexin and calreticulin. EMBO J 15:6921–6930
Rutherford SL (2003) Between genotype and phenotype: protein chaperones and evolvability. Nat Rev Genet 4:263–274
Sandqvist A, Björk JK, Akerfelt M, Chitikova Z, Grichine A et al (2009) Heterotrimerization of heat-shock factors 1 and 2 provides a transcriptional switch in response to distinct stimuli. Mol Biol Cell 20:1340–1347
Sangster TA, Salathia N, Lee HN, Watanabe E, Schellenberg K et al (2008) HSP90-buffered genetic variation is common in *Arabidopsis thaliana*. Proc Natl Acad Sci USA 105(8):2969–2974
Sanmiya K, Suzuki K, Egawa Y, Shono M (2004) Mitochondrial small heat-shock protein enhances thermotolerance in tobacco plants. FEBS Lett 557:265–268
Sarria R, Lyznik A, Vallejos CE, Mackenzie SA (1998) A cytoplasmic male sterility-associated mitochondrial peptide in common bean is post-translationally regulated. Plant Cell 10:1217–1228
Scharf KD, Siddique M, Vierling E (2001) The expanding family of *Arabidopsis thaliana* small heat stress proteins and a new family of proteins containing α-crystallin domains (Acd proteins). Cell Stress Chaperones 6(3):225–237
Schlecht R, Erbse AH, Bukau B, Mayer MP (2011) Mechanics of Hsp70 chaperones enables differential interaction with client proteins. Nat Struct Mol Biol 31:1160–1173
Schlesinger MJ (1990) Heat shock proteins. J Biol Chem 265:12111–12114
Schnell DJ, Kessler F, Blobel G (1994) Isolation of components of the chloroplast protein import machinery. Science 266:1007–1012
Schoffl F, Prändl R, Reindl A (1999) Molecular responses to heat stress. In: Shinozaki K, Yamaguchi-Shinozaki K (eds) Molecular responses to cold, drought, heat and salt stress in higher plants. R.G. Landes Co, Austin, pp 81–98

Schrag JD, Bergeron JJ, Li Y, Borisova S, Hahn M, Thomas DY, Cygler M (2001) The structure of Calnexin, an ER chaperone involved in quality control of protein folding. Mol Cell 8 (3):633–644

Schramm F, Larkindale J, Kiehlmann E, Ganguli A, Englich G et al (2008) A cascade of transcription factor DREB2A and heat stress transcription factor HsfA3 regulates the heat stress response of *Arabidopsis*. Plant J 52(2):264–274

Schulman BA, Harper JW (2009) Ubiquitin-like protein activation by E1 enzymes: the apex for downstream signalling pathways. Nat Rev Mol Cell Biol 5:319–331

Selles B, Jacquot JP, Rouhier N (2011) Comparative genomic study of protein disulfide isomerases from photosynthetic organisms. Genomics 97(1):37–50

Sewelam N, Kazan K, Schenk PM (2016) Global plant stress signaling: reactive oxygen species at the cross-road. Fron Plant Sci. https://doi.org/10.3389/fpls.2016.00187

Shaterian J, Georges F, Hussain A, Waterer D, Jong HD et al (2005) Root to shoot communication and abscisic acid in calreticulin (CR) gene expression and salt-stress tolerance in grafted diploid potato clones. Environ Exp Bot 53:323–332

Shirasu K, Schulze-Lefert P (2003) Complex formation, promiscuity and multi-functionality: protein interactions in disease-resistance pathways. Trends Plant Sci 8(6):252–258

Sigler PB, Xu Z, Rye HS, Burston SG, Fenton WA et al (1998) Structure and function in GroEL-mediated protein folding. Annu Rev Biochem 67:581–608

Silflow CD, Sun X, Haas NA, Foley JW, Lefebvre PA (2011) The Hsp70 and Hsp40 chaperones influence microtubule stability in *Chlamydomonas*. Genetics 189(4):1249–1260

Simon SM, Peskin CS, Oster GF (1992) What drives the translocation of proteins? Proc Natl Acad Sci USA 89(9):3770–3774

Singer MA, Lindquist S (1998) Thermotolerance in *Saccharomyces cerevisiae*: the Yin and Yang of trehalose. Trends Biotechnol 16:460–468

Singh A, Singh U, Mittal D, Grover A (2010) Genome-wide analysis of rice ClpB/HSP100, ClpC and ClpD genes. BMC Genom 11:95

Singla SL, Grover A (1993) Antibodies raised against yeast HSP104 crossreact with heat and abscisic acid regulated polypeptide in rice. Plant Mol Biol 22:1177–1180

Singla SL, Pareek A, Grover A (1997) High temperature stress. In: Prasad MNV (ed) Physiological ecology of plants. Wiley, New York, pp 101–127

Sokolenko A, Lerbs-Mache S, Altschmied L, Herrmann RG (1998) Clp protease complexes and their diversity in chloroplasts. Planta 207:286–295

Sondermann H, Scheufler C, Schneider C, Hohfeld J, Hartl FU et al (2001) Structure of a Bag/Hsc70 complex: convergent functional evolution of Hsp70 nucleotide exchange factors. Science 291:1553–1557

Sousa M, Parodi AJ (1995) The molecular basis for the recognition of misfolded glycoproteins by the UDP-Glc: glycoprotein glucosyltransferase. EMBO J 14:4196–4203

Su W, Liu Y, Xia Y, Hong Z, Li J (2011) Conserved endoplasmic reticulum-associated degradation system to eliminate mutated receptor-like kinases in *Arabidopsis*. Proc Natl Acad Sci USA 108:870–875

Sung DY, Vierling E, Charles L (2001) Guy comprehensive expression profile analysis of the *Arabidopsis* Hsp70 gene family. Plant Physiol 126(2):789–800

Suzuki K, Nakanishi H, Bower J, Yoder DW, Osteryoung KW, Miyagishima S (2009) Plastid chaperonin proteins Cpn60α and Cpn60β are required for plastid division in *Arabidopsis thaliana*. BMC Plant Biol 9:38

Suzuki N, Rivero RM, Shulaev V et al (2014) Abiotic and biotic stress combinations. New Phytol 203:32–43

Taipale M, Jarosz DF, Lindquist S (2010) HSP90 at the hub of protein homeostasis: emerging mechanistic insights. Nat Rev Mol Cell Biol 11:515–528

Tatu U, Helenius A (1997) Interactions between newly synthesized glycoproteins, calnexin and a network of resident chaperones in the endoplasmic reticulum. J Cell Biol 136:555–565

Taylor P, Dornan J, Carrello A, Minchin RF, Ratajczak T et al (2001) Two structures of cyclophilin 40: folding and fidelity in the TPR domains. Structure 9:431–438

Tissieres A, Mitchell HK, Tracey UM (1974) Protein synthesis in salivary grands of *Drosophila melanogaster*, realation to chromosome pufs. J Mol Biol 84:389–398
Tominaga H, Coury DA, Amano H, Miki W, Kakinuma M (2012) cDNA cloning and expression analysis of two heat shock protein genes, Hsp90 and Hsp60, from a sterile *Ulva pertusa* (Ulvales, Chlorophyta). Fish Sci 78:415–429
Tsou PL, Lee SY, Allen NS, Winter-Sederoff N, Robertson D (2012) An ER-targeted calcium-binding peptide confers salt and drought tolerance mediated by CIPK6 in *Arabidopsis*. Planta 235(3):539–552
Tsugeki R, Mori H, Nishimura M (1992) Purification, cDNA cloning and northern blot analysis of mitochondrial chaperonin 60 from pumpkin cotyledons. Eur J Biochem 209:453–458
van Montfort RL, Basha E, Friedrich KL, Slingsby C, Vierling E (2001) Crystal structure and assembly of a eukaryotic small heat shock protein. Nat Struct Biol 8(12):1025–1030
Vasiljev A, Ahting U, Nargang FE, Go NE, Habib SJ et al (2004) Reconstituted TOM core complex and Tim9/Tim10 complex of mitochondria are sufficient for translocation of the ADP/ ATP carrier across membranes. Mol Biol Cell 15:1445–1458
Vasquez-Robinet C, Watkinson JL, Sioson AA, Ramakrishnan N, Heath LS et al (2010) Differential expression of heat shock protein genes in preconditioning for photosynthetic acclimation in water-stressed loblolly pine. Plant Physiol Biochem 48:256–264
Verchot J (2012) Cellular chaperones and folding enzymes are vital contributors to membrane bound replication and movement complexes during plant RNA virus infection. Front Plant Sci 3:275
Verghese J, Abrams J, Wang Y, Morano KA (2012) Biology of the heat shock response and protein chaperones: budding yeast (*Saccharomyces cerevisiae*) as a model system. Microbiol Mol Biol Rev 76(2):115–158
Viner R, Clegg JS (2001) Influence of trehalose on the molecular chaperone activity of p26, a small heat shock/α-crystallin protein. Cell Stress Chaperones 6(2):126–135
Wadahama H, Kamauchi S, Nakamoto Y, Nishizawa K, Ishimoto M et al (2008) A novel plant protein disulfide isomerase family homologous to animal P5—molecular cloning and characterization as a functional protein for folding of soybean seed-storage proteins. FEBS J 275:399–410
Wagner I, Arlt H, Van Dyck L, Langer T, Neupert W (1994) Molecular chaperones cooperate with PIMl protease in the degradation of misfolded proteins in mitochondria. EMBO J 13:5135–5145
Wall D, Zylicz M, Georgopoulos C (1994) The NH2-terminal 108 amino-acids of the *Escherichia coli* DnaJ protein stimulate the ATPase activity of DnaK and are sufficient for λ DNA replication. J Biol Chem 269:5446–5451
Walsh P, Bursac D, Law YC, Cyr D, Lithgow T (2004) The J-protein family: modulating protein assembly, disassembly and translocation. EMBO Rep 5:567–571
Wang D, Luthe DS (2003) Heat sensitivity in a bentgrass variant. Failure to accumulate a chloroplast heat shock protein isoform implicated in heat tolerance. Plant Physiol 133:319–327
Wang W, Vinocur B, Altman A (2003) Plant responses to drought, salinity and extreme temperatures: towards genetic engineering for stress tolerance. Planta 218:1–14
Wimmer B, Lottspeich F, van der Klei I, Veenhuis M, Gietl C (1997) The glyoxysomal and plastid molecular chaperones (70-kDa heat shock protein) of watermelon cotyledons are encoded by a single gene. Proc Natl Acad Sci USA 94:13624–13629
Wu C, Seibert FS, Ko K (1994) Identification of chloroplast envelope proteins in close physical proximity to a partially translocated chimeric precursor protein. J Biol Chem 269(51):32264–32271
Wu C (1995) Heat shock transcription factors: structure and regulation. Annu Rev Cell Dev Biol 11:441–469
Wu Q, Shigaki T, Han JS, Kim CK, Hirschi KD et al (2012) Ectopic expression of a maize calreticulin mitigates calcium deficiency-like disorders in sCAX1-expressing tobacco and tomato. Plant Mol Biol 80:609–619

Xu ZS, Li ZY, Chen Y, Chen M, Li LC et al (2012) Heat shock protein 90 in plants: molecular mechanisms and roles in stress responses. Int J Mol Sci 13(12):15706–15723

Yamada AM, Sekiguchi M, Mimura T, Ozeki Y (2002) The role of plant CCT in salt- and osmotic-stress tolerance. Plant Cell Physiol 43(9):1043–1048

Yang JY, Sun Y, Sun AQ, Yi SY, Qin J et al (2006) The involvement of chloroplast HSP100/ClpB in the acquired thermotolerance in tomato. Plant Mol Biol 62:385–395

Yeh CH, Chen YM, Lin CY (2002) Functional regions of rice heat shock protein, Oshsp16.9, required for conferring thermotolerance in *Escherichia coli*. Plant Physiol 128(2):661–668

Yokotani N, Ichikawa T, Kondou Y, Matsui M, Hirochika H et al (2007) Expression of rice heat stress transcription factor OshsfA2e enhances tolerance to environmental stresses in transgenic *Arabidopsis*. Planta 227(5):957–967

Young JC, Agashe VR, Siegers K, Hartl FU (2004) Pathways of chaperone-mediated protein folding in the cytosol. Nat Rev Mol Cell Biol 5:781–791

Young ME, Keegstra K, Froehlich JE (1999) GTP promotes the formation of early-import intermediates but is not required during the translocation step of protein import into chloroplasts. Plant Physiol 121:237–244

Young TE, Ling J, Geisler-Lee CJ, Tanguay RL, Caldwell C et al (2001) Developmental and thermal regulation of the maize heat shock protein, HSP101. Plant Physiol 27(3):777–791

Yuen CY, Wong K, Christopher DA (2016) Phylogenetic characterization and promoter expression analysis of a novel hybrid protein disulfide isomerase/cargo receptor subfamily unique to plants and chromalveolates. Mol Genet Genom 291:455–469

Zhang Z, Quick MK, Kanelakis KC, Gijzen M, Krishna P (2003) Characterization of a plant homolog of Hop, a cochaperone of Hsp90. Plant Physiol 131(2):525–535

Zhu C, Wang Y, Li Y, Bhatti KH, Tian Y, Wu J (2011) Overexpression of a cotton cyclophilin gene (GhCyp1) in transgenic tobacco plants confers dual tolerance to salt stress and *Pseudomonas syringae* pv. tabaci infection. Plant Physiol Biochem 49:1264–1271

Zolkiewski M, Zhang T, Nagy AM (2012) Aggregate reactivation mediated by the Hsp100 chaperones. Biochem Biophys 520:1–6

Chapter 7
Role of Chromatin Assembly and Remodeling in Water Stress Responses in Plants

Shoib Ahmad Baba, Deepti Jain and Nasheeman Ashraf

Abstract Owing to the sessile lifestyle, plants are exposed to the ever-changing and harsh environmental conditions. To defend themselves and to cope up with multitude of biotic and abiotic aggressors that compromise their development and reproduction, plants have developed various survival strategies. These include a plethora of physiological responses as well as developmental and morphological adaptations. Defense response to environmental stresses largely depends on the plant's capability of stress sensing and transcriptional reprogramming to minimize trade-off between growth and stress. Changes in chromatic organization, chromatin remodeling and action of chromatin modifying enzymes constitute an important phenomenon involved in establishing these transcriptional states. This helps plants to attain higher degree of flexibility and facilitate activation or repression of specific sets of genes in response to environmental stresses. The present chapter provides insights about the events and the mechanisms involved in chromatin reorganization that occurs in plants in response to water stress.

Keywords Chromatin assembly · Chromatin remodelling · Histone modification · Histone acetylation · Water stress

S. Ahmad Baba · N. Ashraf (✉)
Plant Biotechnology Division, CSIR-Indian Institute of Integrative Medicine, Sanat Nagar, Srinagar 190005, Jammu and Kashmir, India
e-mail: nashraf@iiim.ac.in

D. Jain
Department of Plant Molecular Biology, Interdisciplinary Centre for Plant Genomics, Delhi University, South Campus, New Delhi 110020, India

V. R. Rajpal et al. (eds.), *Genetic Enhancement of Crops for Tolerance to Abiotic Stress: Mechanisms and Approaches, Vol. I*, Sustainable Development and Biodiversity 20, https://doi.org/10.1007/978-3-319-91956-0_7

7.1 Introduction

Plants are sessile organisms and, therefore, constantly exposed to changing environmental conditions like drought, high salinity, and low temperature. These conditions have adverse effects on growth and development of plants and subsequently their productivity. Therefore, plants have adapted to undergo cellular reprogramming in order to perceive and respond to these stresses (Han and Wagner 2014). Further, plants may face multiple stresses at the same time and therefore, possess elaborate regulatory mechanisms to coordinate the activation of stress-specific responses. Uncovering the key mechanisms by which plants tailor their responses to different stresses has become a major focus of stress biology. Chromatin reorganization is one of the principal mechanisms involved in regulating abiotic stress responses in plants.

The genetic information encoded by DNA is organized into chromatin. This chromatin is composed of nucleosome units each of which is formed by wrapping 147 bp of DNA around a histone octamer (two copies each of histones H2A, H2B, H3 and H4). This nucleosomal DNA does not allow interaction of proteins with DNA thus regulating gene expression. Therefore, expression of many genes is brought about by reorganizing the chromatin so as to alter these constraints and create accessible genome in response to endogenous and exogenous cues. Three different processes are involved in making genome accessible and these include:

1. Histone modifications which alter histone-DNA interaction and subsequently expose or block protein binding sites
2. Chromatin remodeling in which ATPases utilize the energy derived from ATP hydrolysis to alter position or composition of nucleosomes
3. Cytosine methylation in DNA which interferes with binding of some proteins while facilitates that of others

The regulation of chromatin structure through above-mentioned processes occurs by the dynamic interplay between DNA-binding proteins, histone variants, histone-modifying enzymes, chromatin-associated proteins and ATP-dependent nucleosome remodelers. These factors provide instructions that direct the transcriptional dynamics of a genome in response to developmental or environmental cues.

7.2 Histone Modifications

N-terminal regions of nucleosome core complex histones undergo various post-translational modifications including acetylation, methylation, phosphorylation, ubiquitination, sumoylation, or ADP-ribosylation (Bannister and Kouzarides 2011; Zentner and Henikoff 2013). Further, each histone has variants encoded by different genes. These post-translational modifications and histone variants

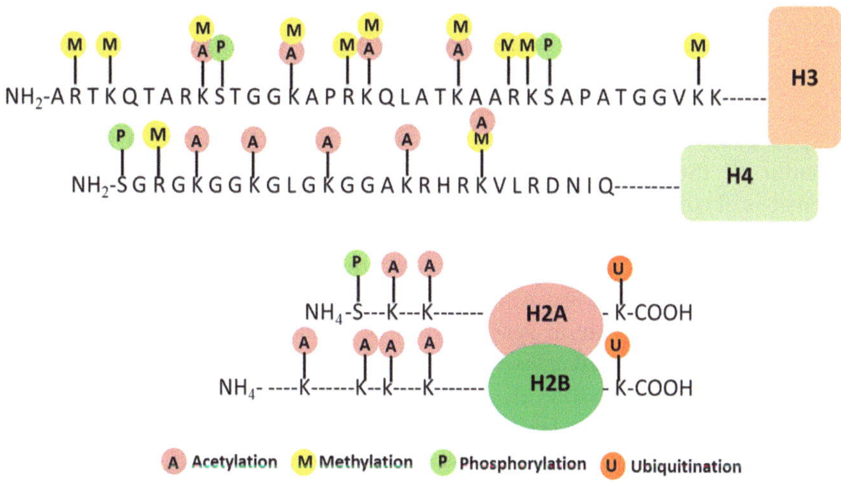

Fig. 7.1 The amino acid terminal of H2A, H2B, H3 and H4 are subject to several posttranslational alterations. Modified amino acid positions in H2A and H2B correspond to human sequences (Redrawn from Lusser et al. 2002)

constitute a 'histone code', which determines the transcriptional state and level of expression of genes. Some histone modifications like acetylation, certain phosphorylation and ubiquitination (Fig. 7.1) (Sridhar et al. 2007; Zhang et al. 2007) result in active transcription state, while biotinylation and sumoylation repress gene expression (Nathan et al. 2006; Camporeale et al. 2007).

7.2.1 *Histone Acetylation*

Acetylation of core histones is a mysterious process. The study of histone acetyltransferases and deacetylases in yeast and vertebrates has advanced our knowledge and understanding about the biological role of histone acetylation. The basics of histone acetylation in plants are other eukaryotes are more or less similar. However, there are some differences, which are reflected in new classes of histone deacetylases. Many new classes or new members of similar classes of histone modifying enzymes have been identified which are specific to plants (Lusser et al. 2001).

There are many explanations for the effects of lysine acetylation on chromatin structure. For examples, initially, acetylation weakens the interaction of the histone octamer with the negatively charged DNA by neutralizing with a positive charge. This in turn destabilizes the nucleosomes thereby allowing transcriptional regulators to gain access to the DNA. Acetylation might also interfere with packing of chromatin and thus modify the availability of chromatin areas for proteins which play regulatory role. Further, acetylation may also act as signal that changes histone–protein interactions (Loidl 1994). This possibility is further supported by the observation that non-histone proteins are also acetylated and deacetylated by histone acetyltransferases (HATs) and histone deacetylases (HDACs), respectively. These non-histone proteins include high mobility group (HMG) proteins, nuclear receptor coactivators, transcriptional activators, general transcription factors and importin α7 (Sterner and Berger 2000). It has been seen that the N-terminal extensions of core histones have sites for multiple modifications such as acetylation, phosphorylation, ubiquitination, ADP-ribosylation and methylation. I has raised the question of how these modifications could cross talk to each other in histone-encoded language (Strahl and Allis 2000). There are evidences which indicate that acetylation has major role in gene expression. Acetylation of H3 and H4 within promoter chromatin coordinates with gene expression as revealed by chromatin immunoprecipitation (Krebs et al. 1999; Parekh and Maniatis 1999). Further, use of antibodies recognizing acetylated isoforms of all the core histones has revealed that highly acetylated histones are not restricted to promoter regions of active genes only (Madisen et al. 1998; Crane-Robinson et al. 1999). These studies suggest that the acetylation of different histones may have different effects in different regions of chromatin. So far several HATs have been discovered and among them many are related to General Control Nonderepressible 5 (Gcn 5) family members, a principal integrator of various signaling pathways; transcriptional regulators such as CREB, Jun/Fos and hormone receptors are associated with CREB binding protein (CBP/p30, CBP/p300) (Davie and Chadee 1998). Acetylation by CBP can either switch on (Parekh and Maniatis 1999) or switch off (Munshi et al. 1998) the transcription of a gene, subjected to the protein modified. It is important to note that histone acetylations generally have a positive effect on transcription but acetylation of non-histone factors may have activating or repressing effects. Most HATs consist of a bromodomain. This domain has a conserved sequence motif with functions yet to be elucidated but includes protein–protein recognition and interaction. These recognize acetyl-lysines (Dhalluin et al. 1999) suggesting that they might function as acetylation receptors (Sterner and Berger 2000). Further, HATs may modulate the binding of factors to chromatin by acetylating transcription factors but not histones. In animals and fungi, H4 is the main target of acetylation while in plants, it is H3 (Waterborg et al. 1990). Recently, it has been reported that the H3 and H4 acetylation patterns of field bean (*Vicia faba*) nuclei change during the cell cycle and it correlates with the replication than with the transcriptional activity. H4 from *Medicago*, *Arabidopsis*, tobacco and carrot was detected in five acetylated isoforms. The core histones and histone H1 are phosphorylated at specific serine/threonine residues. During activation of c-Jun and c-Fos heterodimers in response to stimulation by growth factors phosphorylation of Ser^{10} of H3 has been reported. It plays a vital role in chromosome condensation during mitosis. Further, the pattern of staining in meiotic cells has been reported to differ significantly between plant and animal chromosomes (Manzanero et al. 2000).

Maize (*Zea mays*) is a model monocot organism for molecular and functional characterization of different HAT and HDAC types. Three HAT and four different HDAC activities have been reported in germinating seedlings of maize (Loidl 1994;

Lechner et al. 2000). Two nuclear HATs (HATA1, HATA2) have been biochemically identified in maize apart from cytoplasmic HAT1. An antibody against a maize Gcn5-related protein has shown the presence of *Zm*Gcn5 in both HATA fractions, indicating the presence of complexes with different compositions (Sterner and Berger 2000). HATB of *Zea mays* is responsible for the acetylation of newly synthesized H4 at lysines 5 and 12 before chromatin assembly. H4 acetylation is essential for the transport of newly synthesized H4 into the nucleus and/or correct assembly into the nucleosome (Grunstein 1997). Therefore, HATB activity is needed to provide free H4 with a tag for its subsequent fate. However, deletion of the *Hat1* gene from yeast exhibited no mutant phenotype. Moreover, recently, it was reported that histones need not to be acetylated for interacting with chromatin assembly factor (CAF-1) or to be deposited onto chromatin. Since H3 and H4 N-termini in yeast are functionally redundant, it is believed that acetylation of H3 N-termini by another enzyme could complement in H4 acetylation (Verreault 2000), however, the enzyme is not identified yet.

Histone deacetylases (HDACs) have been grouped into three classes (Davie and Chadee 1998). The enzymes which are related to the yeast proteins Rpd3 and HAD are placed in classes 1 and 2, respectively, and the proteins similar to maize HD2 are placed in class 3. The yeast-silencing information protein (Sir2) has been recently reported to be homologous to the HDA1 family. However, so far none of them has been studied in detail. Further, it has been shown that maize HD1BI, HD1BII and HATB co-fractionate with a protein related to tomato *Le*MSI, a WD repeat containing protein. These WD repeat-containing proteins are responsible for targeting the enzymes to histones (Verreault 1998). In maize, only the HD1B can be responsible for deacetylation pattern introduced by HATB on H4, indicating a possible function for HD1B in the histone deposition process (Kolle et al. 1999). The ZmRpd3/HD1BI can functionally complement a yeast *rpd3* null mutant (Rossi et al. 1998). Also downregulation of *Arabidopsis* AtRPD3A resulted in a delayed flowering phenotype, suggesting that histone acetylation has a role in plant development (Wu et al. 2000). Another HDAC activity was found in germinating maize embryos and loosely chromatin-associated HD1A. The activity of this enzyme is controlled by phosphorylation. Its dephosphorylation leads to an increase in enzyme activity and an alteration in substrate specificity (Verreault 1998). However, it is yet to be established whether HD1A belongs to one of the known HDAC families or represents a different family. HD2-like HDACs (Lusser et al. 1997) form multigene families of highly conserved members across the plant kingdom (Dangl et al. 2001). HD2 was isolated from maize chromatin as a high molecular weight complex and is composed of three identical polypeptides. This enzyme also undergoes phosphorylation but unlike HD1A, its dephosphorylation almost abolishes its activity. The location of HD2 in nucleus of maize cells indicates its potential role in the regulation of rRNA genes (Lusser et al. 1997). *Arabidopsis* homologue *AtHD2A* has been shown to repress transcription when targeted to a reporter gene *in vivo*, (Wu et al. 2000; Guarente 2000), thus defining a fourth class. Three biochemically different HDAC activities have been reported in pea and four in maize (HD1A, HD1BI, HD1BII and HD2) (Loidl 1994; Lechner et al. 2000). Maize HD1BI and HD1BII are class-1 HDACs, and there is at least one additional member of this family in the databases. Several EST clones from maize, *Arabidopsis* and other plant species are also present in the databases that are homologous to the *HDA1* family, however, not much is known about them.

Histone acetylation leads to more open chromatin and hence more active transcription, while reverse is true for histone deacetylation (Zentner and Henikoff 2013). Several reports have shown that in plants drought sensing or treatment with abiotic stress hormone ABA results in activation of histone acetylases resulting in acetylation at coding regions of drought stress-responsive genes, thereby enhancing their expression (Kim et al. 2008). This correlation holds true for expression of stress-responsive genes (Zong et al. 2013). There are many reports to support this. For example, ABA could induce H3S10 phosphorylation and H4K14 acetylation in cultured *Arabidopsis* and tobacco cells (Sokol et al. 2007). Further, in *Arabidopsis* seedlings, acetylation of H3K9, H3K23, and H3K27 was induced at coding regions of drought stress-responsive genes after drought treatment, which in turn resulted in gene activation. Also, HATs have been found to interact with transcription factors involved in stress-responsive expression (Chinnusamy and Zhu 2009).

Various developmental and environmental cues can also repress the target genes by reducing histone acetylation levels. For example, HDA19 interacts with ERF7 (AP2/EREBP-type transcriptional repressor) and SIN3 repressor complex causing repression of abiotic stress-responsive genes (Song et al. 2012). Furthermore, HDA19 also represses transcription activities of WRKY38 and WRKY62 during plant defense responses (Kim et al. 2008). Thus, HDA19 plays important roles in abiotic and biotic stress signaling pathways. HDA6 is also involved in biotic and abiotic stress responses. Mutation in this gene increased the sensitivity of the plants to ABA and salt stresses, which suggests its positive role in these stresses (Chen 2010). Loss-of-function mutants of *HDA6* also exhibited reduced freezing and cold tolerance. This indicates that HDA6 has a critical role in cold acclimation and freezing tolerance (To et al. 2011). Further, expression of four *Arabidopsis* HD2 genes, *HD2A*, *HD2B*, *HD2C*, and *HD2D*, was repressed by ABA and salt (Chinnusamy et al. 2008; Luo et al. 2012). Also, it was observed that overexpression of *AtHD2* causes ABA-insensitive phenotype and leads to enhanced tolerance to salt and drought stresses (Sridha and Wu 2006). Also HD2C protein was found to interact with HDA6 *in vitro* and *in vivo* (Luo et al. 2012), which suggests that these two proteins may work together in plant responses to stresses. To Thus histone acetylation and deacetylation processes coordinate regulation of gene expression in response to various stresses.

7.2.2 Histone Methylation

Methylation of histones is being studied from last five decades or so and studies have revealed that lysines 4, 9, 27 and 36 of histone H3 and lysine 20 in histone H4 can be mono-, di- or tri-methylated (Zhang and Reinberg 2001). The reults of this modification for epigenetic regulation was also revealed (Rea et al. 2000). An important relationship between H3K9^{Me} and heterochromatin was identified by the discovery that the mammalian Su(var)3-9, Enhancer-of-zeste and Trithorax (SET DOMAIN) protein methylates H3 at K9 (Rea et al. 2000). Suv39 h histone methyltransferase (HMT) homologues have been identified in fission yeast, *Neurospora* and in *Arabidopsis* and their mutants show loss of transcriptional silencing. Thus, it shows that this epigenetic mechanism is evolutionary conserved. Further, *in vivo* links between histone and DNA methylation have been reported by examination of *Neurospora* and *Arabidopsis* mutants that are defective in H3K9 and DNA methylation (Nakayama et al. 2001; Tamaru and Selker 2001; Jackson et al. 2002; Malagnac et al. 2002). In *Neurospora*, DIM-5 was shown to encode a H3K9-specific methylase that resulted in trimethylated *K9* and *dim-5* mutants showing abnormal growth and complete loss of DNA methylation (Tamaru et al. 2003).

In *Arabidopsis*, the majority of H3K9Me is present in centromeric and in pericentromeric heterochromatin, which appears as densely 40, 60-diamidino-2-phenylindole (DAPI)-stained chromocenters. Different genetic approaches led to the identification and characterization of the Kryptonite gene (KYP), the first H3K9-specific methylase in plants. DNA methylation was reported to be affected in *kyp* mutants which resembled *dim-5* mutants. However, a reduction of DNA methylation in *Arabidopsis* associated with a depletion of *H3K9Me* occurred predominantly in plant-specific CNG-methylation sites. In *Arabidopsis*, about 29 HMT-related genes have been reported and KYP being the major one because of the drastic loss of H3K9^{Me} observed at *kyp* mutant chromocenters. Methylation of H3K9 is believed to be an important characteristic of heterochromatin and the presence of H3K9^{Me} at specific loci in *Arabidopsis* correlates with heterochromatic silencing. But it is important to note that the loss of H3K9^{Me} does not always alleviate silencing suggesting that increased levels of H3K9^{Me} are not sufficient to maintain silencing in plants. In fission yeast and in mammals, the methylation at H3K9 is recognized by the chromodomain containing heterochromatin protein 1 (HP1), which is a key component of heterochromatin (Gendrel et al. 2002; Johnson et al. 2002; Soppe et al. 2002; Tariq et al. 2003; Probst et al. 2003).

Methylation at lysine and arginine residues of histones is also known to regulate gene expression states. Histone lysine methyltransferases have a characteristic SET domain. In *Arabidopsis* and rice, 31 and 25 SET domain-containing proteins have been identified, respectively (Ng et al. 2006). SET domain proteins in plants are classified into four categories, SU(VAR) 3-9, E(Z) (enhancer of zeste), TRX (trithorax) and ASH1 (absent, small, or homeotic discs 1) based on their homology with animals and yeast SET proteins (Springer et al. 2003). Plant SET proteins play crucial roles in many developmental and stress responses. For example, loss of function of *Arabidopsis* ATX1 results in decreased dehydration tolerance. ATX1 has a direct influence on transcription of *NCED3*, a key gene in ABA biosynthetic pathway. ATX1-mediated H3K4 methylation is required for activating transcription of *NCED3* and subsequent accumulation of ABA in response to water stress (Ding et al. 2011).

Arginine methylation mainly occurs at Arg2, Arg8, Arg17, and Arg26 of histone H3, and Arg3 of histone H4 by means of a small group of protein arginine methyltransferases (PRMTs). It was shown that a PRMT5 designated as SKB1 is involved in salt stress response (Zhang et al. 2011). *SKB1* mutant plants resulted in salt hypersensitivity phenotype and it was demonstrated that SKB1 associates with chromatin and thereby increases the H4R3sme2 leading to suppression of the transcription of stress-responsive genes. During salt stress, the H4R3sme2 level is reduced leading to dissociation of SKB1 from chromatin and subsequent induction of the stress-responsive genes. This shows that SKB1 mediates salt response by altering the methylation status of H4R3sme2 of stress-responsive genes. Taken together, there are strong evidences supporting the view that post-translational modifications of histones play a critical role during water stress response. However, the target genes and the enzymes involved, needs to be identified.

7.2.3 Histone Demethylases

It was known that methylated histones are irreversible because of the more stable nature of the C-N bond, while in case of histone deacetylation simple hydrolysis of an amide bond is involved. The irreversible nature of methylation was based on experiments indicating that the half-life of histone methyl marks was approximately equal to that of the histone itself (Byvoet et al. 1972; Thomas et al. 1972). Other mechanisms of histone demethylation were also identified which included active histone exchange (Ahmad and Henikoff 2002) and proteolytic removal of histone amino-termini (Allis et al. 1980). Furthermore, another potential mechanism was also recognized which involved conversion of methylarginine to citrulline by a peptidylarginine deiminase. This enzyme also worked equally well on arginine (Wang et al. 2004; Cuthbert et al. 2004). However, all of these mechanisms would require passive or active histone exchange to get back to the original unmethylated state.

On the basis of bioinformatics, several groups proposed different mechanisms of direct histone demethylation. For example, S-adenosylmethionine (SAM) was proposed as the source of a reactive radical intermediate, which would target the N-methyl group, create an unstable aminium cation radical which may spontaneously hydrolyzes to form formaldehyde and a demethylated residue (Chinenov 2002). There were many other possibilities like oxidation of the methyl group coupled to reduction of a cofactor, releasing the methyl group as formaldehyde or as another higher oxidative state (Bannister et al. 2002). However, the experimental evidence was not available for many decades. The first experimental evidence of enzymatic demethylation was provided with discovery of lysine-specific demethylase 1 (LSD1). The mechanism of histone demethylation by LSD1 is highly conserved among most eukaryotes. Homologs of LSD1 have been identified and characterized in *Arabidopsis*, *Drosophila*, *Caenorhabditis elegans*, and the fission yeast *S. pombe* (Liu et al. 2007; Rudolph et al. 2007; Opel et al. 2007; Lan et al. 2007a, b; Katz et al. 2009). In all of these examples, each organism contains at least two LSD1 homologs. In humans, a recently characterized LSD1 homolog, LSD2, has also been shown to encode an H3K4me2/1 demethylase (Ciccone et al. 2009; Karytinos et al. 2009). The LSD1 counterparts in plants, flies, and worms that have been characterized also encode H3K4 demethylases. Further, it has been found that *S. pombe* LSD1 homologs demethylate H3K9 and not H3K4 (Opel et al. 2007; Lan et al. 2007a, b), suggesting that this mechanism of demethylation may also be used to target other sites of methylation. After the discovery of LSD1, Zhang et al. (2011) provided the first experimental evidence for an alternative oxidation-reduction mechanism for histone demethylation. This was followed by reports from several groups that were independently pursuing new demethylases (Tsukada et al. 2006; Whetstine et al. 2006; Fodor et al. 2006; Cloos et al. 2006).

There are not evidences so far regarding involvement of plant HDMs in abiotic stresses. However, recent studies suggest that process of histone demethylation occurs and therefore, the histone demethylases might have a possible role in plant stress responses. For example, abscisic acid (ABA) increases methylation of H3K4, while it decreases methylation of H3K9 and of ABA-responsive genes, such as *ABI1*, *ABI2*, and *RD29B* in *Arabidopsis* (Chen 2010). This suggests possible role of HDM in removing the methyl groups from histone $H3K9^{ME2}$. Further, changes in the methylation status of histone residues were observed in *Arabidopsis* in response to dehydration stress, which again supports the involvement of HDM (van Dijk et al. 2010).

7.2.4 Histone Variants

Most of the organisms including plants have genes coding for highly conserved canonical histones (H3, H4, H2A, and H2B). These genes are expressed during the S phase of the cell cycle. There are other less conserved histones which are expressed throughout the cell cycle and they are known as histone variants. The sequence of genes coding for canonical histones and the histone variants is not much different. However, histone variants can bring about change in the nucleosome characteristics. These histone variants may also lead to changes in extent to which various histone modifications can occur (Talbert and Henikoff 2010; Burgess and Zhang 2013; Skene and Henikoff 2013).

In plants, there are reports which show that linker histone (H1) variants are involved in water stress response. For example, *HIS1-3* gene, a linker histone variant is expressed in response to salt, drought, and ABA in *Arabidopsis* (Ascenzi and Gantt 1997; Zhu et al. 2012). Also in tomato, water stress leads to expression of a linker gene *H1-S* (Scippa et al. 2002). Although, the role of these histone variants in plant stress response is known to be conserved, the mechanism of action needs to be established.

7.3 Chromatin Remodeling

Chromatin remodeling is carried out by ATP-dependent chromatin remodeling ATPases, which bring about changes in histone–DNA interactions non-covalently by utilizing the energy derived from ATP hydrolysis. This leads to changes in position and/or composition of nucleosome. This change increases or decreases the accessibility of given DNA segment to *trans*-acting factors and hence induces or represses transcription respectively (Clapier and Cairns 2009; Hargreaves and Crabtree 2011; Narlikar et al. 2013). There are four types of chromatin remodelers: SWI/SNF, ISWI, CHD, and INO80/SWR1. Among these, SWI/SNF and CHD sub-groups have been shown to play roles in water stress responses in plants.

Plant genomes have three types of SWI/SNF ATPases known as BRAHMA (BRM), SPLAYED (SYD), and MINUSCULE (MINU) (Jerzmanowski 2007; Kwon and Wagner 2007; Sang et al. 2012). The functional unit of ATPase consists of catalytic subunit forming a core complex together with SWIRM- and SANT-domain proteins (SWI3) and SNF5-domain proteins. Also, some accessary proteins are present which regulate tissue- and developmental-stage-specific targeting and activity of the complex (Clapier and Cairns 2009; Hargreaves and Crabtree 2011; Kwon and Wagner 2007). Recently, it has been shown that ATP-dependent chromatin remodeling factors play important roles in plant responses to abiotic stresses. In *Arabidopsis*, a gene called *SWI3B* (homolog of yeast *SWI3*) is shown to directly interact with HAB1, a protein phosphatase type 2C, which is a negative regulator of ABA signaling (Saez et al. 2008). The *swi3b* mutants were demonstrated to be less sensitive to ABA–mediated inhibition of seed germination and growth. They were also shown to have reduced expression of ABA-responsive genes suggesting that SWI3B is a negative regulator of ABA signaling and HAB1 modulates the ABA response through the regulation of SWI/SNF chromatin-remodeling complex (Saez et al. 2008). Further, it was shown that a SWI2/SNF2 chromatin remodeling ATPase BRM (BRAHMA) plays an essential role in stress responses in *Arabidopsis*. *Brm* mutants were found to display increased drought tolerance (Han 2012). Moreover, loss of BRM activity led to destabilization of nucleosomes, suggesting that BRM-mediated stress responses occur through the regulation of nucleosome stability. The CHD subgroup chromatin remodeler PKL (PICKLE) has also been shown to be involved in ABA response. CHD chromatin remodelers have two tandem chromodomains, which bind methylated lysines. These domains further couple ATP hydrolysis to remodeling (Hauk et al. 2010).

7.4 DNA Methylation

Methylation of cytosine bases in genome also influences chromatin structure and gene expression (Jones 2012). Methylation on the 5th carbon of cytosine bases is considered as an important epigenetic mark. In plants, cytosine methylation is found in the context of CG, CHG, and CHH (H = A, C, or T). The symmetric CG and CHG methylation is catalyzed by DNA methyltransferase I (MET1) and chromo-methyltransferase 3 (CMT3), respectively. The asymmetric CHH methylation is carried out through de novo methylation by Domains Rearranged Methyltransferase 2 (DRM2) and RNA-directed DNA methylation pathway (Chan et al. 2005; Goll and Bestor 2005; Law and Jacobsen 2010). While CMT3 is a plant specific methyltransferase, MET1 is a homologue of the mammalian methyltransferase DNMT1. DRM2 is also a homologue of mammalian DNMT3A/b gene. In plants many chromosomal regions, repetitive DNA sequences, and transposons are heavily methylated and the extent of methylation is related to the level of gene expression. Different developmental or environmental cues affect the extent of methylation. For example, abiotic stress has been shown to trigger hyper or hypomethylation of different genomic regions, thereby leading to different manifestations. It has been proposed that hypermethylation of coding regions and hypomethylation of promoters leads to stress adaptation in plants (Sahu et al. 2013). Changes in DNA methylation in response to drought was shown to be more pronounced in drought-tolerant plant species. More work needs to be done to explore the in depth effect of DNA methylation on abiotic stress responses in plants.

7.5 Conclusion

With the advancements in our knowledge about the mechanisms that regulate gene activity, it has become quite clear that chromatin is not merely a way of packaging DNA in the nucleus. It signifies a vital regulatory entity that enables maintenance of genome stability and permits the integration of several endogenous and exogenous signals at the gene level. All the post-translational histone modifications as well as DNA methylation play a role through expression or suppression of specific genes in response to various stress stimuli. However, the knowledge about their role in water stress responses in plants is still fragmentary and requires further investigation.

References

Ahmad K, Henikoff, S (2002) The histone variant H3.3 marks active chromatin by replication-independent nucleosome assembly. Mol cell 9(6): 1191–200

Allis CD, Bowen JK, Abraham GN, Glover CV, Gorovsky MA (1980) Proteolytic processing of histone H3 in chromatin: a physiologically regulated event in *Tetrahymena* micronuclei. Cell 20(1):55–64

Ascenzi R, Gantt JS (1997) A drought-stress-inducible histone gene in *Arabidopsis thaliana* is a member of a distinct class of plant linker histone variants. Plant Mol Biol 34(4):629–641

Bannister AJ, Kouzarides T (2011) Regulation of chromatin by histone modifications. Cell Res 21 (3):381–395

Bannister AJ, Schneider R, Kouzarides T (2002) Histone methylation: dynamic or static? Cell 109 (7):801–806
Burgess RJ, Zhang Z (2013) Histone chaperones in nucleosome assembly and human disease. Nat Struct Mol Biol 20(1):14–22
Byvoet P, Shepherd GR, Hardin JM, Noland BJ (1972) The distribution and turnover of labeled methyl groups in histone fractions of cultured mammalian cells. Arch Biochem Biophys 148 (2):558–567
Camporeale G, Oommen AM, Griffin JB, Sarath G, Zempleni J (2007) K12-biotinylated histone H4 marks heterochromatin in human lymphoblastoma cells. J Nutr Biochem 18(11):760–768
Chan SW, Henderson IR, Jacobsen SE (2005) Gardening the genome: DNA methylation in *Arabidopsis thaliana*. Nature Rev Genet 6(5):351–360
Chen LT (2010) Involvement of Arabidopsis histone deacetylase HDA6 in ABA and salt stress response. J Exp Bot 61:3345–3353
Chinenov Y (2002) A second catalytic domain in the Elp3 histone acetyltransferases: a candidate for histone demethylase activity? Trends Biochem Sci 27:115–117
Chinnusamy V, Gong Z, Zhu JK (2008) Abscisic Acid-mediated Epigenetic Processes in Plant Development and Stress Responses. J Integr Plant Biol 50(10):1187–1195
Chinnusamy V, Zhu JK (2009) Epigenetic regulation of stress responses in plants. Curr Opin Plant Biol 12(2):133–139
Ciccone DN, Su H, Hevi S, Gay F, Lei H, Bajko J, Xu G, Li E, Chen T (2009) KDM1B is a histone H3K4 demethylase required to establish maternal genomic imprints. Nature 461 (7262):415–418
Clapier CR, Cairns BR (2009) The biology of chromatin remodeling complexes. Ann Rev Biochem 78:273–304
Cloos PA, Christensen J, Agger K, Maiolica A, Rappsilber J, Antal T, Hansen KH, Helin K (2006) The putative oncogene GASC1 demethylates tri- and dimethylated lysine 9 on histone H3. Nature 442:307–311
Crane-Robinson C, Myers FA, Hebbes TR, Clayton AL, Thorne AW (1999) Chromatin immunoprecipitation assays in acetylation mapping of higher eukaryotes. Methods Enzymol 31 (304):533–547
Cuthbert GL, Daujat S, Snowden AW, Erdjument-Bromage H et al (2004) Histone deimination antagonizes arginine methylation. Cell 118(5):545–553
Dangl M, Brosch G, Haas H, Loidl P, Lusser A (2001) Comparative analysis of HD2 type histone deacetylases in higher plants. Planta 213(2):280–285
Davie JR, Chadee DN (1998) Regulation and regulatory parameters of histone modifications. J Cell Biochem 72(S30–31):203–213
Dhalluin C, Carlson JE, Zeng L, He C, Aggarwal AK, Zhou MM (1999) Structure and ligand of a histone acetyltransferase bromodomain. Nature 399(6735):491–496
Ding Y, Avramova Z, Fromm M (2011) The Arabidopsis trithorax-like factor ATX1 functions in dehydration stress responses via ABA-dependent and ABA-independent pathways. Plant J 66 (5):735–744
Fodor BD, Kubicek S, Yonezawa M, O'Sullivan RJ, Sengupta R, Perez-Burgos L et al (2006) Jmjd2b antagonizes H3K9 trimethylation at pericentric heterochromatin in mammalian cells. Genes Dev 20:1557–1562
Gendrel AV, Lippman Z, Yordan C, Colot V, Martienssen RA (2002) Dependence of heterochromatic histone H3 methylation patterns on the *Arabidopsis* gene DDM1. Science 297(5588):1871–1873
Goll MG, Bestor TH (2005) Eukaryotic cytosine methyltransferases. Ann Rev Biochem 74:481–514
Grunstein M (1997) Histone acetylation in chromatin structure and transcription. Nature 389 (6649):349–352
Guarente L (2000) Sir2 links chromatin silencing, metabolism, and aging. Genes Dev 14(9):1021–1026

Han SK (2012) The SWI2/SNF2 chromatin remodeling ATPase BRAHMA represses abscisic acid responses in the absence of the stress stimulus in *Arabidopsis*. Plant Cell 24(12):4892–4906
Han SK, Wagner D (2014) Role of chromatin in water stress responses in plants. J Exp Bot 65 (10):2785–2799
Hargreaves DC, Crabtree GR (2011) ATP-dependent chromatin remodeling: genetics, genomics and mechanisms. Cell Res 21(3):396–420
Hauk G, McKnight JN, Nodelman IM, Bowman GD (2010) The chromodomains of the Chd1 chromatin remodeler regulate DNA access to the ATPase motor. Mol Cell 39(5):711–723
Jerzmanowski A (2007) SWI/SNF chromatin remodeling and linker histones in plants. Biochim Biophys Acta (BBA)-Gene Struct Express 1769(5): 330–45
Johnson LM, Cao X, Jacobsen S (2002) Interplay between two epigenetic marks: DNA methylation and histone H3 lysine 9 methylation. Curr Biol 12(16):1360–1367
Jackson JP, Johnson L, Jasencakova Z, Zhang X et al (2002) Dimethylation of histone H3 lysine 9 is a critical mark for DNA methylation and gene silencing in *Arabidopsis thaliana*. Chromosoma 112(6):308–315
Jones PA (2012) Functions of DNA methylation: islands, start sites, gene bodies and beyond. Nature Rev Genet 13(7):484–492
Karytinos A, Forneris F, Profumo A, Ciossani G, Battaglioli E, Binda C, Mattevi A (2009) A novel mammalian flavin-dependent histone demethylase. J Biol Chem 284(26):17775–17782
Katz DJ, Edwards TM, Reinke V, Kelly WG (2009) A *C. elegans* LSD1 demethylase contributes to germline immortality by reprogramming epigenetic memory. Cell 137(2): 308–320
Kim JM, To TK, Ishida J, Morosawa T et al (2008) Alterations of lysine modifications on the histone H3 N-tail under drought stress conditions in *Arabidopsis thaliana*. Plant Cell Physiol 49(10):1580–1588
Kolle D, Brosch G, Lechner T, Pipal A, Helliger W, Taplick J, Loidl P (1999) Different types of maize histone deacetylases are distinguished by a highly complex substrate and site specificity. Biochem 38(21):6769–6773
Krebs JE, Kuo MH, Allis CD, Peterson CL (1999) Cell cycle-regulated histone acetylation required for expression of the yeast *HO* gene. Genes Dev 13(11):1412–1421
Kwon CS, Wagner D (2007) Unwinding chromatin for development and growth: a few genes at a time. Trends Genet 23(8):403–412
Lan F, Bayliss PE, Rinn JL, Whetstine JR, Wang JK et al (2007a) A histone H3 lysine 27 demethylase regulates animal posterior development. Nature 449(7163):689–694
Lan F, Zaratiegui M, Villén J, Vaughn MW et al (2007b) *S. pombe* LSD1 homologs regulate heterochromatin propagation and euchromatic gene transcription. Mol Cell 26(1): 89–101
Law JA, Jacobsen SE (2010) Establishing, maintaining and modifying DNA methylation patterns in plants and animals. Nat Rev Genet 11(3):204–220
Lechner T, Lusser A, Pipal A, Brosch G, Loidl A et al (2000) RPD3-type histone deacetylases in maize embryos. Biochem 39(7):1683–1692
Liu F, Quesada V, Crevillén P, Bäurle I, Swiezewski S, Dean C (2007) The *Arabidopsis* RNA-binding protein FCA requires a lysine-specific demethylase 1 homolog to downregulate FLC. Mol Cell 28(3):398–407
Loidl P (1994) Histone acetylation: facts and questions. Chromosoma 103(7):441–449
Luo M (2012) HD2C interacts with HDA6 and is involved in ABA and salt stress response in *Arabidopsis*. J Exp Bot 63(8):3297–3306
Lusser A, Brosch G, Loidl A, Haas H, Loidl P (1997) Identification of maize histone deacetylase HD2 as an acidic nucleolar phosphoprotein. Science 277(5322):88–91
Lusser A, Kölle D, Loidl P (2001) Histone acetylation: lessons from the plant kingdom. Trends Plant Sci 6(2):59–65
Lusser A (2002) Acetylated, methylated, remodeled: chromatin states for gene regulation. Curr Opin Plant Biol 5(5):437–443
Madisen L, Krumm A, Hebbes TR, Groudine M (1998) The immunoglobulin heavy chain locus control region increases histone acetylation along linked c-myc genes. Mol Cell Biol 18 (11):6281–6292

Malagnac F, Bartee L, Bender J (2002) An *Arabidopsis* SET domain protein required for maintenance but not establishment of DNA methylation. EMBO J 21(24):6842–6852
Manzanero S, Arana P, Puertas MJ, Houben A (2000) The chromosomal distribution of phosphorylated histone H3 differs between plants and animals at meiosis. Chromosoma 109 (5):308–317
Munshi N, Merika M, Yie J, Senger K, Chen G, Thanos D (1998) Acetylation of HMG I (Y) by CBP turns off IFNβ expression by disrupting the enhanceosome. Mol Cell 2(4):457–467
Nakayama Rice JC, Strahl BD, Allis CD, Grewal SI (2001) Role of histone H3 lysine 9 methylation in epigenetic control of heterochromatin assembly. Science 292(5514):110–113
Narlikar GJ, Sundaramoorthy R, Owen-Hughes T (2013) Mechanisms and functions of ATP-dependent chromatin-remodeling enzymes. Cell 154(3):490–503
Nathan D, Ingvarsdottir K, Sterner DE, Bylebyl GR, Dokmanovic M (2006) Histone sumoylation is a negative regulator in *Saccharomyces cerevisiae* and shows dynamic interplay with positive-acting histone modifications. Genes Dev 20(8):966–976
Ng DW, Chandrasekharan MB, Hall TC (2006) Ordered histone modifications are associated with transcriptional poising and activation of the phaseolin promoter. Plant Cell 18(1):119–132
Opel M, Lando D, Bonilla C, Trewick SC, Boukaba A (2007) Genome-wide studies of histone demethylation catalyzed by the fission yeast homologues of mammalian LSD1. PLoS ONE 2 (4):e386
Parekh BS, Maniatis T (1999) Virus infection leads to localized hyperacetylation of histones H3 and H4 at the IFN-β promoter. Mol Cell 3(1):125–129
Probst AV, Fransz PF, Paszkowski J, Mittelsten SO (2003) Two means of transcriptional reactivation within heterochromatin. Plant J 33(4):743–749
Rea S, Eisenhaber F, O'Carroll D, Strahl BD, Sun ZW et al (2000) Regulation of chromatin structure by site-specific histone H3 methyltransferases. Nature 406(6796):593–599
Rossi V, Hartings H, Motto M (1998) Identification and characterization of an RPD3 homologue from maize (*Zea mays* L.) that is able to complement an *rpd3* null mutant of *Saccharomyces cerevisiae*. Mol Gen Genet 258(3): 288–296
Rudolph T, Yonezawa M, Lein S, Heidrich K, Kubicek S et al (2007) Heterochromatin formation in Drosophila is initiated through active removal of H3K4 methylation by the LSD1 homolog SU(VAR) 3-3. Mol Cell 26(1):103–115
Saez A, Rodrigues A, Santiago J, Rubio S, Rodriguez PL (2008) HAB1-SWI3B interaction reveals a link between abscisic acid signaling and putative SWI/SNF chromatin-remodeling complexes in *Arabidopsis*. Plant Cell 20(11):2972–2988
Sahu PP, Pandey G, Sharma N, Puranik S, Muthamilarasan M, Prasad M (2013) Epigenetic mechanisms of plant stress responses and adaptation. Plant Cell Rep 32(8):1151–1159
Sang Y, Silva-Ortega CO, Wu S, Yamaguchi N (2012) Mutations in two non-canonical Arabidopsis SWI2/SNF2 chromatin remodeling ATPases cause embryogenesis and stem cell maintenance defects. Plant J 72(6):1000–1014
Scippa GS, Griffiths A, Chiatante D, Bray EA (2002) The H1 histone variant of tomato, H1-S, is targeted to the nucleus and accumulates in chromatin in response to water-deficit stress. Planta 211(2):173–181
Skene PJ, Henikoff S (2013) Histone variants in pluripotency and disease. Development 140 (12):2513–2524
Sokol A, Kwiatkowska A, Jerzmanowski A, Prymakowska-Bosak M (2007) Up-regulation of stress-inducible genes in tobacco and *Arabidopsis* cells in response to abiotic stresses and ABA treatment correlates with dynamic changes in histone H3 and H4 modifications. Planta 227 (1):245–254
Song J, Angel A, Howard M, Dean C (2012) Vernalization–a cold-induced epigenetic switch. J Cell Sci 125(16):3723–3731
Soppe WJ, Jasencakova Z, Houben A, Kakutani T, Meister A et al (2002) DNA methylation controls histone H3 lysine 9 methylation and heterochromatin assembly in *Arabidopsis*. EMBO 21(23):6549–6559

Springer NM, Napoli CA, Selinger DA, Pandey R, Cone KC et al (2003) Comparative analysis of SET domain proteins in maize and *Arabidopsis* reveals multiple duplications preceding the divergence of monocots and dicots. Plant Physiol 132(2):907–925

Sridha S, Wu K (2006) Identification of *AtHD2C* as a novel regulator of abscisic acid responses in *Arabidopsis*. Plant J 46(1):124–133

Sridhar VV, Kapoor A, Zhang K, Zhu J, Zhou T et al (2007) Control of DNA methylation and heterochromatic silencing by histone H2B deubiquitination. Nature 447(7145):735–738

Sterner DE, Berger SL (2000) Acetylation of histones and transcription-related factors. Microbiol Mol Biol Rev 64(2):435–459

Strahl BD, Allis CD (2000) The language of covalent histone modifications. Nature 403(6765):41–45

Talbert PB, Henikoff S (2010) Histone variants-ancient wrap artists of the epigenome. Nat Rev Mol Cell Biol 11(4):264–275

Tamaru H, Selker EU (2001) A histone H3 methyltransferase controls DNA methylation in *Neurospora crassa*. Nature 414(6861):277–283

Tamaru H, Zhang X, McMillen D, Singh PB, Nakayama J et al (2003) Trimethylated lysine 9 of histone H3 is a mark for DNA methylation in *Neurospora crassa*. Nat Genet 34(1):75–79

Tariq M, Saze H, Probst AV, Lichota J, Habu Y, Paszkowski J (2003) Erasure of CpG methylation in *Arabidopsis* alters patterns of histone H3 methylation in heterochromatin. Proc Natl Acad Sci USA 100(15):8823–8827

Thomas G, Lange HW, Hempel K (1972) Relative stability of lysine-bound methyl groups in arginin-rich histones and their subfractions in Ehrlich ascites tumor cells *in vitro*. Hoppe Seylers Z Physiol Chem 353(9):1423–1428

To TK, Nakaminami K, Kim JM, Morosawa T, Ishida J et al (2011) *Arabidopsis* HDA6 is required for freezing tolerance. Biochem Biophys Res Commun 406(3):414–419

Tsukada Y, Fang J, Erdjument-Bromage H, Warren ME, Borchers CH, Tempst P, Zhang Y (2006) Histone demethylation by a family of JmjC domain-containing proteins. Nature 439:811–816

Van Dijk K, Ding Y, Malkaram S, Riethoven JJ, Liu R et al (2010) Dynamic changes in genome-wide histone H3 lysine 4 methylation patterns in response to dehydration stress in *Arabidopsis thaliana*. BMC Plant Biol 10(1):238

Verreault A (2000) *De novo* nucleosome assembly: new pieces in an old puzzle. Genes Dev 14 (12):1430–1438

Verreault A, Kaufman PD, Kobayashi R, Stillman B (1998) Nucleosomal DNA regulates the core-histone-binding subunit of the human Hat1 acetyltransferase. Curr Biol 8(2):96–108

Wang Y, Wysocka J, Sayegh J, Lee YH, Perlin JR et al (2004) Human PAD4 regulates histone arginine methylation levels via demethylimination. Science 306(5694):279–283

Waterborg JH, Harrington RE, Winicov I (1990) Dynamic histone acetylation in alfalfa cells: Butyrate interference with acetate labeling. Biochim Biophys Acta (BBA)-Gene Struct. Express 1049(3): 324–330

Whetstine JR, Nottke A, Lan F, Huarte M, Smolikov S (2006) Reversal of histone lysine trimethylation by the JMJD2 family of histone demethylases. Cell 125(3):467–481

Wu K, Malik K, Tian L, Brown D, Miki B (2000) Functional analysis of a RPD3 histone deacetylase homologue in *Arabidopsis thaliana*. Plant Mol Biol 44(2):167–176

Zentner GE, Henikoff S (2013) Regulation of nucleosome dynamics by histone modifications. Nat Struct Mol Biol 20(3):259–266

Zhang X, Germann S, Blus BJ, Khorasanizadeh S, Gaudin V, Jacobsen SE (2007) The *Arabidopsis* LHP1 protein co-localizes with histone H3 Lys27 trimethylation. Nat Struct Mol Biol 14:869–871

Zhang Y, Reinberg D (2001) Transcription regulation by histone methylation: interplay between different covalent modifications of the core histone tails. Genes Dev 15(18):2343–2360

Zhang Z, Zhang S, Zhang Y, Wang X, Li D et al (2011) *Arabidopsis* floral initiator SKB1 confers high salt tolerance by regulating transcription and pre-mRNA splicing through altering histone H4R3 and small nuclear ribonucleoprotein LSM4 methylation. Plant Cell 23(1):396–411

Zhu Y, Dong A, Shen WH (2012) Histone variants and chromatin assembly in plant abiotic stress responses. Biochim Biophys Acta (BBA)-Gene Regul Mech 1819(3): 343–48

Zong W, Zhong X, You J, Xiong L (2013) Genome-wide profiling of histone H3K4-tri-methylation and gene expression in rice under drought stress. Plant Mol Biol 81 (1–2):175–188

Chapter 8
The 'Omics' Approach for Crop Improvement Against Drought Stress

Deepti Jain, Nasheeman Ashraf, J. P. Khurana and M. N. Shiva Kameshwari

Abstract Plants are sessile organisms and are inevitably exposed to various stress factors during their lifetime. Among abiotic stresses, drought is the most prominent which affects plant growth and yield worldwide. To combat with drought, plants have developed various adaptive strategies. Understanding the mechanisms by which plants perceive and transduce stress signals to initiate adaptive responses is of extreme relevance for rational engineering of hardier crop. Crop improvement against drought stress has been particularly enthralling; consequently, the complex drought stress response has been extensively studied in order to understand tolerance mechanisms thoroughly. As conventional breeding strategies for crop improvement approach their limits, agriculture has to adapt novel approaches to meet the demands of an ever-growing world population. Recent technical advances have led to the emergence of high-throughput tools to explore and exploit plant genomes for crop improvement. In this context, the high-throughput '-Omics' era of research has arisen with most propitious perspectives in developing improved varieties. These omics-based approaches aim to decipher the entire genome for gaining insights into plant molecular responses, which will in turn provide specific strategies for crop improvement. The three main omics technologies—genomics, proteomics and metabolomics are aimed at unraveling the overall expression of genes, proteins and metabolites, respectively, in a functionally relevant context. Advances in this area have provided insights into the molecular basis of various fundamental processes involved in plant stress responses and thus opened up new perspectives and opportunities for improving crop plants. In this chapter, how three core '-omics' techniques can be translated to create new crops that are more efficiently adapted to adverse conditions.

D. Jain (✉) · J. P. Khurana
Department of Plant Molecular Biology, Interdisciplinary Centre for Plant Genomics, Delhi University, South Campus, New Delhi, India
e-mail: deeptivjain@gmail.com

N. Ashraf
Plant Biotechnology Division, CSIR-Indian Institute of Integrative Medicine, Sanat Nagar, Srinagar 190005, Jammu and Kashmir, India

D. Jain · M. N. Shiva Kameshwari
Department of Botany, Bangalore University, Janana Bharati Campus, Bangalore, India

V. R. Rajpal et al. (eds.), *Genetic Enhancement of Crops for Tolerance to Abiotic Stress: Mechanisms and Approaches, Vol. I*, Sustainable Development and Biodiversity 20, https://doi.org/10.1007/978-3-319-91956-0_8

Keywords Drought stress · -Omics · Gene expression · Metabolite profiling Crop breeding

8.1 Introduction

Sustainable agricultural production is an urgent issue to meet the huge challenge of feeding a thriving global population, using fewer resources, in what is likely to be an increasingly difficult climate. As sessile organisms, plants are often exposed to various adverse conditions during their lifetime that affect their growth and yield. Stress responses to these biotic or abiotic factors vary considerably from plant to plant and thus the need to understand these variations prompted us to study underlying regulatory mechanisms. Abiotic stresses comprise various stresses caused by abiotic factors like ultraviolet radiations, high and low temperatures, drought, salinity, heavy metals, hypoxia, etc. Amongst all, drought being the most prominent and widespread has been studied extensively in terms of regulatory mechanisms. Drought stress can occur in any developmental phase of plant and it often occurs in concurrence with other environmental stresses, such as heat stress and salinity (Suzuki et al. 2014). Drought stress induces a range of physiological and biochemical responses in plants, including stomatal closure, repression of cell growth and photosynthesis, and activation of respiration.

For several decades, stress genetics and physiology has been dominated by the view that physiological or biochemical models can be developed for improving stress responses and then strategies can be designed to test these approaches. The advent of high-throughput whole genome, metabolome, proteome and related technologies has offered a valuable tool to further dissect proven adaptive traits, and provided opportunities to develop collections of sequence-based resources for specific organisms. In this chapter, we will briefly discuss the recent resources and advancements in omics technologies with particular emphasis on genomics, transcriptomics, proteomics and metabolomics in plant research.

8.2 Stress-Adaption Mechanisms in Plant Species

Plants have developed adaptive strategies to cope with environmental stresses during their lifecycle. To cope with water deficit, they have evolved three major mechanisms: escape, avoidance and tolerance. Plants escape drought stress by formation of seeds before drought conditions appear. To avoid stress, plants undergo various morphological changes such as development of specialized leaf surfaces to decrease the rate of transpiration, the reduction of leaf area, sunken stomata or an altered root system to use water more efficiently. Stress tolerance is a complex trait of co-ordinating physiological and biochemical alterations at the cellular and molecular level i.e. the accumulation of various osmolytes and proteins specifically involved in stress tolerance coupled with an efficient antioxidant system.

In any living being, one of the fundamental properties of adaptive mechanisms is that they must be able to sense and respond to the external threat, to ensure that the resources are being used when required. Accordingly, plant cells have evolved to perceive differential signals from their environment, integrate them and respond by modulating the gene expression and regulation. Molecular and genomic analyses have shown that several different transcriptional regulatory systems are involved in cumulative stress-responsive gene induction or repression of an assortment of genes with diverse functions (Shinozaki et al. 2003; Yamaguchi-Shinozaki and Shinozaki 2005). Recently, a number of stress-inducible genes have been identified using microarray analysis in various plant species and analyzing the functions of these genes is critical to help our understanding of the molecular mechanisms governing plant stress response and tolerance. More importantly, the expression of such inducible genes has been used as markers and thus an overall scheme of transcriptional regulation has been developed. In the proposed model, transcriptional activation occurs at distinct time points in response to stress stimuli and these variations in induction phases are controlled by different signaling mechanisms and different transcription factors (Hirayama and Shinozaki 2010). Also, genetic screens for mutations that affect the expression of stress inducible genes have allowed the identification of novel components of regulatory system (Chinnusamy et al. 2002). The study of plant species living under extreme environmental conditions has also provided important information on stress tolerance mechanisms. These findings altogether have provided a basic model of gene regulatory networks in abiotic stress responses in plants, ultimately leading to enhancement of stress tolerance in crops through genetic manipulation.

8.3 Strategies in Crop Breeding

The main aim of studying stress responses in plants lies in improving the abiotic stress tolerance of crops by either conventional breeding methods or genetic manipulation. The results of basic research using *Arabidopsis* as a model plant have been applied to other crop plants in improving stress tolerance. Though such strategies are promising to improve stress tolerance, however, improvement of crops will require further research. The biggest challenge in crop improvement is to fill the gap between laboratory and field conditions. For example, in many reports the tolerance levels of genetically manipulated plant against a stress was examined only over short periods, whilst in fields, plants are subjected to various stresses simultaneously. The combinatorial effect of various abiotic stresses have been reported to cause unexpected physiological changes in plant cells (Larkindale et al. 2005; Mittler 2006). To overcome these difficulties, we need to fully understand whole stress-response system of plants (Fig. 8.1).

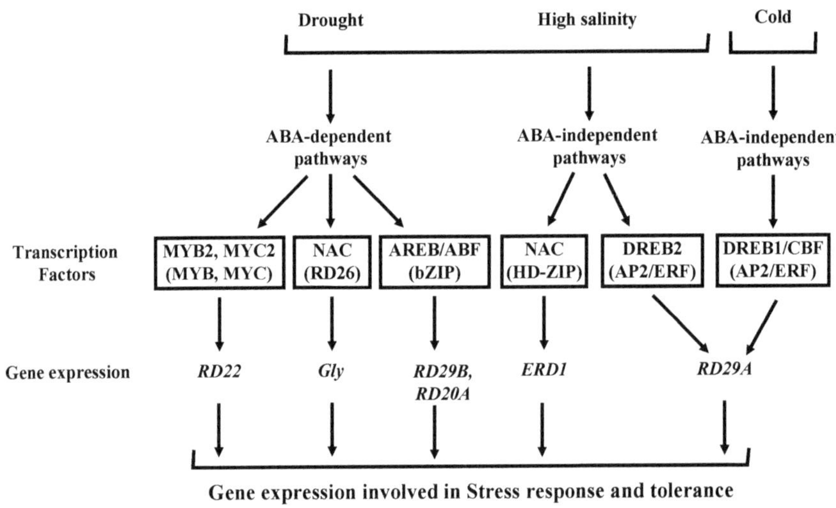

Fig. 8.1 Transcriptional regulatory networks of abiotic stress signaling. Abiotic stress responses involve ABA-independent or -dependent pathways. In the ABA-dependent pathway, AREB/ABFs transcription factors function in ABA-inducible gene expression of *RD20A, RD29B* genes and MYB2/MYC2 is responsible for *RD22* gene expression. In one of the ABA-independent pathways, DREB transcription factors are mainly involved in *RD29A* gene expression not only by drought and salt but also by cold stress. DREB1/CBFs are important transcription factors in cold-responsive gene expression. DREB2s are involved in dehydration and high salinity stress-responsive gene expression. Another ABA-independent pathway involves NAC transcription factors is controlled by drought and salt, but not by cold. NAC regulates *ERD1* gene expression

8.4 Omics for Gene Discovery and Breeding

A significant challenge in gene discovery based on forward genetics or positional cloning approach is the final identification of the gene or regulatory sequence responsible for the phenotype. Though positional cloning helps us to generate very useful genetic data, it does not conclusively identify the key sequence variant associated with the target phenotype. Various omics platforms improve our ability to identify likely candidate genes that control specific traits and elucidate the biological role or process that determines the gene effect. One can generate a series of large ‘omics’ datasets from stress-adapted versus unadapted lines which would further be used in the identification of a number of genes or emerging pathways that could be associated with enhanced stress response (Langridge and Fleury 2011). A schematic representation of each relevant omics resource is shown in Fig. 8.2. A conceptual model is representing each biological element in a corresponding plane with layers ranging from genome to phenome, a model termed ‘omic space’(Toyoda and Wada 2004). Such comprehensive models often provide an excellent boost for designing experiments and generating hypotheses based on the integrated knowledge found in the omic space of a particular organism.

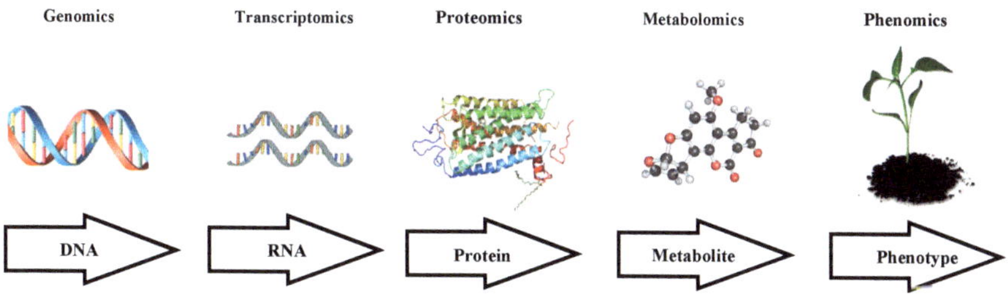

Fig. 8.2 Omics space and its related techniques in plants. Involvement of different -omics techniques in plant genome to phoneme analysis

Furthermore, the comparison of such omic resources and datasets among species promises to be an efficient way to find collateral evidence for conserved gene functions that might be evolutionarily supported.

8.4.1 Genomics Related Platforms and Resources

Evolution of novel technologies like molecular markers, trait/physical mapping, transcriptome/genome sequencing during the last two decades has improved our understanding from genome to gene level to gene networks for plant development and stress management in many model or crop species (Varshney and Dubey 2009). The efficacy of breeding programs can increase by either extending the amount or nature of genetic variation, or by accelerating the selection process to produce varieties more rapidly. Genomics can provide support for both approaches. 'Genomics-assisted breeding' is a revolutionizing breeding technology based on genomics tools (Varshney et al. 2005) that significantly enhances the efficiency of breeding for improvement of agronomical traits. Further, the studies done in the areas of functional, structural and comparative genomics suggested that the information gained from one plant species could be used for the improvement of related species. The genomics-based approaches help in deciphering the entire genome to gain insights into plant molecular responses, which will in turn provide specific strategies for crop improvement.

8.4.1.1 Functional Genomics

Functional genomics tools have been used to study gene functions and the interactions between genes in regulatory networks. RNA (transcriptomics), protein (proteomics) and metabolite (metabolomics) levels can be assessed for parental lines to gain information linked to the trait of interest. This further can be exploited to generate improved varieties. For example, if a quantitative trait locus (QTL) is associated with drought tolerance, transcriptomics of QTL ± lines will reveal genes in the region that are differentially regulated in response to drought stress. Similarly, protein profiling will provide information on changes in protein abundance or modification in response to the stress and metabolomics data might indicate that the region is associated with a major change in levels of any metabolite suggesting those genes to be involved in the biosynthesis of any osmoprotectant (Fleury et al. 2010). Information on loci that control gene expression levels, protein modification or levels of a particular metabolite can directly be mapped onto a segregating population by the help of transcript, protein and metabolite profiles. The QTLs associated with such traits are known as expression (eQTL), protein (pQTL) or metabolite (mQTL). The preliminary information on molecular phenotypes helps elucidate genotypic variation that underlies morphological and physiological traits. Functional genomic studies are perhaps the most readily applicable information for crop improvement. These approaches employ predominantly either sequence- or hybridization-based methodologies.

Sequencing-based approaches include Expressed Sequence Tag (EST) sequencing, Serial Analysis of Gene Expression (SAGE) analysis or Massively Parallel Signature Sequencing (MPSS). Expressed gene catalogue of any species can be easily analyzed by EST sequencing as ESTs have been shown to identify corresponding genes unambiguously in a rapid and cost-effective fashion (Bouchez and Höfte 1998). National Center for Biotechnological Information (NCBI) has the EST database for important crops such as maize, soybean, wheat, and rice, along with several thousands of ESTs for other plants. EST sequencing of different cDNA libraries from various tissues, developmental stages, or stress treatments generally help in revealing differentially expressed genes (Yamamoto and Sasaki 1997). EST sequencing is very useful in crops lacking whole genome sequence information or in crops with large and repetitive genomes. SAGE is an alternative approach to quantitate the abundance of thousands of transcripts simultaneously. In this technique, short sequence tags from transcripts are concatenated and sequenced, giving an absolute measure of gene expression (Velculescu et al. 1995; Vega-Sánchez et al. 2007). The first report of SAGE in plants identified stress-responsive novel genes and also implied novel functions for known genes in rice seedlings (Lee and Lee 2003; Matsumura et al. 1999). Similarly, in MPSS technique, longer sequence tags are ligated to microbeads and sequenced in parallel, enabling analysis of millions of transcripts simultaneously (Brenner et al. 2000). MPSS captures rare transcripts too in species that lack a whole genome sequence (Reinartz et al. 2002). MPSS has also been employed in the expression studies of small RNAs in plants (Meyers et al. 2006; Nobuta et al. 2007).

Hybridization-based approaches are array-based techniques that assess expression by hybridization of the target DNA with cDNA or oligonucleotide probes attached to a solid surface (Lockhart et al. 1996). Substantial microarray expression data exists for model crops like *Arabidopsis thaliana* and rice (Wang et al. 2011; Zimmermann et al. 2004) and other crop species such as wheat (Ergen et al. 2009), barley (Close et al. 2004), maize (Luo et al. 2010), cotton (Ranjan et al. 2012), cassava (Utsumi et al. 2012), and tomato (Loukehaich et al. 2012). The relevance of microarray study depends upon the choice of tissue or genotype. Agricultural gain and stress adaptation are best studied in reproductive tissues and stress-tolerant genotypes, respectively (Deyholos 2010). Generally, abiotic stresses are complex in nature, eliciting intricate mechanisms of responses in plants, so slight differences in the stress conditions may bring significant differences in stress responses. Similarly, pre- and post-translational modifications may affect the result interpretations. Whole genome tiling arrays are a successful expansion of array-based transcript profiling to investigate abiotic stress responses in species with an available whole genome sequence (Rensink and Buell 2005). Candidate genes for desired traits, such as stress tolerance can be detected by genome wide expression profiles whose inactivation or overexpression will further help in their characterization and utilization. The whole idea of high-throughput phenotyping, or 'phenomics', has developed into a highly active research field. The complex nature of drought stress has been proven by the involvement of a number of genes in drought stress. Drought tolerant genotypes revealed the presence of multiple pathways conferring drought stress tolerance. Since, transcription factors are generally the key players for diversifying stress responses; they are often targeted to examine crops for drought stress responses. A comprehensive EST database is a prerequisite for the success of the above-mentioned approaches to identify genes accurately and unambiguously. ESTs are the source for designing 'functional markers' which are the polymorphic sites on causal genes responsible for phenotypic variation of traits in crops (Andersen and Lübberstedt 2003). The importance of functional markers has been well documented in stress tolerance studies (Bagge et al. 2007).

8.4.1.2 Structural Genomics

Functional genomics focuses on the functions of genes and gene networks, whilst structural genomics on the physical structure of the genome to identify, locate, and order genomic features along chromosomes. In the last decade, advances in DNA sequencing technologies have paved the way for the exploitation of plant genomics studies for breeding improved varieties. Through Next Generation Sequencing (NGS) reference or draft genome sequences for a number of species like *Arabidopsis thaliana, Brachypodium distachyon*, rice, sorghum, soybean, and maize, have been published (Morrell et al. 2012). Detailed information on genomic features including coding and noncoding genes, regulatory sequences, repetitive elements, and GC content can be obtained from whole genome sequences which can further be exploited for crop improvement via molecular breeding, particularly for complex traits (Mochida and Shinozaki 2010).

NGS-mediated shotgun sequencing made its impact by contributing in the development of molecular markers. In contrast to morphological markers which once had been the focus of traditional breeding studies, DNA based markers are not affected by the environment (Mohan et al. 1997). Amongst all markers, Single Nucleotide Polymorphisms (SNPs) (Edwards and Batley 2010) are beneficial for use in genomics because of their abundance in genome, codominant nature and amenability to high-throughput genotyping. SNPs are readily identified by comparison of different genotypes in crop species provided reference genome sequences or extensive transcript databases are available and may thus serve as functional markers (Mammadov et al. 2012). Insertion Site-Based Polymorphisms (ISBPs) is a recently developed molecular marker type that utilizes the insertional polymorphisms observed in the repeat junctions of complex genomes (Paux et al. 2010).

The utilization of molecular markers in breeding is referred to as Marker-Assisted Selection (MAS) that helps in improving varieties with respect to desired traits, such as pathogen resistance, abiotic stress tolerance, or high yield (Collard and Mackill 2008). Any trait, if tightly linked to a molecular marker or more preferably flanked by two close markers results in a more efficient and accurate MAS (Edmeades et al. 2004; Edwards and Batley 2010). Additionally, markers should be highly polymorphic in the germplasm used for breeding for efficient MAS. Since, genetic diversity plays a very important role in improving crops through breeding techniques; molecular markers are used in the exploration of the variation among the germplasm to select the best candidate parental lines. Likewise, molecular markers ensure genomic purity or may identify heterotic groups of cultivars to achieve heterosis. Several backcrossing steps are involved in conventional plant breeding to enable transfer of one or a few traits to an elite cultivar while retaining most of the recurrent genome and molecular markers have provided the opportunity for accelerated backcrossing. Genotyping by MAS enables early selection of traits that are labor and/or cost-intensive to score phenotypically, that are under complex genetic control, or that are manifested late in development (Collard and Mackill 2008). Though MAS usually requires the validation of QTLs when applied in different genetic backgrounds, but functional markers, however, may overcome the issue of QTL validation (Edmeades et al. 2004). In recent years, MAS has been successfully utilized to improve crops for abiotic stress tolerance, including drought (Ashraf 2010), salinity (Yamaguchi and Blumwald 2005), and waterlogging (Ahmed et al. 2012).

8.4.1.3 Comparative Genomics

Comparative genomics is a promising tool to gain information for species with largely unexplored genomes by utilizing the conservation between closely related plant species (Guyot et al. 2012). A collinear order of genes and markers shared by genomes of different species has been shown by comparative genome mapping studies. Though plant genomes greatly differ in size, yet these differences generally correspond to intergenic regions. A recent concept of "genome zipper" was significantly contributed by comparative genomics, which enables the determination of a virtual gene order in a partially sequenced genome. Genome zippers predict the gene order and organization in species like *Brachypodium*, sorghum, and rice by comparing the fully sequenced and annotated genomes with various sources of data from less well-studied species (Mayer et al. 2011). Despite of comparative genomics and genome zippers concept, species-specific genomic features, like homoeologous genes with different orthologous relationships, can still only be accessed through a fully annotated reference genome sequences (Rustenholz et al. 2010).

8.4.2 *Transcriptome Resources*

Transcriptome analysis is a useful tool to screen candidate genes, predict gene function and discover cis-regulatory motifs. It comprises the comprehensive and high-throughput analysis of gene expression. Large-scale gene expression profiles for various species have been done by the well-established hybridization-based methods, such as microarrays and gene chips. The multiple data sets thus generated containing large-scale gene expression profiles serve as a common pool of large amounts of information. This pool helps us to access, analyse and disseminate expression data for comprehensive functional genomics studies, such as co-expression and comparative analyses. Furthermore, this data helps in answering biological questions and leveraging existing large-scale expression studies results for developing novel strategies to demonstrate the expression and translation of a particular gene. Recently, deep sequencing of short fragments of expressed RNAs is becoming an efficient tool to study genome-sequenced species (Harbers and Carninci 2005; de Hoon and Hayashizaki 2008). High-throughput expression profiling and analysis systems have been developed and have matured rapidly through the past decade. Broadly, there can be two categories: sequencing-based and hybridization-based approaches. Though, these approaches are based on different principles they should be considered complementary to each other and currently, both are important tools for transcriptome profiling.

8.4.2.1 Sequence-Tag Based Resources

Expressed sequence tag (EST) and complementary DNA (cDNA) sequences are currently the most important resources for transcriptome exploration. Transcriptome profiles of contrasting genotypes are generally acquired by large-scale sequencing of ESTs randomly picked in their unbiased cDNA libraries and then classified into clusters of transcript sequences using sequence-clustering and/or assembling methods. The transcript abundance in each tissue is estimated by counting the number of ESTs with identifiers for each cDNA library and/or each sequence cluster. The formal identification of candidate genes thus proceeds first by gathering information from expressed sequences tags (ESTs) and then by cloning the complete gene. Furthermore, the sequence information from ESTs enable gene discovery, complement genome annotation, guide SNP (single nucleotide polymorphism) characterization and ease proteome analysis. Digital differential display (DDD) tool, part of NCBI's UniGene database uses the same platform by comparing and identifying changes in mRNA transcript levels between two or more samples. This technique is widely used in comprehensive cDNA projects for a number of eukaryotic species (Fei et al. 2004; Mochida et al. 2004; Sterky et al. 2004; Zhang et al. 2004). SAGE is a technique for deep sequencing of short read cDNA tags that further let the user identify a large number of transcripts present in tissues and allows quantitative comparison of transcriptomes (Velculescu et al. 1995). The development of SAGE technique has further expanded its utility. These are MAGE (Multiplex Automated Genomic Engineering), CAGE (Cap Analysis of Gene Expression), SADE (SAGE Adaptation for Down-sized Extracts), microSAGE, miniSAGE, longSAGE, superSAGE, deepSAGE, super SAGE, 5′ SAGE, etc. (Hashimoto et al. 2004; Anisimov 2008). Another sequencing-based technology is Massively Parallel Serial Sequencing (MPSS) that uniquely quantifies gene expression levels (Brenner et al. 2000). This method combines the non-gel based sequencing with *in vitro* cloning of millions ESTs on microbeads. This approach was adopted to perform genome-scale discovery and expression profiling of sRNAs in *Arabidopsis* and rice (Lu et al. 2006; Nobuta et al. 2007).

8.4.2.2 Hybridization-Based Resources

DNA microarray and chip-related technologies are widely used in a variety of life sciences disciplines. The power of DNA microarrays lies in the simultaneous hybridization of mRNA extract from biological samples to a pre-selected mRNA library (probe set), immobilized on a glass slide or on a silicon chip. The expression levels of each transcript are obtained by reading out intensities of hybridization signals. With the rapid increase in the number of sequenced species, a number of DNA microarrays have been developed for transcriptome analysis in various plant species and thus enables users to use a particular DNA microarray design to obtain transcriptome data from many experiments. Gene expression profiles can be linked to other information to gain insight into biological regulatory processes, predicting binding sites, predicting protein interactions, predicting functionally conserved modules, predicting protein functions, etc. Various web tools have been developed for in-depth expression analysis and functional predictions. GenExpress (https://www.genialis.com/tag/genexpress/), a web application tool is a hard earned multinational effort designed to uncover the transcriptome of *A. thaliana* and stands as one of the most comprehensive resources for the *Arabidopsis* transcriptome (Kilian et al. 2007; Goda et al. 2008). DAVID (Database for Annotation, Visualization and Integrated Discovery) is another web-based tool for annotation and functional analysis. Similarly, NCBI's Gene Expression Omnibus (GEO) and the European Bioinformatics Institute (EBI)'s Array Express have been used as the primary sources of transcriptome data in the public domain (Parkinson et al. 2007; Barrett et al. 2008). Xcluster is a cross platform software for expression analysis. Likewise, a number of commercial comprehensive software for microarray data analysis are available that helps in extensive visualization and graphics. Additionally, more databases for transcriptome data are available such as ATTED II, which provides co-expression analysis data calculated from *Arabidopsis*

ATH1 Gene Chip data (Obayashi et al. 2007, 2009). Genevestigator is another database that summarizes hundreds of microarray experiments from various organisms (Zimmermann et al. 2004). Electronic Fluorescent Pictograph (eFP) browser provides gene expression patterns collected from *Arabidopsis*, poplar, *Medicago*, rice and barley (Winter et al. 2007). *Arabidopsis* Gene Expression Database AREX provides data sets of high-resolution gene expression patterns of root tissues in *Arabidopsis* (Birnbaum et al. 2003). RICEATLAS database covers rice transcriptome data covering various types of tissues (Jiao et al. 2009). Moreover, tiling arrays, a subtype of microarray chips, provide a platform for analyzing expressed regions throughout a whole genome and is an effective method to discover novel genes, elucidate their structure and interacting partners through chromatin immunoprecipitation (Matsui et al. 2008). Linking with the immune precipitation method has recently extended the usefulness of tiling arrays. A comprehensive DNA methylation map of the *Arabidopsis* genome was constructed by combining methylcytosine immunoprecipitation (mCIP) method with the *Arabidopsis* tiling array (Zhang et al. 2006). 'ChIP-seq', has also become an alternative but powerful approach to sequence co-precipitated DNAs together with a protein using the next generation sequencer (Park 2009). It helps in identifying the genome wide DNA-binding positions of transcription factors and proteins.

8.4.3 Proteome Resources

Proteome analysis, the detailed investigation of the functions, functional networks and 3D structures of proteins, is increasingly gaining attention after the completion of genome sequencing projects in model and non-model crops. These high-throughput proteome data sets serves as an important resource for understanding protein functions in cellular systems that reflect cell and organ states in terms of growth, development and response to environmental changes. Functional proteomics helps in the high-throughput identification of all of the proteins in cells and/or tissues. Recently, the advanced proteome tools took us to second generation of functional proteomics, including quantitative proteomics, subcellular proteomics, and protein-protein interactions (Rossignol et al. 2006; Jorrín-Novo et al. 2009; Yates et al. 2009).

8.4.3.1 Proteome Profiling

Protein sample preparation, separation, detection and their identification are the key steps for protein profiling of any organism. Protein samples can be fractionated based on solubility, molecular mass and isoelectric point by sequential solubilization. Complex proteins can be fractionated by one-dimensional SDS-PAGE based on their molecular masses. Two-dimensional gel electrophoresis (2-DE), which uses isoelectric focusing (IEF) as the first dimension and SDS-PAGE as the second dimension, is an effective method for high-resolution separation of complex proteins. This technique has been widely used in proteomics for various species (Méchin et al. 2004; Chen and Harmon 2006; Yates et al. 2009). Chromatography-based separation methods are effective in separating proteins based on their physicochemical properties. In recent times, mass spectrometry (MS) plays an important role in proteomics. MS equipment consists of an ionizer and mass spectrometer(s) to detect the ionized samples. The obtained peptide mass fingerprint data are then searched against a database of theoretically predicted masses of known amino acid sequences to identify the target proteins (Hirano et al. 2004; Newton et al. 2004). Moreover, in the gel-free separation method the protein mixture is directly digested into peptides and separated by the multi-dimensional separation method (Yates et al. 2009).

8.4.3.2 Quantitative Proteomics

Comprehensive quantification of each protein's abundance in a cell has become quite important to discover key proteomic changes, including expression, interaction and modification, that may be associated with genetic variations and/or visible phenotypic changes (Gstaiger and Aebersold 2009). Difference Gel Electrophoresis (DIGE) method is popularly used for differential display of proteins for quantitative protein comparison in which protein samples are labeled with different fluorophores before 2-DE, enabling accurate analysis of differences in protein abundance between samples (Rossignol et al. 2006). Other widely used methods for protein differential display using stable isotope labeling are Isotope-Coded Affinity Tags (ICATs), Isobaric Tags for Relative and Absolute Quantitation (iTRAQ) and Stable Isotope Labeling with Amino acids in Cell culture (SILAC) (Jorrín-Novo et al. 2009). Recently, label-free quantitative techniques like (LC)-MS/MS or MS/MS are developed for high-throughput comparisons of proteomic expression.

8.4.3.3 Subcellular Proteomics

The enzymatic inventory of a cell organelle, the compartmentalization of metabolic pathways, protein targeting, trafficking and proteomic dynamics at the organelle level are an integral part of the understanding for a large-scale proteome analysis of cellular systems undergoing changes (Andersen and Mann 2006; Chen and Harmon 2006; Baginsky 2009). The proteome of different organelles of plant cells such as chloroplasts, etioplasts, amyloplasts, chromoplasts, mitochondria, vacuoles, plasma membranes, nucleus, peroxisomes, cytosolic ribosome and cell wall has been studied by a variety of approaches (Baginsky 2009). ICAT and iTRAQ techniques are also effective for acquiring quantitative data on proteomes in each organelle. In *Arabidopsis*, rice and alga, differential proteome profiles of plant plasma membranes were studied to identify differentially expressed proteins in response to environmental factors such as cold, salt and bacterial stress (Benschop et al. 2007;

Katz et al. 2007; Cheng et al. 2009; Minami et al. 2009). Several databases are available that provide subcellular proteome information like, the rice proteome database (Komatsu 2005), the soybean proteome database (Sakata et al. 2009), the Nottingham *Arabidopsis* Stock Centre (NASC) Proteomics database and the SUB-cellular location database for *Arabidopsis* proteins (SUBA) (Dunkley et al. 2006).

8.4.3.4 Post-translational Protein Modifications

Modificome research investigates various kinds of post-translational protein modifications and plays a key role in the current study of proteomics. It elucidates the role of each protein's functional modification with its associated biological event (Kwon et al. 2006). Protein phosphorylation is a most common protein modification affecting most basic cellular processes in eukaryotic organisms. Large-scale *in vivo* phosphorylation site mapping can be done by MS-based technologies accompanied by phosphopeptide enrichment techniques. The Plant Protein Phosphorylation Database (P3DB) (http://www.p3db.org) provides information for phosphoproteomes from multiple plants (Gao et al. 2009). Ubiquitination of protein is another major post-translational modification that controls protein abundance, localization and activity. Large-scale protein ubiquitination analysis has been reported in many plants (Shirasu 2007; Manzano et al. 2008; Igawa et al. 2009).

8.4.3.5 Structural Proteomics

Protein 3D structural information is crucial for elucidating relationships between protein functions and structures or for analyzing molecules in protein complexes. International Structural Genomics Organization (ISGO, http://www.isgo.org) was formed to facilitate global structural genomics research efforts (Stevens et al. 2001). For structural proteomics, production of soluble and folded proteins remains a major limiting step. Mostly *Escherichia coli* cells are used for protein production but advanced *E. coli* cell-free system and wheat germ embryo cell-free system have also been developed (Kigawa et al. 1999; Endo and Sawasaki 2003). NMR spectroscopy and X-ray crystallography has also played an important role in structural proteomics. Bioinformatics and related databases are also necessary tools for advanced structural proteomics.

8.4.4 Metabolome Resources and Related Platforms

Metabolomics aims to understand the metabolic systems and allows us to conduct parallel assessments of multiple metabolites and to undertake quantitative analysis of particular metabolites. The plant metabolome is enormously diversified due to complex set of metabolites produced in each plant species (Bino et al. 2004; von Roepenack-Lahaye et al. 2004). Metabolomics is able to elucidate plant cellular systems and allows us to engineer crops to improve the productivity and functionality of plants under different conditions (Fernie and Schauer 2009; Oksman-Caldentey and Saito 2005; von Roepenack-Lahaye et al. 2004). Various analytical instruments such as GC-MS (Gas Chromatography-Mass Spectrometry), LC-MS (Liquid Chromatography-Mass Spectrometry), FT-MS (Fourier Transform-Mass Spectrometry), FT-IR (Fourier Transform-Infrared Spectroscopy) and NMR (Nuclear Magnetic Resonance) spectroscopy can be used for metabolomics experiments (Roessner et al. 2001; Schripsema 2010). The metabolic pathway maps often combines metabolic profiles with other omics methods, including gene expression profiles of genes encoding enzymes involved in particular pathways (Thimm et al. 2004).

8.4.4.1 Metabolite Profiling

Metabolite profiling is the systematic collection of metabolite profiles and is the prerequisite step in metabolomics. It improves the understanding of the cellular systems responsive to changes in intracellular and extracellular environments. Furthermore, the changes in metabolic profiles helps in identifying genes involved in particular metabolic pathways. Several metabolomics databases are available from various studies of metabolic profiling in plant species, such as the Metabolome Tomato Database (MoTo DB) (Moco et al. 2006), the KOMICS (Kazusa-omics) database for the tomato cultivar (Iijima et al. 2008) and the Golm Metabolome Database (GMD) (Kopka et al. 2004). These databases are the information resources and serve as tools for further integration of metabolic profiles and other omics data (Akiyama et al. 2008).

8.4.4.2 Integrated Metabolomics and Other Omics Resources

Several MS technologies and bioinformatics are applied to extensively analyze the metabolic changes in plants in response to environmental stress factors (Fiehn 2002; Shulaev et al 2008). A combinatorial approach for metabolomics and other omics instances elucidate the gene-to-metabolites molecular networks and has markedly increased our understanding of plants' responses to various stresses. Metabolite profiling has been used to characterize stress responses to various abiotic factors such as drought, cold, high salinity or temperature for comprehensive analyses of stress signal transduction pathways. The integrated approaches have also elucidated regulatory networks, like ABA-dependent or -independent networks that act in response to environmental stresses in plants. The endogenous ABA level significantly increases in response to drought stress to regulate physiological stress responses and gene expression. A combined approach of the metabolome and transcriptome was adopted to analyze the dehydration-stress responses of an

Arabidopsis *NCED3*-knockout mutant and the wild-type plant (Urano et al. 2009). *NCED3* plays a critical role in the dehydration-inducible biosynthesis of ABA (Yamaguchi-Shinozaki and Shinozaki 2006). Metabolite profiling showed that the ABA accumulated during drought regulates the accumulation of various amino acids and sugars such as glucose and fructose. Particularly, the drought-inducible expression of key biosynthetic genes (*BCAT2*, *LKR/SDH*, *P5CS1*, and *ADC2*) is correlated with the drought-inducible accumulations of branch-chain amino acids (BCAAs), saccharopine, proline, and agmatine respectively, which are regulated by endogenous ABA. On the contrary, the accumulation of raffinose and galactinol is not regulated by ABA during drought stress. Moreover, metabolic network analysis showed a strong correlation between dehydration stress and raffinose in the *nced3* mutant. These results altogether revealed the important role of ABA in regulating the metabolic changes that occur during the drought-stress response. Metabolite and transcript profiles were compared between dehydration and salinity stress in grapevine. Higher concentrations of glucose, malate, and proline were found in dehydration-treated plants, compared with salt-stressed plants. The differential levels of metabolites were correlated with those of transcript levels of many genes involved in energy metabolism and nitrogen assimilation. In comparison with salt-stressed plants, dehydration-treated plants were weaker in adjusting osmotically, detoxifying ROS, and ameliorating photoinhibition (Cramer et al. 2007). Metabolic profiling in fact showed that sucrose replaces proline in plants as the major osmoprotectant during the combined dehydration and heat-stress treatment (Rizhsky et al. 2004).

In order to identify genes involved in anthocyanin biosynthesis in *Arabidopsis*, an integrated approach that comprised metabolome and transcriptome analysis was conducted for investigation of an activation-tagged mutant and overexpressors of an MYB TF, PAP1 gene (Tohge et al. 2005). The co-expression data of the *Arabidopsis* transcriptome provided by the ATTED-II database was used to identify novel genes involved in lipid metabolism, leading to identification of a novel gene, *UDP-glucose pyrophosphorylase 3* (*UGP3*) that is required for the first step of sulfolipid biosynthesis (Okazaki et al. 2009). All of the genes related to flavonoid biosynthesis was identified by co-expression analysis (Yonekura-Sakakibara et al. 2008).

8.5 Modeling Plant Responses for Crop Improvement

Molecular plant breeding often revolves around predictions of phenotype based on genotypes. The reliability of these predictions is derived from measurements of phenotypic performance in large segregating populations. Statistical and modelling techniques for phenotypic data, which have been generated from field and controlled environment studies further supported the analysis of complex traits (Hammer et al. 2006). Since phenotypic prediction sometimes becomes difficult because of the large genotype by environment interaction, drought tolerance has been modeled to produce an 'index of the climatic environment'. This index can identify the stages of crop development in which the interaction between the genotype and the environment is strongest which can further be deployed to identify the components of the crop response that offer the greatest response to breeding and selection (Chapman 2008). Physiological and molecular knowledge is linked to the more conventional phenotyping and genetic analysis by the advent of different omics platforms. Genomics and genome analysis provide valuable information regarding structure and behavior of crop genomes. Genotype or environment interaction can be analyzed by the detailed genomics and related omics data. Availability of detailed omics datasets is now helping in interrogating components of the environment that react with particular regions of the genome. This type of analysis offers the potential to target studies to specific environmental issues, like drought stress and identify responses that will result in the greatest genetic gain. The objective is to develop gene-network and ecophysiological models that link agronomy to gene structure and provide information on selection targets for breeding. Any gene responsible for a particular trait can be used to enhance breeding in different ways. Firstly, if the gene is known then diagnostic markers can be developed and used for screening the phenotypic differences, which could be due to variation in the protein product, differential expression levels, gene duplication or due to the presence or absence of the gene. Screening germplasm collections provides the opportunity to look for sequence variants at the locus or can be extended to look at landraces or wild relatives (Barkley and Wang 2008). Discovery of new alleles with increased expression should be given a high priority to determine the most appropriate strategy for crop improvement. High-throughput screening methods such as Targeting Induced Local Lesions in Genomes (TILLING) have made mutant populations a valuable source of variation and have reinvigorated interest in mutation breeding. Site-directed mutagenesis is the basis of a recently designed zinc finger nuclease technology (Townsend et al. 2009) that has been successfully used to generate herbicide-tolerant maize lines by precise insertion in a target gene (Shukla et al. 2009).

8.6 Conclusions and Future Prospects

The aim of molecular plant breeding is to improve crop variety for yield, quality and resistance by means of latest innovations in the fields of genetics and genomics. Genomic tools help in improving our understanding about the association between genotype and phenotype. Recent advances in sequencing and genotyping technologies have made it possible to identify molecular markers associated with the trait of interest for breeders. Such markers have greatly helped the breeding communities in overcoming the constraints of phenotypic selection for several crops. These approaches together with 'omics' analyses are crucial to understand the whole processes of molecular networks in response to drought and related stress and can be used to identify the genomic regions or genes involved in expression of trait(s) that are of interest to the breeders. It is important to validate the functions of newly identified stress-responsive protein-coding and non-coding RNAs to understand the complex drought-related stress responses of plants. An integrated metabolome and transcriptome analysis is necessary to identify the broad function of metabolite regulatory networks during responses to abiotic stresses.

In summary, the presented tools and approaches in this chapter have great potential to impact crop breeding. However, it is really important at this stage to bring integrated technologies/approaches together with conventional breeding methodologies for enhancing the genetic gain leading to crop improvement.

References

Ahmed F, Rafii MY, Ismail MR, Juraimi AS, Rahim HA, Asfaliza R, Latif MA (2012) Waterlogging tolerance of crops: breeding, mechanism of tolerance, molecular approaches, and future prospects. Bio Med Res Int 2013

Akiyama K, Chikayama E, Yuasa H, Shimada Y, Tohge T, Shinozaki K, Hirai MY, Sakurai T, Kikuchi J, Saito K (2008) PRIMe: a web site that assembles tools for metabolomics and transcriptomics. In Silico Biol 8(3–4):339–345

Andersen JR, Lübberstedt T (2003) Functional markers in plants. Trends Plant Sci 8(11):554–560

Andersen JS, Mann M (2006) Organellar proteomics: turning inventories into insights. EMBO Rep 7(9):874–879

Anisimov SV (2008) Serial analysis of gene expression (SAGE): 13 years of application in research. Curr Pharma Biotechnol 9(5):338–350

Ashraf M (2010) Inducing drought tolerance in plants: recent advances. Biotechnol Adv 28 (1):169–183

Bagge M, Xia X, Lübberstedt T (2007) Functional markers in wheat. Curr Opin Plant Biol 10 (2):211–216

Baginsky S (2009) Plant proteomics: concepts, applications, and novel strategies for data interpretation. Mass Spectrom Rev 28(1):93–120

Barkley NA, Wang ML (2008) Application of TILLING and EcoTILLING as reverse genetic approaches to elucidate the function of genes in plants and animals. Curr Genomics 9(4):212–226

Barrett T, Troup DB, Wilhite SE, Ledoux P, Rudnev D, Evangelista C, Kim IF, Soboleva A et al (2008) NCBI GEO: archive for high-throughput functional genomic data. Nucleic Acids Res 37:D885–D890

Benschop JJ, Mohammed S, O'Flaherty M, Heck AJ, Slijper M, Menke FL (2007) Quantitative phosphoproteomics of early elicitor signaling in *Arabidopsis*. Mol Cell Proteomics 6(7):1198–1214

Bino RJ, Hall RD, Fiehn O, Kopka J, Saito K, Draper J, Nikolau BJ et al (2004) Potential of metabolomics as a functional genomics tool. Trends Plant Sci 9(9):418–425

Birnbaum K, Shasha DE, Wang JY, Jung JW, Lambert GM, Galbraith DW, Benfey PN (2003) A gene expression map of the *Arabidopsis* root. Science 302(5652):1956–1960

Bouchez D, Höfte H (1998) Functional genomics in plants. Plant Physiol 118(3):725–732

Brenner S, Johnson M, Bridgham J, Golda G, Lloyd DH, Johnson D, Luo S et al (2000) Gene expression analysis by massively parallel signature sequencing (MPSS) on microbead arrays. Nature Biotechnol 18(6):630–634

Chapman SC (2008) Use of crop models to understand genotype by environment interactions for drought in real-world and simulated plant breeding trials. Euphytica 161(1–2):195–208

Chen S, Harmon AC (2006) Advances in plant proteomics. Proteomics 6(20):5504–5516

Cheng Y, Qi Y, Zhu Q, Chen X, Wang N, Zhao X, Chen H, Cui X, Xu L, Zhang W (2009) New changes in the plasma-membrane-associated proteome of rice roots under salt stress. Proteomics 9(11):3100–3114

Chinnusamy V, Stevenson B, Lee BH, Zhu JK (2002) Screening for gene regulation mutants by bioluminescence imaging. Sci STKE 140:1–10

Close TJ, Wanamaker SI, Caldo RA, Turner SM, Ashlock DA, Dickerson JA et al (2004) A new resource for cereal genomics: 22 K barley gene chip comes of age. Plant Physiol 134(3):960–968

Collard BCY, Mackill DJ (2008) Marker-assisted selection: an approach for precision plant breeding in the twenty-first century. Philos Trans R Soc Lond B Biol Sci 363(1491):557–572

Cramer GR, Ergül A, Grimplet J, Tillett RL, Tattersall EA, Bohlman MC et al (2007) Water and salinity stress in grapevines: early and late changes in transcript and metabolite profiles. Func Integr Genomics 7(2):111–134

de Hoon M, Hayashizaki Y (2008) Deep cap analysis gene expression (CAGE): genome-wide identification of promoters, quantification of their expression, and network inference. Biotechniques 44(5):627

Deyholos MK (2010) Making the most of drought and salinity transcriptomics. Plant Cell Environ 33(4):648–654

Dunkley TP, Hester S, Shadforth IP, Runions J, Weimar T, Hanton SL et al (2006) Mapping the *Arabidopsis* organelle proteome. Proc Natl Acad Sci USA 103(17):6518–6523

Edmeades GO, McMaster GS, White JW, Campos H (2004) Genomics and the physiologist: bridging the gap between genes and crop response. Field Crops Res 90(1):5–18

Edwards D, Batley J (2010) Plant genome sequencing: applications for crop improvement. Plant Biotechnol J 8(1):2–9

Endo Y, Sawasaki T (2003) High-throughput, genome-scale protein production method based on the wheat germ cell-free expression system. Biotechnol Adv 21(8):695–713

Ergen NZ, Thimmapuram J, Bohnert HJ, Budak H (2009) Transcriptome pathways unique to dehydration tolerant relatives of modern wheat. Func Integr Genomics 9(3):377–396

Fei Z, Tang X, Alba RM, White JA, Ronning CM, Martin GB, Tanksley SD, Giovannoni JJ (2004) Comprehensive EST analysis of tomato and comparative genomics of fruit ripening. Plant J 40(1):47–59

Fernie AR, Schauer N (2009) Metabolomics-assisted breeding: a viable option for crop improvement? Trends Genet 25(1):39–48

Fiehn O (2002) Metabolomics–the link between genotypes and phenotypes. Plant Mol Biol 48(1–2):155–171

Fleury D, Jefferies S, Kuchel H, Langridge P (2010) Genetic and genomic tools to improve drought tolerance in wheat. J Exp Bot 61(12):3211–3222

Gao J, Agrawal GK, Thelen JJ, Xu D (2009) P3DB: a plant protein phosphorylation database. Nucleic Acids Res 37:D960–D962

Goda H, Sasaki E, Akiyama K, Maruyama-Nakashita A, Nakabayashi K, Li W et al (2008) The AtGenExpress hormone and chemical treatment data set: experimental design, data evaluation, model data analysis and data access. Plant J 55(3):526–542

Gstaiger M, Aebersold R (2009) Applying mass spectrometry-based proteomics to genetics, genomics and network biology. Nat Rev Genet 10(9):617–627

Guyot R, Lefebvre-Pautigny F, Tranchant-Dubreuil C, Rigoreau M, Hamon P et al (2012) Ancestral synteny shared between distantly-related plant species from the asterid (*Coffea canephora* and *Solanum* Sp.) and rosid (*Vitis vinifera*) clades. BMC Genom 13(1):103

Hammer G, Cooper M, Tardieu F, Welch S, Walsh B, van Eeuwijk F, Chapman S, Podlich D (2006) Models for navigating biological complexity in breeding improved crop plants. Trends Plant Sci 11(12):587–593

Harbers M, Carninci P (2005) Tag-based approaches for transcriptome research and genome annotation. Nat Methods 2(7):495–502

Hashimoto S et al (2004) 5′-end SAGE for the analysis of transcriptional start sites. Nat Biotechnol 22(9):1146–1149

Hirano H, Islam N, Kawasaki H (2004) Technical aspects of functional proteomics in plants. Phytochem 65(11):1487–1498

Hirayama T, Shinozaki K (2010) Research on plant abiotic stress responses in the post-genome era: Past, present and future. Plant J 61(6):1041–1052

Igawa T, Fujiwara M, Takahashi H, Sawasaki T, Endo Y, Seki M, Shinozaki K et al (2009) Isolation and identification of ubiquitin-related proteins from *Arabidopsis* seedlings. J Exp Botany 60(11):3067–3073

Iijima Y, Nakamura Y, Ogata Y, Tanaka KI, Sakurai N, Suda K, Suzuki T et al (2008) Metabolite annotations based on the integration of mass spectral information. Plant J 54(5):949–962

Jiao Y, Tausta SL, Gandotra N, Sun N, Liu T, Clay NK, Ceserani T, Chen M, Ma L et al (2009) A transcriptome atlas of rice cell types uncovers cellular, functional and developmental hierarchies. Nat Genet 41(2):258–263

Jorrín-Novo JV, Maldonado AM, Echevarría-Zomeño S, Valledor L, Castillejo MA et al (2009) Plant proteomics update (2007–2008): second-generation proteomic techniques, an appropriate experimental design, and data analysis to fulfill MIAPE standards, increase plant proteome coverage and expand biological knowledge. J Proteomics 72(3):285–314

Katz A, Waridel P, Shevchenko A, Pick U (2007) Salt-induced changes in the plasma membrane proteome of the halotolerant alga *Dunaliella salina* as revealed by blue native gel electrophoresis and nano-LC-MS/MS analysis. Mol Cell Proteomics 6(9):1459–1472

Kigawa T, Yabuki T, Yoshida Y, Tsutsui M, Ito Y, Shibata T, Yokoyama S (1999) Cell-free production and stable-isotope labelling of milligram quantities of proteins. FEBS Lett 442 (1):15–19

Kilian J, Whitehead D, Horak J, Wanke D, Weinl S, Batistic O, D'Angelo C et al (2007) The AtGenExpress global stress expression data set: protocols, evaluation and model data analysis of UV-B light, drought and cold stress responses. Plant J 50(2):347–363

Komatsu S (2005) Rice proteome database: a step toward functional analysis of the rice genome. Plant Mol Biol 59(1):179–190

Kopka J, Schauer N, Krueger S, Birkemeyer C, Usadel B, Bergmüller E, Dörmann P et al (2004) GMD@CSB.DB: the Golm metabolome database. Bioinformatics 21(8):1635–1638

Kwon SJ, Choi EY, Choi YJ, Ahn JH, Park OK (2006) Proteomics studies of post-translational modifications in plants. J Exp Bot 57(7):1547–1551

Langridge P, Fleury D (2011) Making the most of 'omics' for crop breeding. Trends Biotechnol 29 (1):33–40

Larkindale J, Hall JD, Knight MR, Vierling E (2005) Heat stress phenotypes of *Arabidopsis* mutants implicate multiple signaling pathways in the acquisition of thermotolerance. Plant Physiol 138(2):882–897

Lee JY, Lee DH (2003) Use of serial analysis of gene expression technology to reveal changes in gene expression in *Arabidopsis* pollen undergoing cold stress. Plant Physiol 132(2):517–529

Lockhart DJ, Dong H, Byrne MC, Follettie MT, Gallo MV, Chee MS, Mittmann M et al (1996) Expression monitoring by hybridization to high-density oligonucleotide arrays. Nat Biotechnol 14(13):1675–1680

Loukehaich R, Wang T, Ouyang B, Ziaf K, Li H, Zhang J, Lu Y, Ye Z (2012) SpUSP, an annexin-interacting universal stress protein, enhances drought tolerance in tomato. J Exp Bot 63(15):5593–5606

Lu C, Kulkarni K, Souret FF, MuthuValliappan R, Tej SS, Poethig RS (2006) MicroRNAs and other small RNAs enriched in the *Arabidopsis* RNA-dependent RNA polymerase-2 mutant. Genome Res 16(10):1276–1288

Luo M, Liu J, Lee RD, Scully BT, Guo B (2010) Monitoring the expression of maize genes in developing kernels under drought stress using oligo-microarray. J Integ Plant Biol 52 (12):1059–1074

Mammadov J, Aggarwal R, Buyyarapu R, Kumpatla S (2012) SNP markers and their impact on plant breeding. Int J Plant Genomics 2012:728398

Manzano C, Abraham Z, López-Torrejón G, Del Pozo JC (2008) Identification of ubiquitinated proteins in *Arabidopsis*. Plant Mol Biol 68(1–2):145–158

Matsui A, Ishida J, Morosawa T, Mochizuki Y, Kaminuma E, Endo TA, Okamoto M et al (2008) *Arabidopsis* transcriptome analysis under drought, cold, high-salinity and ABA treatment conditions using a tiling array. Plant Cell Physiol 49(8):1135–1149

Matsumura H, Nirasawa S, Terauchi R (1999) Transcript profiling in rice (*Oryza sativa* L.) seedlings using serial analysis of gene expression (SAGE). Plant J 20(6):719–726

Mayer KF, Martis M, Hedley PE, Šimková H, Liu H, Morris JA, Steuernagel B (2011) Unlocking the barley genome by chromosomal and comparative genomics. Plant Cell 23(4):1249–1263

Méchin V, Balliau T, Château-Joubert S, Davanture M, Langella O, Négroni L et al (2004) A two-dimensional proteome map of maize endosperm. Phytochem 65(11):1609–1618

Meyers BC, Souret FF, Lu C, Green PJ (2006) Sweating the small stuff: microRNA discovery in plants. Curr Opin Biotechnol 17(2):139–146

Minami A, Fujiwara M, Furuto A, Fukao Y, Yamashita T, Kamo M (2009) Alterations in detergent-resistant plasma membrane microdomains in *Arabidopsis thaliana* during cold acclimation. Plant Cell Physiol 50(2):341–359

Mittler R (2006) Abiotic stress, the field environment and stress combination. Trends Plant Sci 11 (1):15–19

Mochida K, Shinozaki K (2010) Genomics and bioinformatics resources for crop improvement. Plant Cell Physiol 51(4):497–523

Mochida K, Yamazaki Y, Ogihara Y (2004) Discrimination of homoeologous gene expression in hexaploid wheat by SNP analysis of contigs grouped from a large number of expressed sequence tags. Mol Genet Genomics 270(5):371–377

Moco S, Bino RJ, Vorst O, Verhoeven HA, de Groot J, van Beek TA, Vervoort (2006) A liquid chromatography-mass spectrometry-based metabolome database for tomato. Plant Physiol 141 (4):1205–1218

Mohan M, Nair S, Bhagwat A, Krishna TG, Yano M, Bhatia CR, Sasaki T (1997) Genome mapping, molecular markers and marker-assisted selection in crop plants. Mol Breed 3(2):87–103

Morrell PL, Buckler ES, Ross-Ibarra J (2012) Crop genomics: advances and applications. Nat. Rev Genet 13(2):85

Newton RP, Brenton AG, Smith CJ, Dudley E (2004) Plant proteome analysis by mass spectrometry: principles, problems, pitfalls and recent developments. Phytochemistry 65 (11):1449–1485

Nobuta K, Venu RC, Lu C, Beló A, Vemaraju K, Kulkarni K et al (2007) An expression atlas of rice mRNAs and small RNAs. Nat Biotech 25(4):473–477

Obayashi T, Kinoshita K, Nakai K, Shibaoka M, Hayashi S, Saeki M et al (2007) ATTED-II: a database of co-expressed genes and cis elements for identifying co-regulated gene groups in *Arabidopsis*. Nucleic Acids Res 35:D863–D869

Obayashi T, Hayashi S, Saeki M, Ohta H, Kinoshita K (2009) ATTED-II provides coexpressed gene networks for *Arabidopsis*. Nucleic Acids Res 37:D987–D991

Okazaki Y, Shimojima M, Sawada Y, Toyooka K, Narisawa T, Mochida K et al (2009) A chloroplastic UDP-glucose pyrophosphorylase from *Arabidopsis* is the committed enzyme for the first step of sulfolipid biosynthesis. Plant Cell 21(3):892–909

Oksman-Caldentey KM, Saito K (2005) Integrating genomics and metabolomics for engineering plant metabolic pathways. Curr Opin Biotechnol 16(2):174–179

Park PJ (2009) ChIP-seq: advantages and challenges of a maturing technology. Nat Rev Genet 10 (10):669

Parkinson H, Kapushesky M, Shojatalab M, Abeygunawardena N, Coulson R et al (2007) ArrayExpress—a public database of microarray experiments and gene expression profiles. Nucleic Acids Res 35:D747–D750

Paux E, Faure S, Choulet F, Roger D, Gauthier V, Martinant JP, Sourdille P et al (2010) Insertion site-based polymorphism markers open new perspectives for genome saturation and marker assisted selection in wheat. Plant Biotechnol J 8(2):196–210

Ranjan A, Pandey N, Lakhwani D, Dubey NK, Pathre UV, Sawant SV (2012) Comparative transcriptomic analysis of roots of contrasting *Gossypium herbaceum* genotypes revealing adaptation to drought. BMC Genom 13(1):680

Reinartz J, Bruyns E, Lin JZ, Burcham T, Brenner S, Bowen B, Kramer M, Woychik R (2002) Massively parallel signature sequencing (MPSS) as a tool for in-depth quantitative gene expression profiling in all organisms. Brief Funct Genomics 1(1):95–104

Rensink WA, Buell CR (2005) Microarray expression profiling resources for plant genomics. Trends Plant Sci 10(12):603–609

Rizhsky L, Liang H, Shuman J, Shulaev V, Davletova S, Mittler R (2004) When defense pathways collide. The response of *Arabidopsis* to a combination of drought and heat stress. Plant Physiol 134(4):1683–1696

Roessner U, Luedemann A, Brust D, Fiehn O, Linke T, Willmitzer L, Fernie AR (2001) Metabolic profiling allows comprehensive phenotyping of genetically or environmentally modified plant systems. Plant Cell 13(1):11–29

Rossignol M, Peltier JB, Mock HP, Matros A, Maldonado AM, Jorrín JV (2006) Plant proteome analysis: a 2004–2006 update. Proteomics 6(20):5529–5548

Rustenholz C, Hedley PE, Morris J, Choulet F, Feuillet C, Waugh R, Paux E (2010) Specific patterns of gene space organisation revealed in wheat by using the combination of barley and wheat genomic resources. BMC Genom 11(1):714

Sakata K, Ohyanagi H, Nobori H, Nakamura T, Hashiguchi A, Nanjo Y, Mikami Y et al (2009) Soybean proteome database: a data resource for plant differential omics. J Proteome Res 8 (7):3539–3548

Schripsema J (2010) Application of NMR in plant metabolomics: techniques, problems and prospects. Phytochem Anal 21(1):14–21

Shinozaki K, Yamaguchi-Shinozaki K, Seki M (2003) Regulatory network of gene expression in the drought and cold stress responses. Curr Opin Plant Biol 6(5):410–417

Shirasu K (2007) Multidimensional protein identification technology (MudPIT) Analysis of ubiquitinated proteins in plants. Mol Cell Proteomics 6:601–610

Shukla VK, Doyon Y, Miller JC, DeKelver RC, Moehle EA, Worden SE et al (2009) Precise genome modification in the crop species *Zea mays* using zinc-finger nucleases. Nature 459 (7245):437

Shulaev V, Cortes D, Miller G, Mittler R (2008) Metabolomics for plant stress response. Physiol Plant 132(2):199–208

Sterky F, Bhalerao RR, Unneberg P, Segerman B, Nilsson P, Brunner AM et al (2004) A populus EST resource for plant functional genomics. Proc Natl Acad Sci USA 101(38):13951–13956

Stevens RC, Yokoyama S, Wilson IA (2001) Global efforts in structural genomics. Science 294 (5540):89–92

Suzuki N, Rivero RM, Shulaev V, Blumwald E, Mittler R (2014) Abiotic and biotic stress combinations. New Phytol 203(1):32–43

Thimm O, Bläsing O, Gibon Y, Nagel A, Meyer S, Krüger P, Selbig J (2004) Mapman: a user-driven tool to display genomics data sets onto diagrams of metabolic pathways and other biological processes. Plant J 37(6):914–939

Tohge T, Nishiyama Y, Hirai MY, Yano M, Nakajima JI, Awazuhara M (2005) Functional genomics by integrated analysis of metabolome and transcriptome of Arabidopsis plants over-expressing an MYB transcription factor. Plant J 42(2):218–235

Townsend JA, Wright DA, Winfrey RJ, Fu F, Maeder ML, Joung JK, Voytas DF (2009) High frequency modification of plant genes using engineered zinc finger nucleases. Nature 459 (7245):442

Toyoda T, Wada A (2004) Omic space: coordinate-based integration and analysis of genomic phenomic interactions. Bioinformatics 20(11):1759–1765

Urano K, Maruyama K, Ogata Y, Morishita Y, Takeda M, Sakurai N (2009) Characterization of the ABA-regulated global responses to dehydration in Arabidopsis by metabolomics. Plant J 57 (6):1065–1078

Utsumi Y, Tanaka MA, Morosawa T, Kurotani A, Yoshida T (2012) Transcriptome analysis using a high-density oligomicroarray under drought stress in various genotypes of cassava: an important tropical crop. DNA Res 19(4):335–345

Varshney RK, Dubey (2009) Novel genomic tools and modern genetic and breeding approaches for crop improvement. J Plant Biochem Biotechnol 18(2):127–138

Varshney RK, Graner A, Sorrells ME (2005) Genomics-assisted breeding for crop improvement. Trends Plant Sci 10(12):621–630

Vega-Sánchez ME, Gowda M, Wang GL (2007) Tag-based approaches for deep transcriptome analysis in plants. Plant Sci 173(4):371–380

Velculescu VE, Zhang L, Vogelstein B, Kinzler KW (1995) Serial analysis of gene expression. Science 270(5235):484

von Roepenack-Lahaye E, Degenkolb T, Zerjeski M, Franz M, Roth U, Wessjohann L et al (2004) Profiling of *Arabidopsis* secondary metabolites by capillary liquid chromatography coupled to electrospray ionization quadrupole time-of-flight mass spectrometry. Plant Physiol 134(2):548–559

Wang D, Pan Y, Zhao X, Zhu L, Fu B, Li Z (2011) Genome-wide temporal-spatial gene expression profiling of drought responsiveness in rice. BMC Genom 12(1):149

Winter D, Vinegar B, Nahal H, Ammar R, Wilson GV, Provart NJ (2007) An "electronic fluorescent pictograph" browser for exploring and analyzing large-scale biological data sets. PLoS ONE 2(8):e718

Yamaguchi T, Blumwald E (2005) Developing salt-tolerant crop plants: challenges and opportunities. Trends Plant Sci 10(12):615–620

Yamaguchi-Shinozaki K, Shinozaki K (2005) Organization of cis-acting regulatory elements in osmotic-and cold-stress-responsive promoters. Trends Plant Sci 10(2):88–94

Yamaguchi-Shinozaki K, Shinozaki K (2006) Transcriptional regulatory networks in cellular responses and tolerance to dehydration and cold stresses. Annu Rev Plant Biol 57:781–803

Yamamoto K, Sasaki T (1997) Large-scale EST sequencing in rice. Plant Mol Biol 35(1–2):135–144

Yates JR, Ruse CI, Nakorchevsky A (2009) Proteomics by mass spectrometry: approaches, advances, and applications. Ann Rev Biomed Eng 11:49–79

Yonekura-Sakakibara K, Tohge T, Matsuda F, Nakabayashi R, Takayama H et al (2008) Comprehensive flavonol profiling and transcriptome coexpression analysis leading to decoding gene–metabolite correlations in *Arabidopsis*. Plant Cell 20(8):2160–2176

Zhang H, Sreenivasulu N, Weschke W, Stein N, Rudd S, Radchuk V et al (2004) Large-scale analysis of the barley transcriptome based on expressed sequence tags. The Plant J 40(2):276–290

Zhang X, Yazaki J, Sundaresan A, Cokus S, Chan SW, Chen H et al (2006) Genome-wide high-resolution mapping and functional analysis of DNA methylation in *Arabidopsis*. Cell 126 (6):1189–1201

Zimmermann P, Hirsch-Hoffmann M, Hennig L, Gruissem W (2004) GENEVESTIGATOR. *Arabidopsis* microarray database and analysis toolbox. Plant Physiol 136(1):2621–2632

Chapter 9
Genomic Strategies for Improving Abiotic Stress Tolerance in Crop Plants

Jyoti Taunk, Asha Rani, Richa Singh, Neelam R. Yadav and Ram C. Yadav

Abstract Abiotic stresses which adversely affect agricultural production are a serious global concern for food security. Environmental stresses such as drought, salinity, high temperature and frost are predicted to worsen with the anticipated climate change. Hence, production of stress tolerant crops is urgently required for ensuring future food security. Traditional crop improvement approaches have almost reached their limits and will not provide any further gain. Use of genomics approaches is gaining importance and leading to a new revolution of plant breeding for abiotic stress tolerance. Advances in genomics technologies have allowed an in-depth analysis of crop genomes and have enhanced our understanding of the complexity of the mechanisms governing abiotic stress tolerance. For example, next generation sequencing (NGS) technologies are allowing the mass sequencing of genomes and transcriptomes, thus producing a vast array of genomic information. The analysis of NGS data by means of bioinformatics developments have allowed discovery of new genes and regulatory sequences controlling important traits. Also, with the generation of innumerable number of markers and their use in genome-wide association studies, many genomic regions associated with important traits related to abiotic stress tolerance have been identified. This review provides an overview of the various genomics approaches available for crop improvement against abiotic stresses and their successes and failures in different crops. The chapter will be useful particularly for the students and scientists in genomics research and also to the larger community of researchers who have recognized the potential of genomics research and are beginning to explore the technologies involved.

Keywords Association mapping · Marker assisted selection · QTL mapping
Proteomics · Metabolomics

J. Taunk · A. Rani · R. Singh · N. R. Yadav (✉) · R. C. Yadav
Department of Molecular Biology, Biotechnology and Bioinformatics,
Chaudhary Charan Singh Haryana Agricultural University, Hisar 125004, India
e-mail: nryadav58@gmail.com

V. R. Rajpal et al. (eds.), *Genetic Enhancement of Crops for Tolerance to Abiotic Stress: Mechanisms and Approaches, Vol. I*, Sustainable Development and Biodiversity 20, https://doi.org/10.1007/978-3-319-91956-0_9

9.1 Introduction

Abiotic stress, one of the most serious constraints to global food production, are projected to worsen with the anticipated climate change. Researchers have been trying to understand and dissect the mechanisms of plant tolerance to abiotic stresses using a variety of approaches. However, success has been limited as most of these stresses are complex in nature being controlled by networks of genetic and environmental factors (Sinclair 2011). As traditional methods have reached their limits for crop improvement; there is a need to adopt novel approaches to meet the demands of an ever growing world population. Modern genomics and genetics approaches coupled with advances in precise phenotyping and breeding methodologies are expected to effectively unravel the genes and metabolic pathways that confer abiotic stress tolerance in crops. Here, we will discuss the most recent advances in the genetic and genomics strategies to unravel the complex multilayered abiotic stress tolerance mechanisms and their exploitation in crop improvement. Emphasis has been given to molecular dissection of abiotic stress tolerance by quantitative trait loci (QTL) mapping or gene discovery through linkage and

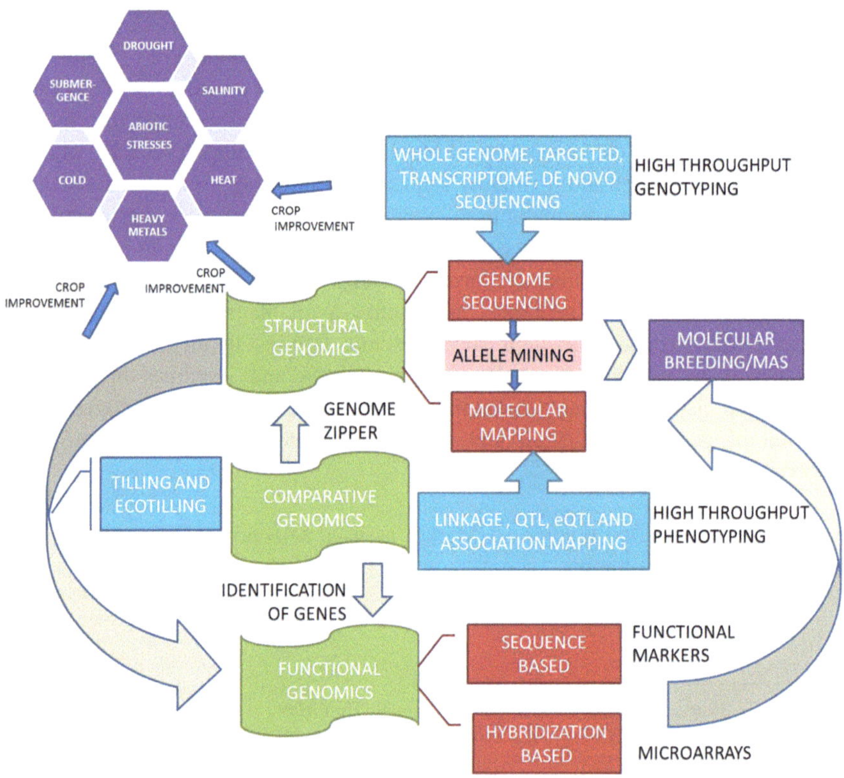

Fig. 9.1 Genomic strategies for improving abiotic stress tolerance in plants

association mapping (AM), QTL cloning, candidate gene identification, transcriptomics and functional genomics. The aim of utilizing these genomics-based approaches is to screen the entire genome including genic and intergenic regions, and to gain insights into plant molecular responses for developing effective strategies for crop improvement (Fig. 9.1).

9.2 Molecular Genetic Approaches and QTL Mapping

The identification of genes (responsible for important agricultural traits) has been mostly conducted by traditional forward and reverse genetic approaches for discrete traits and by QTL mapping for complex traits (Takeda and Matsuoka 2008). The traditional methods are important to reveal the genes that are involved in a particular function, whereas QTL mapping reveals the effects of genetic variants on complex traits, which include most agronomic traits. Molecular genetic approaches are relatively straight forward in model plants such as rice and *Arabidopsis thaliana*, where whole genome sequence information is available and transformation techniques are well established (Alonso and Ecker 2006; Jung et al. 2008). Insertion lines generated by the use of transfer DNA (T DNA) (the modified, transferable DNA of some species of bacteria) and mutant libraries (prepared by chemical mutagenesis and transposon insertions) are available in *A. thaliana* and rice. Moreover, systematic transgenic approaches have been enabled recently in *A. thaliana* and rice by using enormous complementary DNA (cDNA) collections. For example, the full-length cDNA overexpressor (FOX)-hunting system can help screening of several transgenic lines in which full-length cDNAs are overexpressed randomly (Ichikawa et al. 2006; Nakamura et al. 2007). However, transgenic approaches for both forward and reverse genetic studies are not yet practical in many crop species (such as barley, wheat and sorghum), in which gene manipulation technologies are either inefficient or not available.

The important goals of QTL mapping in plants are to increase our biological knowledge of the inheritance and genetic architecture of quantitative traits, both within a species and across related species, and to identify genetic markers that can be used as indirect selection tools in breeding (Bernardo 2008). QTL mapping relies on statistical linkage analyses among quantitative traits of interest and genetic markers, using a population such as second generation (F_2) plants, recombinant inbred lines (RILs), backcross plants (BC), near isogenic lines (NILs), and double haploids (DH). Such mapping populations are developed by crossing two inbred parents with clear contrasting differences in phenotypic traits of interest. Each mapping population has its own advantages and disadvantages and the researchers need to decide the appropriate population depending on project's objectives, available time, trait complexity and whether molecular markers to be used for genotyping are dominant or codominant (Semagn et al. 2010). QTL mapping requires (1) selection and/or development of appropriate mapping population; (2) phenotyping of the population for the trait(s) of interest (morphological characters, agronomic traits, abiotic stress resistance, etc.) under field conditions; (3) genotyping by molecular markers [entire population, selective genotyping or bulk segregant analysis (BSA)] to generate molecular data with adequate and uniformly spaced polymorphic markers; and (4) identification of molecular markers linked to the trait of interest using statistical programs. QTL mapping studies have led to the identification of QTLs in many crops for tolerance to abiotic stresses like drought (Sanchez et al. 2002; Harris et al. 2007; Landi et al. 2007), salinity (Lindsay et al. 2004; Huang et al. 2006; Fan et al. 2015), cold (Juan et al. 2010; Liu et al. 2013; Zhang et al. 2014), submergence (Xu et al. 2000; Gutterson and Reuber 2004; Xu et al. 2006) and heat (Shanmugavadivel et al. 2017).

9.2.1 QTLs for Drought Tolerance

Sorghum has been studied as a model among crop species for drought resistance, because of its adaptation to hot and dry environments. 'Stay green' trait is particularly a relevant trait conferring drought tolerance towards improvement of sorghum, characterized by delayed leaf senescence during grain ripening under water-limited conditions. Sanchez et al. (2002) identified four genomic regions associated with the stay-green trait using a RIL population developed from B35 × T × 7000. These four major stay-green QTLs were consistently identified in all field trials and accounted for 53.5% of the phenotypic variance. In another QTL-mapping study, four major QTLs (Stg1, Stg2, Stg3, and Stg4) for this trait were identified, which account for approximately 20, 30, 16 and 10% of the phenotypic variation, respectively (Harris et al. 2007). In maize, Landi et al. (2007) developed back-cross-derived lines (BDLs) homozygous either for the (+) or for the (−) allele increasing or decreasing, respectively, root size and leaf ABA concentration. They reported a major QTL for root-ABA1 (involved in root architecture, Abscisic acid (ABA) concentration and other traits according to water availability), which accounts for 32% of the phenotypic variation in ABA concentration in the leaf. They found that the (+) root-*ABA1* allele confers not only a consistently lower susceptibility to root lodging but also a lower grain yield, especially when root lodging does not occur. In case of upland rice, QTL (qtl12.1) with a large effect on grain yield under drought stress was detected on chromosome 12. Under stress conditions, the locus also increased harvest index, biomass yield, and plant height while reducing the number of days to flowering (Bernier et al. 2007), whereas in case of lowland rice, drought tolerance QTL was detected on chromosome 1 near *sd1* that explained 32% of the genetic variation for yield under stress (Kumar et al. 2007). In pearl millet, a major QTL for terminal drought tolerance has been identified on linkage group 2 (LG 2) using segregating populations derived from two independent crosses between ICMB 841 and 863B, and H 77/833-2 and PRLT 2/89-33 (Serraj et al. 2005; Bidinger et al. 2007). This QTL on LG 2 has been considered a major target for marker-assisted selcction (MAS) for improving grain yield under variable terminal stress conditions in pearl millet (Yadav et al. 2011).

9.2.2 QTLs for Salinity Tolerance

In order to identify QTL associated with salt tolerance in soybean, lines from the cross of S-100 (salt tolerant) x Tokyo (salt sensitive) were evaluated in field under saline condition. Each line was characterized using restriction fragment length polymorphism (RFLP) markers and an initial QTL single-factor analysis was completed. Finally, these results were used to saturate the selected genomic regions with simple sequence repeat (SSR) markers to improve mapping precision and to identify genomic regions associated with the desired trait. Subsequently, on LG N, a salt tolerant major QTL was identified near the Sat_091 SSR marker. The strong relationship between the SSR marker alleles and salt tolerant character suggested that these markers could be used for MAS in commercial breeding programme (Lee et al. 2004). In durum wheat, the QTL Nax1 was observed as a genetic component that confers lower Na^+ and higher K^+ concentrations in the leaf blade (Lindsay et al. 2004). By using NILs, Nax1 was shown to have a role in salt tolerance through higher levels of Na^+ exclusion from the xylem in the roots and leaf sheath, thereby reducing Na^+ concentration in the leaf blades. By comparative mapping of wheat and rice chromosomes, *HKT7-A2* (encoding a sodium transporter), was suggested to be a strong candidate gene for Nax1 (Davenport and Munns 2006; Huang et al. 2006). Fan (2015) used 72 DH lines from a cross between T X 9425 (a Chinese landrace variety with superior drought and salinity tolerance) and a sensitive variety to identify QTL for drought and salinity tolerance, based on a range of developmental and physiological traits. Two QTLs for drought tolerance (leaf wilting under drought stress) and one QTL for salinity tolerance (plant survival under salt stress) were identified from this population.

9.2.3 QTLs for Submergence Tolerance

A major QTL, Submergence1 (Sub1), was found linked to the submergence tolerance character in FR13A cultivar of the indica rice variety. This locus is a cluster of three genes (Sub1A, Sub1B, and Sub1C) that encode putative ethylene response factors (ERFs). The gene specific for submergence tolerance has been identified as *Sub1A*. Introgression of the *Sub1* genes into the widely grown Indian variety Swarna, (lacking *Sub1A*), confers strong submergence tolerance without affecting plant height, yield, harvest index and grain quality (Xu et al. 2000; Gutterson and Reuber 2004; Xu et al. 2006). Gonzaga et al. (2016) developed rice RILs derived from IR42/FR13A mapping population and identified 5 QTLs complementary to *Sub1* gene on chromosomes 1, 4, 8, 9 and 10, where four were from FR13A and one was from IR42. They found that lines without *Sub1* were still tolerant, with a maximum survival rate up to 95%. The non-*Sub1* QTLs identified have great potential to enhance tolerance as evidenced by the superior tolerance of FR13A compared with that of the developed *Sub1* lines.

9.2.4 QTLs for Heat Tolerance

High temperature (heat) stress during grain filling is a major problem in most of the wheat growing areas. Through composite interval mapping, Talukder et al. (2014) identified five QTL regions significantly associated with response to heat stress. Associations were identified for plasma membrane damage on chromosomes 7A, 2B and 1D; SPAD chlorophyll content on 6A, 7A, 1B and 1D; thylakoid membrane damage on 6A, 7A and 1D. To map the QTLs for heat tolerance in rice, Shanmugavadivel et al. (2017) used 272 F_8 RILs derived from a cross between Nagina22, a well-known heat tolerant *Aus* cultivar and IR64, a heat sensitive popular *Indica* rice variety. They identified two high effect QTLs, one novel (*qSTIPSS9*.1) and one known (*qSTIY5.1/qSSIY5*.2) for heat tolerance in rice in narrow physical intervals, which can be employed for crop improvement by MAS after development of suitable markers. In case of tomato, F_2 mapping population from two contrasting cultivars, i.e. Nagcarlang and NCHS-1, was generated and phenotyped under continuous mild heat conditions for a number of traits underlying reproductive success, i.e. pollen viability, pollen number, style length, anther length, style protrusion, female fertility and flowering characteristics like inflorescence number and flowers per inflorescence. QTLs were identified for most of these traits, including a single, highly significant one for pollen viability, which accounted for 36% of phenotypic variation in the population and modified pollen viability under high temperature with around 20% (Xu et al. 2017).

9.2.5 QTLs for Cold Tolerance

QTLs for cold tolerance have also been found by several scientists at seedling stage in rice, viz. qSPA-1 and qCTS-1 on chromosome 1 (Juan et al. 2010; Liu et al. 2013; Park et al. 2013); qCTS-2 on chromosome 2 (Lou et al. 2007; Liu et al. 2013); qCTS-12 on chromosome 12 (Andaya and Tai 2006; Zhang et al. 2014; Suh et al. 2013). For QTL mapping of cold tolerance in tomato, Liu et al. (2016) derived a population of 146 RILs from the cross between a cold sensitive cultivated *Solanum lycopersicum* L. XF98-7 and a cold tolerant wild *Solanum pimpinellifolium* LA2184. Five QTLs controlling relative germination ratio and four QTLs for chilling index were detected with genetic contribution ranging from 0.95 to 19.55%.

Most of the above mentioned studies identified QTLs that explained a significant proportion of the phenotypic variance of the desired trait, and therefore, gave rise to an optimistic assessment of the prospects of MAS and/or fine mapping. However, many studies have reported multiple small-effect QTLs on every chromosome for various abiotic stress traits (Tuberosa et al. 2002; Arriagada et al. 2017). Such QTLs, which cannot be fine mapped or used for MAS become dead end for any future investigation.

9.3 Association Mapping

Association mapping (AM) is another QTL mapping method based on linkage disequilibrium (LD). AM is one of the important applications of LD. LD refers to non-random association between two markers (alleles at different loci), between two genes or QTLs, between a gene/QTL and a marker locus (Gupta et al. 2005), while AM refers to significant association of a molecular marker with a phenotypic trait.

LD has a central role in AM. The distance over which LD persists will determine the number and density of markers, and experimental design for association analysis, therefore, it is important to determine the extent of LD in the species under investigation. There are many factors contributing to the increase of LD such as inbreeding, small population size, genetic isolation between lineages, population subdivision, low recombination rate, population admixture, genetic drift and epistasis. On the other hand, factors like high recombination rate, outcrossing, high mutation rate, gene conversion, etc., lead to a decrease/disruption in LD which have been extensively discussed in a number of papers (Weiss and Clark 2002; Gaut and Long 2003; Gupta et al. 2005; Kim et al. 2007). LD will tend to decay with genetic distance between the loci, because genetically distant loci are more likely to have recombined in the past as compared to tightly linked loci. Several studies on rates of decay of LD have been reported in various plant taxa (Flint-Garcia et al. 2003) such as *Arabidopsis thaliana* (Nordborg et al. 2005), maize (Palaisa et al. 2003), barley (Caldwell et al. 2006), sorghum (Hamblin et al. 2005) and durum wheat (Maccaferri et al. 2005) which indicate tremendous variation in the extent of LD that is mostly due to founder effect followed by genetic drift.

AM falls into two broad categories; candidate-gene AM and genome-wide AM (GWAM).The first one relates polymorphisms in selected candidate genes with phenotypic traits and second surveys genetic variation in the whole genome to find signals of association for various complex traits (Risch and Merikangas 1996).

9.3.1 Candidate Gene Association Mapping

The candidate genes are selected based on the prior information related to the location or function of the gene involved in genetic, biochemical or physiological pathways that lead to final trait variation (Risch and Merikangas 1996; Mackay 2001). Yu et al. (2013) conducted candidate gene AM of drought tolerance traits in 192 diverse perennial ryegrass (*Lolium perenne* L.) accessions from 43 countries. They identified significant associations between a putative *LpLEA3* encoding late embryogenesis abundant group 3 protein and a putative *LpFeSOD* encoding iron superoxide dismutase and leaf water content, as well as between a putative *LpCyt Cu-ZnSOD* encoding cytosolic copper-zinc superoxide dismutase and chlorophyll fluorescence under drought conditions.

Yu et al. (2015) conducted candidate gene association mapping for winter survival (WS) and spring regrowth in perennial ryegrass. They found significant association between C-repeat binding factor LpCBF1b and WS. Also, significant association of *LpLEA3* (encoding a late embryogenesis abundant group 3 protein) and *LpCAT* (encoding a catalase) with percentage of canopy green cover and chlorophyll index (Chl), and of *LpMnSOD* (encoding a magnesium superoxide dismutase) and *LpChl Cu-ZnSOD* (encoding a chlorophyll copper–zinc superoxide dismutase) with normalized difference vegetation index and Chl. Jespersen et al. (2017) subjected colonial bentgrass plants to heat stress in controlled-environmental growth chambers for phenotypic evaluation and determination of genetic variation in candidate gene expression. They developed molecular markers for genes involved in protein degradation (cysteine protease), antioxidant defense (catalase and glutathione-S-transferase), energy metabolism (glyceraldehyde-3-phosphate dehydrogenase), cell expansion (expansin) and stress protection (heat shock proteins HSP26, HSP70, and HSP101). Through Kruskal-Wallis analysis, they found that the physiological traits of chlorophyll content, electrolyte leakage, normalized difference vegetative index, and turf quality were associated with all candidate gene markers with the exception of HSP101.

Sehgal et al. (2015) phenotyped pearl millet inbred germplasm association panel (PMiGAP) for yield and yield components and morpho-physiological traits under both well-watered and drought conditions and genotyped with SNPs and insertions/deletions (InDels) from seventeen genes underlying a major validated drought tolerance QTL. They obtained significant associations for 22 SNPs and 3 InDels from 13 genes under different treatments. A SNP in putative acetyl CoA carboxylase gene showed constitutive association with grain yield, grain harvest index and panicle yield under all treatments. An InDel in putative chlorophyll a/b binding protein gene was found to be significantly associated with both stay-green and grain yield traits under drought stress. Sehgal et al. (2015) suggested that this InDel can be used as a functional marker for selecting high yielding genotypes with 'stay green' phenotype under drought stress.

9.3.2 Genome Wide Association Mapping

For whole-genome association scans, high-density genome-wide markers are required that accurately reflect genome-wide LD structure and haplotype diversity. Due to decrease in sequencing and genotyping costs, GWAM studies have grown rapidly in many crops, even for complex traits (Rostoks et al. 2005; Bastien et al. 2014; Kumar et al. 2015; Thudi et al. 2014).

Wan et al. (2017) conducted GWAM to identify salt tolerance-related QTL. They identified 75 SNPs distributed across 14 chromosomes, which were associated with 4 salt tolerance-related traits. These SNPs were integrated into 25 QTLs that explained 4.21–9.23% of the phenotypic variation in the cultivars. Additionally, 38 possible candidate genes were also identified which fell into several functional groups associated with plant salt tolerance, including transcription factors, aquaporins, transporters, and enzymes.

Liu and Yu (2017) used a diverse panel of 198 alfalfa accessions for mapping loci associated with plant growth and forage production under salt stress using GWAM. Forty-two markers were found to be significantly associated with salt tolerance, of which 13 were associated with multiple traits. Nineteen putative candidate genes were linked to 24 significant markers. Among them, B3 DNA-binding protein, Thiamine pyrophosphokinase and IQ calmodulin-binding motif protein were identified among multiple traits.

To dissect the genetic basis of heat tolerance in sorghum, Chen et al. (2017) performed GWAM for traits responsive to heat stress at the vegetative stage in an association panel. They identified 9 SNPs that were significantly associated with leaf firing and five SNPs that were associated with leaf blotching. Fourteen candidate genes were found to be directly linked to biological pathways involved in plant stress responses including heat stress.

A joint linkage–LD mapping approach that takes advantage of each approach (linkage mapping and LD mapping) (Wu et al. 2002) was used in maize for detecting QTL for drought tolerance (Lu et al. 2010). Nearly 2000 SNP markers, including 659 SNPs developed from drought-response candidate genes, were screened across 3 RIL populations and 305 diverse inbred lines (ILs). Anthesis-silking interval (ASI), an important trait for maize drought tolerance, was used for mapping. Eighteen QTL were identified, with the sum of phenotypic variation explained (PVE) increasing from 5.4 to 23.3% for single SNP-based analysis (Lu et al. 2010). This combined approach (linkage analysis and LD) for QTL analysis has been extended for fine mapping of multi-trait QTLs (Meuwissen and Goddard 2004; Gupta et al. 2005).

9.4 Marker-Assisted Selection

Following the discovery of promising QTLs and identification of molecular markers, MAS has been used to transfer single genes or QTL in various species. For instance, MAS has been very successful for development of first drought tolerant aerobic rice variety (MAS 946-1), which was released in India. Scientists at the University of Agricultural Sciences (UAS), Bangalore, crossed a deep rooted upland japonica rice variety from the Philippines with a high yielding indica variety that consumed up to 60% less water than other traditional varieties (Gandhi 2007). At International Rice Research Institute (IRRI), MAS was used for submergence tolerance using the *sub1* gene on chromosome 9. Molecular markers that were tightly linked with *Sub1*, flanking *Sub1*, and unlinked to *Sub1* were used to apply foreground, recombinant, and background selection, respectively, in backcrosses between a submergence-tolerant donor and the widely grown recurrent parent Swarna. The mega variety Swarna was efficiently converted to a submergence tolerant variety in three backcross generations within a time frame of 2–3 years

(Neeraja et al. 2007). Through MAS, four QTL alleles for deeper roots from 'Azucena' (a japonica upland cultivar that is well adapted to rainfed conditions) have been transferred into 'IR64', a rice cultivar with shallow root system. Under drought-stressed field conditions, the MAS-generated lines had a root mass greater than that of 'IR64' (Courtois et al. 2003). MAS has been widely used for improvement of rice and due to the prevalence of several rice 'mega varieties' it is likely to continue being a successful approach (Mackill et al. 2005).

Until recently, only limited genomic resources were available for legume crops, so MAS adoption has been slow (Kumar et al. 2011). However, for these crops, advances in DNA sequencing and genotyping technologies have recently delivered large-scale transcriptome sequence data sets (Kaur et al. 2012), that can be exploited for the design of DNA-based genetic markers such as SSRs and SNPs, supporting linkage mapping, analysis of genetic diversity (Fondevilla et al. 2011) as well as gene tagging for MAS (Collard et al. 2005). For physiologically complex traits such as salinity tolerance (Ashraf and Foolad 2013), molecular markers implementation has rarely been achieved. Breeders, therefore, need to select for varying and multiple genomic regions or response mechanisms found in different germplasm and different screening environments. It may therefore be necessary to quantify the adaptive nature of different QTLs according to varying salinity stress.

Molecular markers also assist in backcrossing i.e. marker assisted backcrossing (MABC). This is an effective and precise method for introgression of a single locus controlling trait of interest while retaining the essential characteristics of the recurrent parent (RP) (Collard and Mackill 2008). It is effective for QTLs with large variations in phenotype. Conventional method of plant breeding requires several backcrossing steps to enable transfer of one or a few traits to an elite cultivar while retaining most of the recurrent genome. In general, at least six backcrossing steps are required to achieve the desired homozygosity, particularly for the selection of traits with low heritability. MAS can greatly accelerate this process by utilizing both the flanking markers linked to the desired trait for selecting this particular trait and a set of unlinked markers for tracking the RP genome. Flanking markers and selection for recombination also reduces "linkage drag", which is the main cause of reduction in crop performance due to the co-transfer of undesirable traits located near the trait of interest (Collard and Mackill 2008). Furthermore, MAS enables early selection of traits that are labor and cost-intensive to score phenotypically and are under complex genetic control. This dramatically reduces the number of plants to be screened for genotyping in further steps (Collard and Mackill 2008; Edmeades et al. 2004). The three selection steps of MABC were reported: foreground selection, recombinant selection and background selection. In foreground selection, breeder selects plants having the marker allele of the donor parent at the target locus. Its objective is to maintain the target locus of trait of interest in a heterozygous state (one donor allele and one RP allele) until the final backcross is completed. Those markers tightly linked to the target gene or QTL are used to select the target locus of donor parent in early (BC) progenies for the selection of plants having the target gene. 'Foreground selection' is also referred as 'positive selection'

(Takeuchi et al. 2006). The purpose of second level i.e. recombinant selection is to reduce the size of the donor chromosome segment containing the target locus (i.e. size of the introgression). Recombinant selection is usually performed for at least two backcross generations because double recombination events occurring on both sides of a target locus are extremely rare (Hasan et al. 2015). Background selection is important in order to reduce unnecessary genes (linkage drag) introduced from donor. Its aim is to select the backcross progeny with the greatest proportion of recurrent parent genome. Background selection is also referred as 'negative selection' (Takeuchi et al. 2006). This type of selection is done by using markers that are unlinked to the target gene/QTL on all other chromosome. This is very useful approach because the RP recovery can be greatly accelerated. The use of background selection to accelerate the development of an RP with an additional one or more genes has been referred to as 'variety development or enhancement' (Mackill 2006) and 'complete line conversion' (Ribaut et al. 2002).

Marker-mediated backcrossing approach has been used to generate series of maize NILs derived from an elite recipient line (the recurrent line) and an exotic donor line (Stuber et al. 1999). For foreground and background selection, marker-facilitated selfing and marker-facilitated backcrossing were used. Two BCs and one selfing (to fix the introgressed segment) generations were sufficient to generate different NILs (each with different introgressed genomic regions). At International Crop Research Institute for Semi-Arid Tropics (ICRISAT), MAS was used in chickpea to introgress QTL alleles associated with a large root size into elite germplasm where terminal drought stress increased yield from about 20% to more than 50%. Hence, a deep root system was found capable of extracting additional soil moisture that positively impacted chickpea yield under terminal drought stress conditions (Crouch and Serraj 2002). Successful introgression of *Sub1*, derived from donor rice variety IR64, has been done into popular rice variety AS996 through MABC. Fifty-three polymorphic markers, out of 460 markers, were used for assessment on BC_1F_1, BC_2F_1 and BC_3F_1 generations. RP genome recovery was 87.5, 93.75 and 96.15% in BC_1F_1, BC_2F_1 and BC_3F_1 generations, respectively (Cuc et al. 2012). All mega- varieties [Samba Mahsuri and CR1009 from India, Thadokkham 1 (TDK1) from Laos, IR64 from the Philippines (IRRI), and BR11 from Bangladesh] with *Sub1* introgression were observed with significantly higher survival rate than the original parents by using the MABC strategy (Septiningsih et al. 2009).

'Saltol QTL' obtained from the salt tolerant rice variety FL478 has been transferred into the high-yielding and widely grown cultivar, ASS996 by MABC strategy (Huyen et al. 2012). In each backcross generations, QTL-linked markers were used for screening heterozygous plants and 63 polymorphic markers (distributed on 12 chromosomes) were used to check RP genome recovery. In this study, two plants P284 and P307 with the highest recipient alleles, up to 89.06 and 86.36%, were used to develop BC_2F_1 populations. In another study, 'Saltol QTL' derived from FL478 was introgressed in genetic background of Bacthom 7 cultivar. The background analysis in the introgression line revealed the recovery of up to 96.8–100% of RP alleles after three generations (Vu et al. 2012). NILs were developed by introgression of three drought tolerance root QTLs through MABC procedure from CT9993 (an upland japonica) into IR20, a lowland indica cultivar (Suji et al. 2012).

The major drawbacks of MAS in breeding are its high cost of implementation and the risk of recombination between the marker and the trait. These problems reduce the reliability of MAS to predict phenotype via genotype. The high cost of MAS is particularly relevant in cases where an effective phenotyping method is already established using conventional method of breeding. Despite its drawbacks, MAS has been successfully utilized to improve crops for abiotic stress tolerance, given that the genetic element responsible for the high tolerance is accurately defined and delineated.

In addition, MAS can provide specific advantages in marker-assisted pyramiding, the process of combining several genes together into a single genotype. Pyramiding may be possible using conventional method of breeding but it is very difficult to identify the plants containing more than one gene while DNA markers can greatly facilitate such selection. Marker-assisted pyramiding was successfully done for pyramiding salt tolerance QTLs in rice. Two major QTLs; one on chromosome 7 for shoot Na^+ concentration (called as qSNC-7) and another on chromosome 1 for shoot K^+ concentration (called as qSKC-1) (Lin et al. 2004), along with two others, were pyramided in a salt-tolerant variety. Three F_3 lines were observed with enhanced level of seedling survival in salt stress. A locus conferring submergence tolerance was introgressed from cultivar 'FR13A' into the variety 'Swarna' for the development of strong submergence tolerance lines (Xu et al. 2006).

The successful examples of MAS pyramiding suggest that it could facilitate transfer of combination of QTLs in elite cultivar for development of several abiotic stress tolerance varieties. Hence, the research dealing with such strategies will be able to take advantage of the results being gathered from other approaches such as omics technologies. Nevertheless, current advances in omics technologies together with advances in transgenic technology and MAS will prove useful in improving the present scenario.

9.5 Allele Mining

Allele mining utilizes the DNA sequence of a genotype to isolate useful alleles from related genotypes (Latha et al. 2004). It helps in tracing the evolution of alleles, identification of new haplotypes and development of allele specific markers for use in MAS (Kumar et al. 2010). Latha et al. (2004) used rice calmodulin gene and salt-inducible rice gene for allele mining of stress tolerance genes in related rice germplasm. Platten et al. (2013) conducted allele mining of *HKT1;5* for salinity tolerance in *Oryza sativa* and *O. glaberrima* and identified seven major and three minor alleles of *OsHKT1;5*.

9.6 Ecotype Targeting Induced Local Lesions in Genomes (EcoTILLING)

EcoTILLING is a reverse genetics approach, which utilizes mutation detection strategy to elucidate gene function and for finding desired genotypes. This approach generally characterizes SNP and/or InDels in natural mutation populations (Al-Yassin and Khademian 2015). This is a modification of TILLING strategy, which utilizes induced mutation instead of natural mutation populations (Comai et al. 2004). This technology is cost effective which allows screening of 15–20 kb large gene regions for rare mutations with high sensitivity (Coassin et al. 2008). Also, EcoTILLING could determine heterozygosity levels within a gene fragment. Utilizing EcoTILLING technique for allele mining and haplotype discovery in 9 candidate genes in barley, Cseri et al. (2011) were able to identify 185 SNPs and 46 InDels for drought tolerance. Similarly, Negrão et al. (2011) studied 375 rice accessions for salt tolerance and identified a total of 15 SNPs and 23 InDels in *OsCPK17* and *SalT* genes, respectively. Yu et al. (2012) analyzed diversity in promoter sequences of 24 transcription factor families using 95 diverse rice landraces. Association between the promoters' sequence diversity and drought tolerance index (DTI)/level (DTL), association of three genes with DTI and five genes with DTL was found. In sugarbeet, EcoTILLING revealed polymorphism in the *BvFL1* gene associated with winter hardiness (Frerichmann et al. 2013).

9.7 Functional Markers

Functional markers are derived from sequence polymorphisms found in the allelic variants of a functional gene (Salgotra et al. 2014). In contrast to random DNA markers, functional markers are completely linked to the trait of interest, therefore they are also called "perfect markers" (Akpınar et al. 2013). An et al. (2011) developed functional SSR markers on a large scale across *Brassica* species through functional annotation of publically available PlantGDB-assembled unique transcripts, which showed good transferability among *Brassica* species. Garg et al. (2012) studied the role of *TaMYB2* gene in dehydration tolerance in common wheat. They identified synonymous SNPs associated with dehydration tolerance and developed an allele specific marker for the same. Liu et al. (2012) summarized functional markers being developed and currently in use in common wheat.

9.8 Expression Quantitative Trait Locus (EQTL)

An eQTL is a chromosomal region that accounts for the proportion of the variation in abundance of a mRNA transcript observed between individuals in a genetic mapping population (Druka et al 2010). The concept was first recognized by Jansen and Nap (2001), who coined the term "genetical genomics" in which the combination of a genotyped segregating population and genome-wide expression profiling is used to formulate hypothetic regulatory pathways and unravel complex traits in a higher throughput manner. Level of gene expression varies in response to environmental changes (E) and among individuals of a species due to natural genetic variation (G). Some genes may further exhibit genetic variation in their expression in response to the environment (GxE interactions), which can be studied through eQTL mapping (Snoek et al. 2012). The eQTLs are of two types, cis and trans. When the sequence variation controlling transcript levels is assumed to be determined by the sequence variation that lies within or in the close proximity of the gene, they are of cis types whereas, in trans-eQTL, the observed location of the eQTL does not coincide with the location of the gene. Jiang et al. (2011) identified 76 eQTLs for nine cold-related traits on 12 chromosomes of rice. These eQTLs showed significant interactions of QTLs and environment. Lowry et al. (2013) explored the genetic architecture underlying expression responses to soil drying using eQTL mapping in the Tsu-1 (Tsushima, Japan) x Kas-1 (Kashmir, India) cross-based RIL population of *A. thaliana* and found some statistically significant eQTLs that interacted with soil drying treatments.

9.9 Proteomics

Proteomics is a powerful tool for investigating the molecular mechanisms of the responses of plants to stresses, and it provides a path toward increasing the efficiency of indirect selection for inherited traits (Nouri et al. 2011). Advances in protein profiling methodologies, mass spectrometry instrumentation and bioinformatics tools have paved the way for high throughput analysis. Previously, two-dimensional gel electrophoresis techniques were used (Grimplet et al. 2009; Giribaldi and Giuffrida 2010), but these are now being replaced by shotgun proteomics techniques including isobaric Tag for Relative and Absolute Quantitation (iTRAQ) and Tandem Mass Tag (TMT) (Martinez-Esteso et al. 2011; Liu et al. 2014; Li et al. 2015), or label-free quantitation methods (Cramer et al. 2013). Proteomics allow global investigation of structural, functional, abundance and interactions of proteins at a given time point and it can detect translational and post-translational regulations, thereby providing new insights into complex biological phenomena (Ghosh and Xu 2014). Proteomic studies have led to the identification of various abiotic stress-responsive proteins in a wide range of crops (Abreu et al. 2013; Barkla et al. 2013; Ngara and Ndimba 2014). Following differential expression proteomics approach in soluble chloroplast of

Arabidopsis, Uberegui et al. (2015) revealed the participation of the 'Executer pathway' in response to increased light conditions. A number of light- and genotype-responsive proteins were detected and mass-spectrometry identification showed changes in several abundant photosynthesis- and carbon metabolism-related proteins as well as proteins involved in plastid messenger RNA (mRNA) processing. Li et al. (2015) employed an isobaric tag for iTRAQ-based proteomic technique to identify the early differentially expressed proteins (DEPs) from salt-treated cotton roots and identified 128 differentially expressed proteins, 76 of which displayed increased abundance and 52 decreased under salt stress conditions. A few proteomic studies dealing with combined stress treatments have shown that plant response to a combined stress treatment is specific when compared to the individual stress factors applied separately (Kosová et al. 2015). For example, the effects of drought and salinity proteome response when compared in wheat revealed that salinity induced significant alterations in a higher number of proteins than drought as a consequence of an ionic effect of salinity stress (Peng et al. 2009). Similarly, Li et al. (2014) observed differences between the individual treatments and combined treatments when effects of a spring freezing in combination with either drought or waterlogging were studied in winter wheat. A comparison of drought and flooding in soybean seedlings revealed an increase in enzymes involved in regulation of redox homeostasis in drought-stressed plants while an increase in anaerobic metabolism-related enzymes in flooded plants (Oh and Komatsu 2015). Some proteomic studies have specifically characterized posttranslation modification in crops under abiotic stress like analysis of phosphorylation during salt and water stresses in maize (Zörb et al. 2010; Bonhomme et al. 2012; Hu et al. 2013) and characterization of protein glycosylation in soybean roots under flooding (Mustafa and Komatsu 2014).

9.10 Metabolomics

Under abiotic stress conditions, plant metabolism is disturbed either because of inhibition of metabolic enzymes, shortage of substrate, excess demand for specific compounds or several other factors. Therefore, the metabolic network is reconfigured to maintain essential metabolism and to acclimate adopting a new steady state (Obata and Fernie 2012). Metabolomics is a powerful tool, which elucidates regulation of metabolic networks along with gene functions as part of functional genomics and system biology. The term 'metabolomics' is defined as comprehensive and quantitative analysis of all small molecules in a biological system at a given developmental stage, and in a given tissue or cell type (Fiehn 2001). Amongst all—*omics* technologies, metabolomics is the most transversal and can be applied to different organisms with little or no modifications (Arbona et al. 2013). Several techniques including gas chromatography-mass spectrometry (GC-MS), liquid chromatography (LC)-MS, capillary electrophoresis (CE)-MS, nuclear magnetic resonance spectroscopy (NMR) and fourier transform infrared spectroscopy (FTIR) are commonly used in plant metabolomics research (Obata and Fernie 2012). Plant metabolomics has been used for several purposes including evaluation of the impact of stress/treatment on plant metabolism, tracking of a certain compound or compound category within a particular biosynthetic/degradation pathway and classification of samples. Using metabolic changes as a 'map' or 'marker', factors regulating metabolic movements were investigated in combination with other 'omic' analyses (Hirayama and Shinozaki 2010). Among all primary metabolites, sugars, sugar alcohols and amino acids are the most important metabolites whose concentration in plant tissues is affected by stress.

In maize, an NMR-based metabolite profiling study confirmed that early effects of salt stress are related to the osmotic component of salinity (Gavaghan et al. 2011). Non-targeted metabolomics studies revealed effect on pantothenate/CoA pathways in acclimation of plants to heat stress (Guy et al. 2008). Secondary metabolities like phenolics and carotenoids provide protection against excess light and UV irradiation, glucosinolates and alkaloid glycosides act as semiotic compounds during abiotic stresses. In response to flooding, more than 40 flavonoids in leaves of two citrus rootstock species differing in stress tolerance were identified (Munns and Tester 2008). In *Arabidopsis*, drought stress was found to induce accumulation of aliphatic glucosinolates and flavonoids but repressed accumulation of the phytoalexin camalexin (Grubb and Abel 2006). Dhuique-Mayer et al. (2009) found that the compound triterpenoidlimonin occurs in juice sacs of citrus as a result of physical damage or field freeze. Metabolomic analysis of *Punica granatum* under drought stress found 12 volatile compounds in leaf profiles, mainly aldehydes, alcohols, and organic acids. The study has evidenced a possible role of the oxylipin pathway in response to water stress (Catola et al. 2015). Studies of metabolomics conducted for different abiotic stresses have identified metabolites specific to each abiotic stress like dehydration, salinity, light, heat and low temperature (Cramer et al. 2007; Wienkoop et al. 2008; Caldana et al. 2011). Scalabrin et al. (2015) exposed *Nicotiana langsdorffii* plants, wild and transgenic for the *rol C* gene and the rat glucocorticoid receptor (*GR*) gene, to different abiotic stresses like high temperature, water deficit and high chromium concentrations. Through untargeted metabolomic analysis, they investigated the metabolic effects of the inserted genes in response to the applied stresses. The plants exposed to heat stress showed a unique set of induced secondary metabolites along with changes in lipid composition and induction of both acylsugars and glykoalkaloids. Water deficit and high chromium stresses resulted in enhanced antioxidants [for example, dihydrocoumarin-apocynin derivative (HCAs), polyamine] levels. Metabotypes or the genetic determinants of metabolic phenotypes have helped in development of biomarkers through metabotype QTL (mQTL) mapping and metabolomic genome-wide association studies (mGWAS) (Fernie and Schauer 2009). Schauer et al. (2006) analyzed metabolite profiles of tomato interspecific introgression lines between wild *Solanum pennelli* and *S. lycopersicon* cv. M82 and mapped specific fruit metabolite fingerprints to whole-plant phenotypes. In a population of nine *Arabidopsis* accessions acclimated to different environments and subjected to cold stress, it was found that particular transcript and metabolite profiles correlated with the ability to cold acclimate (Hannah et al. 2006).

9.11 Comparative Genomics

Comparative stress genomics scores various commonalities and differences in expression patterns of different genes relative to populations that differ in stress tolerance (Bressan et al. 2001). The availability of plant genomes, along with the accumulation of expression data and an increasing number of stress-related cDNA libraries, represent valuable resources for comparative genomics-based discovery of stress-related genes and pathways (Ma et al. 2012). Comparative analysis among genotypes within the same species and between species will enable us to identify species-specific genes underlying stress responses. Using this approach, stress-responsive transcription factors (TFs) were predicted in soybean, maize, sorghum, barley and wheat using a comparative analysis of known stress-responsive TFs in *Arabidopsis* and rice (Mochida et al. 2009, 2011; Tran and Mochida 2010). Taji et al. (2004) applied the full-length *Arabidopsis* cDNA microarray to reveal the differences in the regulation of salt tolerance mechanisms between a glycophyte, *Arabidopsis*, and a halophyte, salt cress. Salt cress was found to accumulate proline at much higher levels than *Arabidopsis*, which corresponded to a higher expression of AtP5CS in salt cress, a key enzyme of proline biosynthesis. Also, salt cress was found to be more tolerant to oxidative stress than *Arabidopsis*. Comparative analysis of genotype-dependent expressed sequence tags (EST) and stress-responsive transcriptome of chickpea have revealed 209 gene families and 262 genotype-specific SNPs (Ashraf et al. 2009). Using comparative genomics, Sanchez et al. (2011) analyzed the responses to salinity of three model and three cultivated species of the legume genus *Lotus*. Transcriptome analysis showed that about 60% of expressed genes were responsive to salt treatment in one or more species, but less than 1% was responsive in all.

9.12 High Throughput Sequencing

Next-generation sequencing (NGS) technology along with new complementary computational tools have intensified genome projects like whole genome re-sequencing for diversity analysis and RNA sequencing for transcriptome and non-coding RNAome analysis (Wang et al. 2013). These technologies offer several advantages compared with existing technologies such as EST sequencing and microarrays (Wang et al. 2009; Haas and Zody 2010). NGS technologies have provided important genome-wide insights on the evolution of organisms for which genomic information is lacking (Yang et al. 2015). Miao et al. (2015) revealed novel insights into mechanisms underlying abiotic stress-responsive pathways in *Medicago falcate* grown under standard, dehydration, high salinity, and cold conditions through de novo transcriptome analysis. Mechanisms underlying the metabolism and core signaling components of major phytohormones were revealed and nod factor signaling pathways modified by abiotic stresses were identified.

Further, comparison of homology between the *M. falcate and M. truncatula* transcriptomes, along with five other leguminous species, revealed a high level of global sequence conservation within the family. Wang et al. (2013) employed RNA-seq technology to characterize the de novo transcriptome of radish roots and identified 4,614 differentially expressed genes (DEGs) during lead (Pb) stress. The upregulated DEGs under Pb stress are predominately involved in defense responses in cell walls and glutathione metabolism-related processes, while down regulated DEGs were mainly involved in carbohydrate metabolism-related pathways. Also, many candidate genes, which were involved in defense and detoxification mechanisms including signaling protein kinases, transcription factors, metal transporters and chelate compound biosynthesis-related enzymes were also identified. Kohli et al. (2014) identified and characterized salt stress-responsive micro RNA (miRNA) in chickpea through Illumina Solexa sequencing. A total of 12,135,571 unique reads were obtained. In addition to 122 conserved miRNAs belonging to 25 different families, 59 novel miRNAs along with their star sequences were identified. Four legume-specific miRNAs, including miR5213, miR5232, miR2111 and miR2118, were also found. Yadav et al. (2016) identified a set of novel and known dehydration-responsive miRNAs in foxtail millet, where 32 were found to be upregulated in tolerant cultivar and 22 miRNAs were downregulated in sensitive cultivar. Identified miRNAs were found to encode various TFs and functional enzymes, indicating their involvement in broad spectrum regulatory functions and biological processes. Using RNA-Seq, Digital gene expression (DGE) and sRNA-Seq technologies, Zheng et al. (2015) performed an integrative analysis of miRNA and mRNA expression profiling and their regulatory network of tea plants under chilling (4°C) and freezing (−5°C) stress. They found that karrikins, a new group of plant growth regulators, and β-primeverosidase (BPR), a key enzyme functionally relevant with the formation of tea aroma might play an important role in both early chilling and freezing response of tea plants. More than 600 DEGs after chilling stress were revealed through transcriptome analysis of sugarcane hybrid CP72-1210 (cold susceptible) and *Saccharum spontaneum* TUS05-05 (cold tolerant). Further, to investigate the relevance of transmembrane transporter activity against abiotic stress tolerance, a *S. spontaneum* homolog of a NOD26-like major intrinsic protein gene (*SspNIP2*) was functionally analyzed, which revealed that some degree of tolerance to salt stress was conferred by *SspNIP2* (Park et al. 2015).

9.13 Conclusion

Genomic approaches significantly contribute to design breeding strategies aimed at improved crop production and yields against abiotic stress tolerance. This becomes more pertinent in the light of meeting challenges of climate change. The integration of different "omics" techniques and functional genetics can provide novel insights into genetic and biochemical aspects of cellular function and metabolic network regulation involved in abiotic stress tolerance. The full elucidation of biochemical and genetic mechanisms underlying plant stress-responsive biology depends largely on the comprehensive investigations using systematic omics. Although, different genomic approaches are very useful to build catalogs, but linking abiotic stress genes with phenotypic variation is still a challenging task. The combined power of different omics technologies and post-genomics tools are expected to accelerate the selection process and will considerably shorten the time required for the production of elite lines. Much progress has been demonstrated, but interpretation of the complex information generated and its application needs further investigation. The use of bioinformatics tools in facilitating the management of big data and its integration in genomic approaches is desirable. Collaborations among public and private crop breeding institutions, research centers and academia would play a key role in the success of improving abiotic stress tolerance in crops.

References

Abreu IA, Farinha AP, Negrão S et al (2013) Coping with abiotic stress: proteome changes for crop improvement. J Proteom 93:145–168

Akpınar BA, Lucas SJ, Budak H (2013) Genomics approaches for crop improvement against abiotic stress. Sci World J. https://doi.org/10.1155/2013/361921

Alonso JM, Ecker JR (2006) Moving forward in reverse: genetic technologies to enable genome-wide phenomic screens in *Arabidopsis*. Nature Rev Genet 7:524–536

Al-Yassin A, Khademian R (2015) Allelic variation of salinity tolerance genes in barley ecotypes (natural populations) using EcoTILLING: A review article. Am-Eurasian J Agric Environ Sci 15(4):563–572

An Z, Gao C, Li J et al (2011) Large-scale development of functional markers in *Brassica* species. Genome 54(9):763–770

Andaya VC, Tai TH (2006) Fine mapping of the qCTS12 locus, a major QTL for seedling cold tolerance in rice. Theor Appl Genet 113(3):467–475

Arbona V, Manzi M, de Ollas C et al (2013) Metabolomics as a tool to investigate abiotic stress tolerance in plants. Int J Mol Sci 14:4885–4911

Arriagada O, Mora F, Quitral Y et al (2017) Identification of QTL underlying agronomic, morphological and physiological traits in barley) under rainfed conditions using SNP markers. Acta Sci Agron. https://doi.org/10.4025/actasciagron.v39i3.32612

Ashraf M, Foolad MR (2013) Crop breeding for salt tolerance in the era of molecular markers and marker-assisted selection. Plant Breed 132:10–20

Ashraf N, Ghai D, Barman P et al (2009) Comparative analyses of genotype dependent expressed sequence tags and stress-responsive transcriptome of chickpea wilt illustrate predicted and unexpected genes and novel regulators of plant immunity. BMC Genom 10:415. https://doi.org/10.1186/1471-2164-10-415

Barkla BJ, Vera-Estrella R, Pantoja O (2013) Progress and challenges for abiotic stress proteomics of crop plants. Proteomics 13:1801–1815

Bastien M, Sonah H, Belzile F (2014) Genome wide association mapping of *Sclerotini asclerotiorum* resistance in soybean with a genotyping-by-sequencing approach. Plant Genome. https://doi.org/10.3835/plantgenome2013.10.0030

Bernardo R (2008) Molecular markers and selection for complex traits in plants: learning from the last 20 years. Crop Sci 48(5):1649–1664

Bernier J, Kumar A, Ramaiah V et al (2007) A Large-effect QTL for grain yield under reproductive-stage drought stress in upland rice. Crop Sci 47:507–516

Bidinger FR, Nepolean T, Hash CT et al (2007) Quantitative trait loci for grain yield in pearl millet under variable post flowering moisture conditions. Crop Sci 47:969–980

Bonhomme L, Benoît V, Tardieu F et al (2012) Phosphoproteome dynamics upon changes in plant water status reveal early events associated with rapid growth adjustment in maize leaves. Mol Cell Proteomics 11:957–972

Bressan RA, Zhang C, Zhang H et al (2001) Learning from the *Arabidopsis* experience. The next gene search paradigm. Plant Physiol 127:1354–1360

Caldana C, Degenkolbe T, Cuadros-Inostroza A et al (2011) High-density kinetic analysis of the metabolomic and transcriptomic response of *Arabidopsis* to eight environmental conditions. Plant J 67(5):869–884

Caldwell KS, Russell J, Langridge P et al (2006) Extreme population-dependent linkage disequilibrium detected in an inbreeding plant species, *Hordeum vulgare*. Genetics 172:557–567

Catola S, Marino G, Emiliani G et al (2015) Physiological and metabolomic analysis of *Punica granatum* (L.) under drought stress. Planta 243(2):441–449

Chen J, Chopra R, Hayes C et al (2017) Genome-wide association study of developing leaves' heat tolerance during vegetative growth stages in a sorghum association panel. Plant Genome. https://doi.org/10.3835/plantgenome2016.09.0091

Coassin S, Brandstätter A, Kronenberg F (2008) An optimized procedure for the design and evaluation of Ecotilling assays. BMC Genom 9:510–521

Collard B, Jahufer MZZ, Brouwer JB et al (2005) An introduction to markers, quantitative trait loci (QTL) mapping and marker-assisted selection for crop improvement: the basic concepts. Euphytica 142(1–2):169–196

Collard BCY, Mackill DJ (2008) Marker-assisted selection: an approach for precision plant breeding in the twenty-first century. Phil Trans R Soc B Biol Sci 363:557–572

Comai L, Young K, Till BJ et al (2004) Efficient discovery of DNA polymorphisms in natural populations by Ecotilling. Plant J 37:778–786

Courtois B, Shen L, Petalcorin W et al (2003) Locating QTLs controlling constitutive root traits in the rice population IAC 165 × Co39. Euphytica 134:335–345

Cramer GR, Ergul A, Grimplet J et al (2007) Water and salinity stress in grapevines: early and late changes in transcript and metabolite profiles. Funct Integr Genomics 7(2):111–134

Cramer GR, VanSluyter SC, Hopper DW et al (2013) Proteomic analysis indicates massive changes in metabolism prior to the inhibition of growth and photosynthesis of grapevine (*Vitis vinifera* L.) in response to water deficit. BMC Plant Biol 13:49

Crouch JH, Serraj R (2002) DNA marker technology as a tool for genetic enhancement of drought tolerance at ICRISAT, in field screening for drought tolerance in crop plants with emphasis on rice: international workshop on field screening for drought tolerance in rice, ICRISAT. ICRISAT, Patancheru, India

Cseri A, Cserhati M, Korff MV et al (2011) Allele mining and haplotype discovery in barley candidate genes for drought tolerance. Euphytica 181:341–356

Cuc LM, Huyen LTN, Hien PTM et al (2012) Application of marker assisted backcrossing to introgress the submergence tolerance QTL SUB1 into the Vietnam elite rice variety-AS996. Am J Plant Sci 3:528–536

Davenport RJ, Munns R (2006) Physiological characterization of two genes for Na^+ exclusion in durum wheat, Nax1 and Nax2. Plant Physiol 142:1537–1547

Dhuique-Mayer C, Fanciullino AL, Dubois C et al (2009) Effect of genotype and environment on citrus juice carotenoid content. J Agric Food Chem 57:9160–9168

Druka A, Potokina E, Luo Z et al (2010) Expression quantitative trait loci analysis in plants. Plant Biotechnol J 8(1):10–27

Edmeades GO, McMaster GS, White JW et al (2004) Genomics and the physiologist: bridging the gap between genes and crop response. Field Crops Res 90(1):5–18

Fan Y, Shabala S, Ma Y et al (2015) Using QTL mapping to investigate the relationships between abiotic stress tolerance (drought and salinity) and agronomic and physiological traits. BMC Genom 5:16–43

Fernie AR, Schauer N (2009) Metabolomics-assisted breeding: a viable option for crop improvement? Trends Genet 25:39–48

Fiehn O (2001) Combining genomics, metabolome analysis, and biochemical modelling to understand metabolic networks. Comp Funct Genomics 2:155–168

Flint-Garcia SA, Thornsberry JM, Edward S et al (2003) Structure of linkage disequilibrium in plants. Annu Rev Plant Biol 54:357–374

Fondevilla S, Küster H, Krajinski F et al (2011) Identification of genes differentially expressed in a resistant reaction to *Mycosphaerella pinodes* in pea using microarray technology. BMC Genom 13:12–28

Frerichmann SL, Kirchhoff M, Müller AE et al (2013) EcoTILLING in *Beta vulgaris* reveals polymorphisms in the FLC like gene BvFL1 that are associated with annuality and winter hardiness. BMC Plant Biol 13:52

Gandhi D (2007) UAS scientist develops first drought tolerant rice. The Hindu. www.thehindu.com/2007/11/17/stories/2007111752560500.htm. Verified 20 March 2009

Garg B, Lata C, Prasad M (2012) A study of the role of gene TaMYB2 and an associated SNP in dehydration tolerance in common wheat. Mol Biol Rep 39(12):10865–10871

Gaut BS, Long AD (2003) The lowdown on linkage disequilibrium. Plant Cell 15:1502–1506

Gavaghan CL, Li JV, Hadfield ST et al (2011) Application of NMR-based metabolomics to the investigation of salt stress in maize (*Zea mays*). Phytochem Anal 22:214–224

Ghosh D, Xu J (2014) Abiotic stress responses in plant roots: a proteomics perspective. Front in Plant Sci. https://doi.org/10.3389/fpls.2014.00006

Giribaldi M, Giuffrida MG (2010) Heard it through the grapevine: proteomic perspective on grape and wine. J Proteomics 73:1647–1655

Gonzaga ZJC, Carandang J, Sanchez DL et al (2016) Mapping additional QTLs from FR13A to increase submergence tolerance in rice beyond SUB1. Euphytica 209:627–636

Grimplet J, Wheatley MD, Jouira HB et al (2009) Proteomic and selected metabolite analysis of grape berry tissues under well-watered and water-deficit stress conditions. Proteomics 9:2503–2528

Grubb CD, Abel S (2006) Glucosinolate metabolism and its control. Trends Plant Sci 11:89–100

Gupta PK, Rustgi S, Kulwal PL (2005) Linkage disequilibrium and association studies in higher plants: present status and future prospects. Plant Mol Biol 57:461–485

Gutterson N, Reuber TL (2004) Regulation of disease resistance pathways by AP2/ERF transcription factors. Curr Opin Plant Biol 7:465–471

Guy C, Kaplan F, Kopka J et al (2008) Metabolomics of temperature stress. Physiol Plant 132:220–235

Haas BJ, Zody MC (2010) Advancing RNA-seq analysis. Nat Biotechnol 28(5):421–423

Hamblin M, Salas Fernandez MG, Casa AM et al (2005) Equilibrium processes cannot explain high levels of short- and medium-range linkage disequilibrium in the domesticated grass Sorghum bicolor. Genetics 171:1247–1256

Hannah MA, Wiese D, Freund S et al (2006) Natural genetic variation of freezing tolerance in *Arabidopsis*. Plant Physiol 142:98–112

Harris K, Subudhi PK, Borrell A et al (2007) Sorghum stay-green QTL individually reduce post-flowering drought-induced leaf senescence. J Exp Bot 58:327–338

Hasan MM, Rafii MY, Ismail MR et al (2015) Marker-assisted backcrossing: a useful method for rice improvement. Biotechnol Biotechnol Equip 29(2):237–254

Hirayama T, Shinozaki K (2010) Research on plant abiotic stress responses in the post-genome era: past, present and future. Plant J 61:1041–1052

Hu Y, Guo S, Li X et al (2013) Comparative analysis of salt-responsive phosphoproteins in maize leaves using Ti4+-IMAC enrichment and ESI-Q-TOFMS. Electrophoresis 34:485–492

Huang S, Spielmeyer W, Lagudah ES et al (2006) A sodium transporter (HKT7) is a candidate for Nax1, a gene for salt tolerance in durum wheat. Plant Physiol 142:1718–1727

Huyen LTN, Cuc LM, Ismail AM et al (2012) Introgression the salinity tolerance QTLs Saltol into AS996, the elite rice variety of Vietnam. Am J Plant Sci 3:981–987

Ichikawa T, Nakazawa M, Kawashima M et al (2006) The FOX hunting system: an alternative gain-of-function gene hunting technique. Plant J 48:974–985

International Rice Genome Sequencing Project (2005) The map-based sequence of the rice genome. Nature 436:793–800

Jansen RC, Nap JP (2001) Genetical genomics: the added value from segregation. Trends Genet 17(7):388–391

Jespersen D, Belanger FC, Huang B (2017) Candidate genes and molecular markers associated with heat tolerance in colonial Bentgrass. PLoS ONE 12:e0171183

Jiang W, Jin YM, Lee J et al (2011) Quantitative trait loci for cold tolerance of rice recombinant inbred lines in low temperature environments. Mol Cells 32:579–587

Juan JZ, Xiang ZY, Li ZD et al (2010) Identifiation of QTL for rice cold tolerance at plumule and 3-leaf seedling stages by using QTL network software. Rice Sci 17:282–287

Jung KH, An G, Ronald PC (2008) Towards a better bowl of rice: assigning function to tens of thousands of rice genes. Nat Rev Gene 9:91–101

Kaur S, Pembleton LW, Cogan NO et al (2012) Transcriptome sequencing of field pea and faba bean for discovery and validation of SSR genetic markers. BMC Genom 12:265–276

Kim S, Plagnol V, Hu TT et al (2007) Recombination and linkage disequilibrium in *Arabidopsis thaliana*. Nat Genet 39:1151–1155

Kohli D, Joshi G, Deokar AA et al (2014) Identification and characterization of wilt and salt stress-responsive microRNAs in chickpea through high-throughput sequencing. PLoS ONE 9 (10):e108851. https://doi.org/10.1371/journal.pone.0108851

Kosová K, Vítámvás P, Urban MO et al (2015) Biological networks underlying abiotic stress tolerance in temperate crops—A proteomic perspective Int. J Mol Sci 16:20913–20942

Kumar GR, Sakthivel K, Sundaram RM et al (2010) Allele mining in crops: prospects and potentials. Biotechnol Adv 28(4):451–461

Kumar J, Choudhary AK, Solanki RK et al (2011) Towards marker-assisted selection in pulses: a review. Plant Breed 130:297–313

Kumar R, Venuprasad R, Atlin GN (2007) Genetic analysis of rainfed lowland rice drought tolerance under naturally occurring stress in eastern India: heritability and QTL effects. Field Crops Res 103:42–52

Kumar V, Singh A, Mithra SVA et al (2015) Genome-wide association mapping of salinity tolerance in rice (*Oryza sativa*). DNA Res 22:133–145

Landi P, Sanguineti MC, Liu C et al (2007) Root-ABA1 QTL affects root lodging, grain yield, and other agronomic traits in maize grown under well-watered and water-stressed conditions. J Exp Bot 8:319–326

Latha R, Rubia L, Bennett J et al (2004) Allele mining for stress tolerance genes in *Oryza* species and related germplasm. Mol Biotechnol 27(2):101–108

Lee GJ, Boerma HR, Villagarcia MR et al (2004) A major QTL conditioning salt tolerance in S-100 soybean and descendent cultivars. Theor Appl Genet 109:1610–1619

Li W, Zhao F, Fang W (2015) Identification of early salt stress responsive proteins in seedling roots of upland cotton (*Gossypium hirsutum* L.) employing iTRAQ-based proteomic technique. Front Plant Sci 6:732. https://doi.org/10.3389/fpls.2015.00732

Li X, Cai J, Liu F et al (2014) Physiological, proteomic and transcriptional responses of wheat to combination of drought or waterlogging with late spring low temperature. Funct Plant Biol 41:690–703

Lin HX, Zhu MZ, Yano M et al (2004) QTLs for Na+ and K+ uptake of the shoots and roots controlling rice salt tolerance. Theor Appl Genet 108:253–260

Lindsay MP, Lagudah ES, Hare RA et al (2004) A locus for sodium exclusion (Nax1), a trait for salt tolerance, mapped in durum wheat. Funct Plant Biol 31:1105–1114

Liu FX, Xu WY, Song Q et al (2013) Microarray-assisted fie-mapping of quantitative trait loci for cold tolerance in rice. Mol Plant 6:757–767

Liu GT, Ma L, Duan W et al (2014) Differential proteomic analysis of grapevine leaves by iTRAQ reveals responses to heat stress and subsequent recovery. BMC Plant Biol 14:110. https://doi.org/10.1186/1471-2229-14-110

Liu XP, Yu XL (2017) Genome-Wide association mapping of loci associated with plant growth and forage production under salt stress in alfalfa (*Medicago sativa* L.). Front Plant Sci https://doi.org/10.3389/fpls.2017.00853

Liu Y, Zhou T, Ge H et al (2016) SSR mapping of QTLs conferring cold tolerance in an interspecific cross of tomato. Int J Genomics. https://doi.org/10.1155/2016/3219276

Liu Y, He Z, Appels R et al (2012) Functional markers in wheat: current status and future prospects. Theor Appl Genet 125(1):1–10

Lou Q, Chen L, Sun Z et al (2007) A major QTL associated with cold tolerance at seedling stage in rice (*Oryza sativa* L.). Euphytica 158:87–94

Lowry DB, Logan TL, Santuari L et al (2013) Expression quantitative trait locus mapping across water availability environments revealsca with genomic features in *Arabidopsis*. Plant Cell 25:3266–3279

Lu Y, Zhang S, Shah S et al (2010) Joint linkage linkage disequilibrium mapping is a powerful approach to detecting QTL underlying drought tolerance in maize. Proc Natl Acad Sci USA 107(45):19585–19590

Ma Y, Qin F, Tran LSP (2012) Contribution of genomics to gene discovery in plant abiotic stress responses. Mol Plant 5(6):1176–1178

Maccaferri M, Sanguineti MC, Noli E et al (2005) Population structure and long range disequilibrium in a durum wheat elite collection. Mol Breed 15:271–290

Mackay TF (2001) The genetic architecture of quantitative traits. Annu Rev Genet 35:303–339

Mackill DJ (2006) Breeding for resistance to abiotic stresses in rice: the value of quantitative trait loci. In: Lamkey KR, Lee M (eds) Plant breeding: the Arnel R Hallauer international symposium. Blackwell, Ames, IA, pp 201–212

Mackill DJ, Collard BCY, Neeraja CN et al (2005) QTLs in rice breeding: examples for abiotic stresses. In: Brar DS, Mackill DJ, Hardy B (eds) Proceedings of the 5th international rice genetics symposium; 2005 November 1923; International Rice Research Institute, Manila, Philippines. World Scientific Publishing, Singapore

Martinez-Esteso MJ, Casado-Vela J, Selles-Marchart S et al (2011) iTRAQ-based profiling of grape berry exo-carp proteins during ripening using a parallel mass spectrometric method. Mol BioSyst 7:749–765

Meuwissen THE, Goddard ME (2004) Mapping multiple QTL using linkage disequilibrium and linkage analysis information and multitrait data. Genet Select Evo 36(2):261–279

Miao Z, Xu W, Li D et al (2015) *De novo* transcriptome analysis of *Medicago falcata* reveals novel insights about the mechanisms underlying abiotic stress responsive pathway. Planta 16:818

Mochida K, Yoshida T, Sakurai T et al (2009) *In silico* analysis of transcription factor repertoire and prediction of stress responsive transcription factors in soybean. DNA Res 16:353–369

Mochida K, Yoshida T, Sakurai T et al (2011) *In silico* analysis of transcription factor repertoires and prediction of stress-responsive transcription factors from six major gramineae plants. DNA Res 18:321–332

Munns R, Tester M (2008) Mechanisms of salinity tolerance. Ann Rev Plant Biol 59:651–681

Mustafa G, Komatsu S (2014) Quantitative proteomics reveals the effect of protein glycosylation in soybean root under flooding stress. Front Plant Sci 5. https://doi.org/10.3389/fpls.2014.00627

Nakamura H, Hakata M, Amano K et al (2007) A genome-wide gain-of function analysis of rice genes using the FOX-hunting system. Plant Mol Biol 65:357–371

Neeraja CN, Maghirang-Rodriguez R, Pamplona A et al (2007) A marker-assisted backcross approach for developing submergence-tolerant rice cultivars. Theor Appl Genet 115:767–776

Negrão S, Almadanim C, Pires I et al (2011) Use of EcoTILLING to identify natural allelic variants of rice candidate genes involved in salinity tolerance. Plant Genet Resour 9:300–304

Ngara R, Ndimba BK (2014) Understanding the complex nature of salinity and drought-stress response in cereals using proteomics technologies. Proteomics 14:611–621

Nordborg M, Hu TT, Ishino Y et al (2005) The pattern of polymorphism in *Arabidopsis thaliana*. PLoS Biol 3:e196

Nouri MZ, Toorchi M, Komatsu S (2011) Proteomics approach for identifying abiotic stress responsive proteins in soybean. In: Sudaric A (ed) Soybean—Molecular Aspects of Breeding. ISBN: 978-953-307-240-1

Obata T, Fernie AR (2012) The use of metabolomics to dissect plant responses to abiotic stresses. Cell Mol Life Sci 69:3225–3243

Oh M, Komatsu S (2015) Characterization of proteins in soybean roots under flooding and drought stresses. J Proteom 114:161–181

Palaisa KA, Morgante M, Williams M et al (2003) Contrasting effects of selection on sequence diversity and linkage disequilibrium at two phytoene synthase loci. Plant Cell 15:1795–1806

Park IK, Oh CS, Kim DM et al (2013) QTL Mapping for cold tolerance at the seedling stage using introgression lines derived from an intersubspecifi cross in rice. Plant Breed Biotechnol 1:1–8

Park JW, Benatti TR, Marconi T et al (2015) Cold responsive gene expression profiling of sugarcane and *Saccharum spontaneum* with functional analysis of a cold inducible *Saccharum* homolog of NOD26- like intrinsic protein to salt and water Stress. PLoS ONE 10(5):e0125810. https://doi.org/10.1371/journal.pone.0125810

Peng Z, Wang M, Li F et al (2009) A proteomic study of the response to salinity and drought stress in an introgression strain of bread wheat. Mol Cell Proteom 8:2676–2686

Platten JD, Egdane JA, Ismail AM (2013) Salinity tolerance, Na+ exclusion and allele mining of HKT1;5 in *Oryza sativa* and *O. glaberrima*: many sources, many genes, one mechanism? BMC Plant Biol 13:32–48

Ribaut JM, Jiang C, Hoisington D (2002) Simulation experiments on efficiencies of gene introgression by backcrossing. Crop Sci 42:557–565

Risch N, Merikangas K (1996) The future of genetic studies of complex human diseases. Science 273:1516–1517

Rostoks N, Mudie S, Cardle L et al (2005) Genome-wide SNP discovery and linkage analysis in barley based on genes responsive to abiotic stress. Mol Genet Genomics 274(5):515–527

Salgotra RK, Gupta BB, Stewart CN Jr (2014) From genomics to functional markers in the era of next-generation sequencing. Biotechnol Lett 36(3):417–426

Sanchez AC, Subudhi PK, Rosenow DT et al (2002) Mapping QTLs associated with drought resistance in sorghum (*Sorghum bicolor* L. Moench). Plant Mol Biol 48:713–726

Sanchez DH, Pieckenstain FL, Szymanski J et al (2011) Comparative functional genomics of salt stress in related model and cultivated plants identifies and overcomes limitations to translational genomics. PLoS ONE 6(2):e17094. https://doi.org/10.1371/journal.pone.0017094

Scalabrin E, Radaelli M, Rizzato G et al (2015) Metabolomic analysis of wild and transgenic *Nicotiana langsdorffii* plants exposed to abiotic stresses: unraveling metabolic responses. Anal Bioanal Chem 407(21):6357–6368

Schauer N, Semel Y, Roessner U et al (2006) Comprehensive metabolic profiling and phenotyping of interspecific introgression lines for tomato improvement. Nat Biotechnol 24:447–454

Sehgal D, Skot L, Singh R et al (2015) Exploring potential of pearl millet germplasm association panel for association mapping of drought tolerance traits. PLoS ONE 10(5):e0122165

Semagn K, Bjørnstad A, Xu Y (2010) The genetic dissection of quantitative traits in crops. Electron J Biotechnol 5:1–45

Septiningsih EM, Pamplona AM, Sanchez DL et al (2009) Development of submergence-tolerant rice cultivars: the Sub1 locus and beyond. Ann Bot 103:151–160

Serraj R, Hash CT, Rizvi SMH et al (2005) Recent advances in marker-assisted selection for drought tolerance in pearl millet. Plant Prod Sci 8:334–337

Shanmugavadivel PS, Amitha Mithra SV, Prakash C et al (2017) High resolution mapping of QTLs for heat tolerance in rice using a 5 K SNP array. Rice (N Y) 10:28

Sinclair TR (2011) Challenges in breeding for yield increase for drought. Trends Plant Sci 16 (6):289–293

Snoek LB, Terpstra IR, Dekter R et al (2012) Genetical genomics reveals large scale genotype-by-environment interactions in *Arabidopsis thaliana*. Front Genet 3:317

Stuber CW, Polacco M, Lynn MS (1999) Synergy of empirical breeding, marker-assisted selection, and genomics to increase crop yield potential. Crop Sci 39:1571–1583

Suh JP, Cho YC, Lee JH et al (2013) SSR analysis of genetic diversity and cold tolerance in temperate rice germplasm. Plant Breed Biotechnol 1:103–110

Suji KK, Prince KSJ, Mankhar PS et al (2012) Evaluation of rice (*Oryza sativa* L.) near isogenic lines with root QTLs for plant production and root traits in rainfed target populations of environment. Field Crops Res 137:89–96

Taji T, Seki M, Satou M et al (2004) Comparative genomics in salt tolerance between *Arabidopsis* and *Arabidopsis*-related halophyte Salt cress using *Arabidopsis* microarray. Plant Physiol 135:1697–1709

Takeda S, Matsuoka M (2008) Genetic approaches to crop improvement: responding to environmental and population changes. Nature Rev Genet 9:444–457

Takeuchi N, Ebitani T, Yamamato T et al (2006) Development of isogenics of rice cultivar Koshihikari with early and late heading by marker-assisted selection. Breed Sci 56:405–413

Talukder SK, Babar MA, Vijayalakshmi O et al (2014) Mapping QTL for the traits associated with heat tolerance in wheat (*Triticum aestivum* L.). BMC Genet 15:97

Thudi M, Upadhyaya HD, Rathore A et al (2014) Genetic dissection of drought and heat tolerance in chickpea through genome-wide and candidate gene-based association mapping approaches. PLoS ONE 12:e0175609

Tran LS, Mochida K (2010) A platform for functional pre-diction and comparative analyses of transcription factors of legumes and beyond. Plant Signal Behav 5:550–552

Tuberosa R, Salvi S, Sanguineti MC et al (2002) Mapping QTLs regulating morphophysiological traits and yield: case studies, shortcomings and perspectives in drought stressed maize. Ann Bot 89(7):941–963

Uberegui E, Hall M, Lorenzo Ó et al (2015) An *Arabidopsis* soluble chloroplast proteomic analysis reveals the participation of the Executer pathway in response to increased light conditions. J Exp Bot 66(7):2067–2077

Vu HTT, Le DD, Ismail AM et al (2012) Marker-assisted backcrossing (MABC) for improved salinity tolerance in rice (*Oryza sativa* L.) to cope with climate change in Vietnam. Aus J Crop Sci 6(12):1649–1654

Wan H, Chen L, Guo J et al (2017) Genome-Wide Association Study Reveals the Genetic Architecture Underlying Salt Tolerance-Related Traits in Rapeseed (*Brassica napus* L.). Front Plant Sci 8:593

Wang Y, Xu L, Chen Y et al (2013) Transcriptome profiling of radish (*Raphanus sativus* L.) root and identification of genes involved in response to lead (Pb) stress with next generation sequencing. PLoS ONE 8(6):e66539. https://doi.org/10.1371/journal.pone.0066539

Wang Z, Gerstein M, Snyder M (2009) RNA-Seq: a revolutionary tool for transcriptomics. Nature Rev Genet 10:57–63

Weiss KM, Clark AG (2002) Linkage disequilibrium and the mapping of complex human traits. Trends Genet 18:19–24

Wienkoop S, Morgenthal K, Wolschin F et al (2008) Integration of metabolomic and proteomic phenotypes: analysis of data covariance dissects starch and RFO metabolism from low and high temperature compensation response in *Arabidopsis thaliana*. Mol Cell Proteomics 7 (9):1725–1736

Wu R, Chang-Xing MA, Painter I et al (2002) Simultaneous maximum likelihood estimation of linkage and linkage phases in outcrossing populations. Theor Popul Biol 61(3):349–363

Xu K, Xia X, Takeshi F et al (2006) Sub1A is an ethylene-response-factor like gene that confers submergence tolerance to rice. Nature 442:705–708

Xu K, Xu X, Ronald PC et al (2000) A high-resolution linkage map of the vicinity of the rice submergence tolerance locus Sub1. MGG 263:681–689

Xu J, Driedonks N, Rutten MJM et al (2017) Mapping quantitative trait loci for heat tolerance of reproductive traits in tomato (*Solanum lycopersicum*). Mol Breed 37:58

Yadav A, Khan Y, Prasad M (2016) Dehydration-responsive miRNAs in foxtail millet: genome-wide identification, characterization and expression profiling. Planta 243(3):749–766

Yadav RS, Sehgal D, Vadez V (2011) Using genetic mapping and genomics approaches in understanding and improving drought tolerance in pearl millet. J Exp Bot 62:397–408
Yang Y, Li X, Kong X et al (2015) Transcriptome analysis reveals diversified adaptation of *Stipa purpurea* along a drought gradient on the Tibetan Plateau. Funct Integr Genomics 15(3):295–307
Yu S, Liao F, Wang F et al (2012) Identification of rice transcription factors associated with drought tolerance using the Ecotilling method. PLoS ONE 7(2):e30765. https://doi.org/10.1371/journal.pone.0030765
Yu X, Bai G, Liu S et al (2013) Association of candidate genes with drought tolerance traits in diverse perennial rye grass accessions. J Exp Bot 64:1537–1551
Yu X, Pijut PM, Byrnec S et al (2015) Candidate gene association mapping for winter survival and spring regrowth in perennial rye grass. Plant Sci 235:37–45
Zhang S, Zheng J, Liu B et al (2014) Identifiation of QTL for cold tolerance at seedling stage in rice (*Oryza sativa* L.) using two distinct methods of cold treatment. Euphytica 195:95–104
Zheng C, Zhao L, Wang Y et al (2015) Integrated RNA-Seq and sRNA-Seq analysis identifies chilling and freezing responsive key molecular players and pathways in tea plant (*Camellia sinensis*). PLoS ONE 10(4):e0125031. https://doi.org/10.1371/journal.pone.0125031
Zörb C, Schmitt S, Mühling KH (2010) Proteomic changes in maize roots after short-term adjustment to saline growth conditions. Proteomics 10:4441–4449

Chapter 10
Genomics of Arsenic Stress Response in Plants

Smita Kumar and Prabodh Kumar Trivedi

Abstract Heavy metal stress severely affects plant growth, development and reduces crop yield and productivity. Among different heavy metals, arsenic (As) is a toxic metalloid and a potent carcinogen. It not only hampers plant development but also causes severe health hazards to mankind once entered into the food chain. Infact, As contamination in the groundwater has turned out to be an epidemic in many regions of South and Southeast Asia. Naturally, As is present in trace amounts in the environment, however, geogenic sources and anthropogenic activities have tremendously increased the level of As in the soil. Epidemiological studies have reported that As poses severe health risk in humans. In plants, As affects several physiological and molecular processes, therefore, it is prerequisite to understand As uptake, translocation, accumulation and detoxification. As a part of detoxification mechanism, As undergoes chemical modifications including reduction, methylation, and glutathione conjugation. Alleviation of As phytotoxicity is important and attempts are being made in exploring the molecular components associated with As detoxification and tolerance in plants. In this context, it is important to understand the genetic control of As uptake and accumulation, which might help in protecting the food crops from contamination. In the past decade, significant knowledge has been generated at the level of "omics" which includes genomics, proteomics and metabolomics. Studies have demonstrated that plants' response to stress is associated with profound changes at the level of transcriptome. Modulation in the expression of genes involved in plants' stress response significantly assists in unravelling the pathways and networks providing tolerance towards stress. The information available through transcriptome studies led to the functional characterization of genes and development of plant varieties resistant to As stress. This chapter summarizes the current knowledge on As contamination, transcriptional regulation and biotechnological advances in the functional genomics of As uptake, transport, accumulation and detoxification in plants.

S. Kumar · P. K. Trivedi (✉)
CSIR-National Botanical Research Institute, Council of Scientific and Industrial Research (CSIR-NBRI), Rana Pratap Marg, Lucknow 226001, India
e-mail: prabodht@nbri.res.in

V. R. Rajpal et al. (eds.), *Genetic Enhancement of Crops for Tolerance to Abiotic Stress: Mechanisms and Approaches, Vol. I*, Sustainable Development and Biodiversity 20, https://doi.org/10.1007/978-3-319-91956-0_10

Keywords Accumulation · Arsenic · Detoxification · Genomics
Methylation · Phytotoxicity · Transcriptional regulation

10.1 Introduction

Arsenic is a ubiquitous element and is present naturally in the bedrock, soil, water and plants. Arsenic is further released into the environment through natural processes such as volcanic eruption, forest fires and erosion of rocks (Kumar et al. 2015a). Unrestricted manufacture of industrial chemicals and increased use of arsenicals in the pesticides or defoliants, the fallout from the ore-smelting operations, mining, and coal power stations have tremendously contaminated the soil and groundwater (Neumann et al. 2010; Zhao et al. 2010). Notably, As levels tend to be high in the ground water sources as compared to the surface water sources of drinking water such as lakes and rivers (Christen 2001). Unfortunately, As contamination in drinking water has globally threatened the lives of millions of people (Fendorf et al. 2010; Rodriguez-Lado et al. 2013). Many countries including Bangladesh, Cambodia, China, India and Vietnam have been reported to contain 10–100 fold levels of As in comparison to uncontaminated areas (Abedin et al. 2002; Santra et al. 2013). Arsenic is extremely toxic elemental pollutant endangering human health and ecological integrity. In nature, As occurs in both inorganic and organic forms and has diverse chemical behaviour with different oxidation states (Tripathi et al. 2007; Kumar et al. 2015a). Organic forms, which include arsenobetaine, arsenocholine, tetramethylarsonium salts and arsenosugars are relatively less toxic and occurs in ocean fish and sea food (Newcombe et al. 2010). Historically, organic arsenicals were used as antimicrobial agents to treat infectious diseases (Jones 2007). Inorganic As forms, mainly trivalent and pentavalent oxidation states such as arsenite As(III) and arsenate As(V), are more toxic than the organic forms monomethylarsonic acid (MMA), dimethylarsinic acid (DMA), trimethylarsinic acid (TMA), and trimethylarsine oxide (TMO). The symptoms of As exposure are subtle in onset and are directly related to the degree of dose and duration of exposure (Bernstam and Nriagu 2000). Numerous health hazards such as keratosis, melanosis, bladder, skin, lung and prostate cancers have been reported in humans due to chronic As exposure (Banerjee et al. 2013). In humans, As biotransformation takes place in the liver and the inorganic form is methylated in the body by alternate reduction of As(V) to As(III) and the addition of a methyl group from S-adenosylmethionine (Vahter 2002).

The element is immutable, and cannot be degraded by biological processes. Arsenic chemistry has indicated that As(V) ion is alike orthophosphate. In plants, As(V) enters via phosphate transporters and thus alters phosphate metabolism and interrupts phosphorylation reactions (Meharg and Macnair 1992). Under anaerobic conditions, As(III) is predominant, whereas As(V) is mainly present under aerobic conditions. As(III) reacts with the sulfhydryl group of proteins and leads to inactivation of enzymes (Mosa et al. 2012). Arsenic causes growth inhibition, chlorophyll degradation, nutrient deficiency and induces the production of Reactive Oxygen Species (ROS) leading oxidation of proteins and membrane lipids which causes severe membrane damage (Ahsan et al. 2008; Kumar and Trivedi 2016a; Shukla et al. 2015). To subsist As-induced oxidative burst, plants have evolved biochemical strategies, which enhance the activity of enzymes involved in the antioxidant responses (Shri et al. 2009). In plants, the role of sulphur metabolism in As detoxification has been implicated (Dixit et al. 2015a, b, 2016). Arsenic is chelated to sulphur containing compounds; phytochelatins (PCs) and glutathione (GSH), and gets sequestered into the vacuoles. The sulphate transporters and the genes associated with sulphur assimilation and GSH biosynthesis help in maintaining the sulphur pool inside the cell (Kumar et al. 2011; Khare et al. 2017; Nocito et al. 2006). *In silico* analysis of rice sulphate transporter gene family suggests the role of sulphate transporters in conferring tolerance towards an array of stresses including As stress (Kumar et al. 2011). In addition, comprehensive analysis of the regulatory elements of the rice sulphate transporter promoters and the functional characterization of *OsSul1;1* promoter also affirms the role of sulphate transporter in As stress response (Kumar et al. 2015b).

In recent years, several studies have described As metabolism in different plant species. However, significant research has been focussed on rice as it accumulates more As compared to other crops due to its physiological properties and anaerobic growing conditions (Abedin et al. 2002; Kumar and Trivedi 2016a; Zhao et al. 2010). Under normal conditions, rice takes up large amount of silicon (Si) from the soil, which is used to strengthen plant parts including stems and husks. Also, Si is important in protecting grains against pathogens. Chemically, As and Si are very similar and so the efficient uptake pathway of Si in rice allows the inadvertent uptake of As (Ma et al. 2008). Rice is also the dietary staple for almost half of the world's population. Therefore, it is a prerequisite to avidly understand and unravel the complex chemistry of assimilation and metabolization of As for the development of mitigation strategies to reduce As uptake in plants.

In recent years, several groups have carried out extensive studies and identified genes involved in As uptake, accumulation and detoxification (Tuli et al. 2010). Comparative studies using contrasting germplasm have identified myriad of genes involved in conferring tolerance towards As stress. Association studies and QTL mapping using natural variation have also opened up a new avenue for the identification and characterization of genes crucial in As stress response. Exploration of transcriptional regulation via gene expression profiling, and deep transcriptome sequencing have created a reservoir of information of the genes and pathways involved in As stress response in different plant species. In addition, functional genomics approaches have assisted in developing transgenic plants resilient towards As stress. In this chapter, we aim to provide an overview of the genomics of As stress response and the diverse molecular strategies that have been employed to develop As resistant crops with the purpose to enhance the sustainable growth of plants and improve global health.

10.2 Arsenic Uptake and Accumulation in Plants

Due to geogenically elevated As levels and phytotoxicity, the environmental fate and behaviour of As have gained attention in recent years. Many regions throughout the world rely on the As contaminated groundwater for the irrigation of staple crops including paddy rice. Consequently, there is a disturbing condition of As accumulation in the rice plants grown in these areas and exposure of people to As not only through drinking water but also through vegetation (Zhao et al. 2010). In the recent past, significant advancement has been made in understanding the uptake, accumulation and transport of As in the plants using high throughput technologies and functional genomics. The different forms of As and their uptake mechanism have been described briefly below.

10.2.1 Inorganic As

During the uptake of As from soil, different nutrients such as Iron (Fe), Phosphorus (P), Silicon (Si) and Sulphur (S) interact with it (Zhao et al. 2010). Chemically, As (V) is analogous to phosphate and exhibits similar physicochemical behaviour in soil and thus competes with it for entry inside the plant cell (Abedin et al. 2002; Catarecha et al. 2007; González et al. 2005; Shin et al. 2004). As(V) has an affinity for the iron oxides/hydroxides in soil and so its concentration is less. It alters the phosphorylation reactions and thus severely affects phosphate metabolism in plants. As(III), which is the most toxic and predominant form in the reducing environment, enters into the plants by a number of aquaporin nodulin26-like intrinsic proteins (NIPs) and binds to the sulfhydryl groups of proteins disrupting their structure and catalytic function (Bienert et al. 2008; Zhao et al. 2010). It renders enzymes inactive and thus increases toxicity. The undissociated methylated As species are also taken up by the NIP aquaporin channels (Li et al. 2009). In the past decade, understanding at the level of genomics and transcriptional regulation of As uptake and metabolism in plants has made headway.

In plants, phosphate transporter gene family has been well described and a number of phosphate transporters from different plant species have been functionally characterized by overexpression and mutant analysis. *Arabidopsis* and rice genome encodes 13 and 9 phosphate transporter genes, respectively (Mudge et al. 2002; Paszkowski et al. 2002). Among other phosphate transporters, Pht1;1 and Pht1;4 express in root epidermis and function in phosphate acquisition from both low and high phosphorus environments (Shin et al. 2004). These transporters significantly mediate As(V) uptake inside the plants. Studies have reported that the suppression of phosphate transporters subsequently decreases As(V) uptake (Logoteta et al. 2009; Shukla et al. 2015). Although progress has been made in identifying the genes through genomics, the genetics of As uptake and transport has not been studied in detail in plants. Screening of number of inbred lines and QTL mapping has identified an As tolerant gene on chromosome 6 in rice (Dasgupta et al. 2004). It would be intriguing to understand the mechanistic details underlying relative selectivity of phosphate and As(V) using natural variation in different plant species including rice and *Arabidopsis*. Therefore, quantitative and qualitative data need to be generated using diverse germplasms available globally.

In rice, Nodulin 26-like Intrinsic Proteins (NIPs) have been reported to be involved in As(III) uptake. OsNIP2;1 and OsNIP3;2, bi-directional As(III) channels, have been shown to be involved in transport As(III) across the membrane (Bienert et al. 2008). Apart from NIPs, As(III) is also taken up by the Si uptake pathway in rice. Mutant studies have shown that mutations in *Lsi1* and *Lsi2* significantly decreased As(III) uptake and transport to xylem, respectively (Ma et al. 2008). Recently, a novel aquaporin tonoplast intrinsic protein (TIP), *PvTIP4*, was found to take up As(III) in *Pteris vittata* (He et al. 2016). These studies have been useful in depicting the influx and efflux systems of As, which are the key components in causing toxicity in plants. Further studies on such types of channels would assist in developing strategies for improving tolerance and remediation by plants.

10.2.2 Arsenic Methylation and Volatilization

Studies suggest that several microorganisms transform inorganic species of As to organic species and vice versa by the process of methylation or demethylation (Bentley and Chasteen 2002). Genome sequencing of bacteria and archaea have identified the presence of As resistance (ars) operons, which provide tolerance towards As stress. In a study, a soil bacterium *Arsenicibacter rosenii* gene, *arsM,* was shown to encode protein for As methylation and volatilization (Huang et al. 2016). Expression of *Rhodopseudomonas palustris arsM* gene in *E. coli* increased tolerance towards As(III) stress due to formation of methylated species from As (Qin et al. 2006). This microbial-mediated conversion of As to less toxic species may contribute to the cycling of As wordwide. In a different study, *arsM* from the bacterium *R. palustris* was expressed in rice, which depicted increased As volatilization with overall low As methylation efficiency (Meng et al. 2011). Further, studies have reported that overexpression of As(III)-S-adenosyl methyl transferase (arsM) methylates As and maintains low As levels in the grains, and organic As species methylarsonic acid (MMA(V) and dimethylarsinic acid (DMA (V) in the roots and shoots of the transgenic rice in comparison to wild type plants (Meng et al. 2011). In addition, overexpression of Met synthase and AdoMet synthetase has been shown to enhance As(III) methylation and could be a strategy for the phytoremediation. Recent study by Zhang et al. (2017) revealed that recombinant Rhiobium-legume symbiont expressing *Chlamydomonas reinhardtii* arsM gene methylates and subsequently volatilizates As more efficiently. In addition, genetically engineered *Pseudomonas putida* with the alga *Chlamydomonas reinhardtii* arsM gene (*CrarsM*) depicted methylation and volatilization of As by the engineered bacterium (Chen et al. 2013a, b). In a study, overexpression of an arsM gene from the thermophilic alga *Cyanidioschyzon merolae* in *Bacillus subtilis* showed As volatilization (Huang et al. 2015). Above studies by different groups suggest an important role of legume-rhizobia symbionts and recombinant bacterium in As bioremediation. Contrary, higher plants lack the potential of methylation of inorganic As (Lomax et al. 2012). It has been reported that uptake efficiency of inorganic As species in the roots is higher as compared to the methylated species, however, the translocation efficiency of methylated As species is higher in the shoots (Raab et al. 2007).

10.3 Transcriptional Regulation of As Uptake and Accumulation

With the advent of global gene expression profiling including microarray and next-generation sequencing, number of studies have been carried out to understand the differential gene expression pattern in As stress response (Kumar and Trivedi 2016b; Tripathi et al. 2012). Studies have reported that As stress alters metabolic pathways and affects molecular and physiological processes in plants (Chakrabarty et al. 2009; Norton et al. 2008). A number of defense and stress-responsive genes including metallothioneins, heat shock proteins, multi drug resistance proteins have been found to be up-regulated in plants grown under As stress. In addition, many transporters such as sulphate transporter, multidrug and toxic compound extrusion (MATE) transporters, glutathione-conjugated transporters, and metal transporter including Natural Resistance Associated Macrophage Protein 1 (NRAMP1) were up-regulated under As(V) stress in rice (Chakrabarty et al. 2009). Also, genome-wide transcriptome and miRNA analyses under As(III) stress in rice have shown up-regulation of transporters, transcription factors, and genes related to lipid metabolism and phytohormone pathways (Yu et al. 2012; Sharma et al. 2015). As(V) stress induces the expression of antioxidant system including superoxide dismutase, transcription factors, vacuolar proteins and genes associated with plant growth and development (Abercrombie et al. 2008). Moreover, differential expression of genes is observed in different rice genotypes in response to As stress (Rai et al. 2011). In a study, large-scale screening of rice germplasm showed contrasting responses towards As sensitivity. This study also revealed that expression of genes associated with sulphur assimilation pathway and antioxidant defence system are modulated in the tolerant and sensitive rice cultivars (Rai et al. 2011).

Differential expression of rice sulphate transporter gene family has been observed in response to As stress in different rice cultivars (Kumar et al. 2011). Interestingly, a unique pattern of alternative splicing for one member of high affinity sulphate transporter *OsSul1;1* was observed in response to As stress in rice (Kumar et al. 2011). These observations significantly advance our knowledge of the post-transcriptional regulatory mechanisms operating to regulate sulphur demand by the plants under As stress. In addition, a comparative biochemical and transcriptional analysis of contrasting varieties of *Brassica juncea* showed differential regulation of the sulphate transporters and phytohormone pathways (Srivastava et al. 2009). Furthermore, expression analysis of *Crambe abyssinica* in response to As(V) stress identified a set of novel genes and regulatory networks under As stress. Differential expression profiling of the genes encoding members of GST gene family and those associated with defence and sulphur metabolism, transporters, heat shock proteins as well as encoding enzymes of the ubiquitination pathway of protein degradation was observed (Paulose et al. 2010). These studies have opened up the possibility of unravelling the genetic architecture and the targeted identification of molecular pathways involved in As stress response. Further studies will help in underpinning the As tolerance mechanisms. Remarkably, the integrated studies utilizing genomics, proteomics and metabolomics intend to elucidate complex mechanisms, which can bridge the gap between current knowledge about As stress resistance in plants. The transcriptional regulation of As uptake and translocation has now well defined that As tolerance is regulated by several genes and pathways. The identified genes enable plant to withstand As stress, and thus are prospective candidates for the wide application in crop breeding. Apart from this, in the past few years, several mitigation processes including genetic modifications have been employed to reduce As uptake and enhance detoxification mechanism in plants to increase agronomic productivity.

Apart from As stress, nutrient deficiency leads tomodulation in several physiological and biological processes and causes reduction in productivity and yield (Kumar et al. 2017). Recently, studies have been carried out to understand the response of natural variation under combined stress of As and nutrient deficiency. The expression analysis of the genes related to regulation and detoxification of As (V) and As(III) stress under limiting phosphate and sulphate conditions, respectively, were studied (Shukla et al. 2015; Khare et al. 2017). Results suggested that the genetic variation-dependent regulatory mechanisms might be the plausible reason for the differential response of *Arabidopsis* natural variants towards As stress under limiting nutrient conditions. These studies have offered a deeper understanding of the genes and the pathways involved in providing As resilience and the genetic basis of *Arabidopsis* response to nutrient deficiency and heavy metal stress.

10.4 Detoxification Mechanism of As in Plants

Arsenic intoxication affects biochemical, physiological and molecular processes in plants. To overcome As stress, plants have developed a sophisticated mechanism of detoxification. Studies have unravelled detoxification mechanisms in mammals (Aposhian 1997), fungi and algae (Cullen and Reimer 1989), suggesting methylation and biotransformation including incorporation of As into organic molecules such as arsenocholine, arsenobetaine and arsenosugars. An enzyme arsenate reductase converts As(V) into As(III), which is then extruded from the plant cell via specific As(III) efflux transporters (Pickering et al. 2000). *ACR2* from yeast *Saccharomyces cerevisiae* was reported to provide resistance towards As stress (Mukhopadhyay et al. 2000). Similarly, heterologous expression of bacterial arsenate reductase (ArsC) in *Arabidopsis thaliana* conferred tolerance towards As stress and enhanced its accumulation (Dhankher et al. 2002). Studies have functionally characterized arsenate reductases from different plant species including *Arabidopsis* (Dhankher et al. 2006), rice (Duan et al. 2007), *Holcus lanatus* (Bleeker et al. 2006), and *Pteris vittata* (Ellis et al. 2006). Genome-wide association (GWA) mapping of loci has identified a gene in *Arabidopsis,* termed as High Arsenic Content 1 (HAC1), which functions as arsenate reductase and regulates As accumulation (Chao et al. 2014; Sánchez-Bermejo et al. 2014). Very recently, a new gene *HAC4* has been identified, which encodes a rhodanase-like protein having As(V) reductase activity. Overexpression of *OsHAC4* enhanced tolerance towards As(V) stress and reduced As accumulation in rice. In contrast, mutation in *OsHAC4* showed decreased As(V) reduction in roots, and As(III) efflux to the external medium with remarkably enhanced accumulation of As in rice shoots (Xu et al. 2017). Thus, such new genes showing As tolerance have emerged as key players in detoxification and regulation of As accumulation in plants.

Studies have shown that As stress induces antioxidant system and increases the activities of the isozymes superoxide dismutase, ascorbate peroxidase, peroxidase and glutathione reductase (Rai et al. 2011; Shri et al. 2009). Number of genes associated with glutathione synthesis and metabolism have been found to be up-regulated under As stress. Differential expression of members of GST gene family is found in response to As stress (Kumar et al. 2013a). Also, overexpression of one member of Lambda class GST showed tolerance towards abiotic stress including As stress (Kumar et al. 2013b). As a part of detoxification mechanism, As induces the production of sulphur containing molecules such as GSH and PCs (Srivastava et al. 2007; Schulz et al. 2008). These PCs are heavy-metal-binding peptides derived from GSH and have the general structure (γ-Glu-Cys) n-Gly (n=2–11). Their biosynthesis is due to the transpeptidation of γ-glutamylcysteinyl dipeptides from GSH by the action of a constitutively present PC synthase. Other heavy metals including Cd^{2+}, Cu^{2+}, Ag^{+}, Hg^{2+}, and Pb^{2+} also activate PC synthase. Subsequently, these ions are complexed by the induced PCs via thiolate coordination. Inhibition of PC synthesis by treatment with a potent inhibitor of γ-glutamylcysteine synthetase, L-buthionine-sulphoxime (BSO), causes hypersensitivity towards As stress (Schat et al. 2002). Studies have been performed to understand As chelation by cysteine-rich peptides and intracellular compartmentalization. Expression of *CdPCS* from aquatic macrophyte *Ceratophyllum demersum* in tobacco has been observed to enhance PC content, precursor non-protein thiols, and increased As and Cd accumulation without significant decrease in the plant growth (Shukla et al. 2012). Similarly, heterologous expression of phytochelatin synthase gene *(AtPCS1)* in tobacco showed increased As and Cd accumulation and detoxification in the transgenic plants due to enhanced levels of phytochelatins (Zanella et al. 2016). Interestingly, in a study, synthetic PCs have been proposed as potential candidates for enhancing the metal accumulation capacity of the plants (Shukla et al. 2013).

In terms of As translocation from roots to shoots, it is observed that As(V) has low mobility as compared to phosphate and also most of the As(V) is converted to As(III) by arsenate reductase. This As(III) is further chelated by PCs and sequestered into the vacuoles (Dhankher et al. 2006). Hence, the predominant form of As in the xylem sap is the As(III), which is not complexed and sequestered into the root vacuoles. Therefore, for different plant species, variation in the ratio of the As present in the xylem sap and the external medium has been observed. Studies have determined that the translocation ratio is less than 1 in the non-hyperaccumulating plants, whereas it is more than 1 in the hyperaccumulating plants such as *Pteris vittata* (Wang et al. 2002).

10.5 Arsenic Hyperaccumulating Plants

The metal(oid) hyperacccumulators are efficient in extracting As from the soil and translocating it to the aboveground parts. Arsenic hyperaccumulators have developed different strategies to accumulate large amount of As in their roots and fronds (Ghori et al. 2016). Several plant species have been identified to accumulate As, but generally the hyperaccumulators belong to the family *Pteridaceae*. *Pteris vittata* has been explored in detail to study the efficient system of uptake, translocation and sequestration of As in vacuoles (Danh et al. 2014). Recently, phytate-induced As uptake and growth of *P. vittata* was reported (Liu et al. 2017). Phytate, which is a large pool of unavailable phosphorous in the soil, acts as an energy source for the seeds. It is predominant in the root exudates of *P. vittata* and has been observed to enhance As and phosphorous uptake. Thus, it could be proposed that the use of phytate or similar molecules might help in developing strategies for the efficient phytoremediation of As contaminated soils (Liu et al. 2017). It has been reported that hyperaccumulators have low ROS and possess a strong antioxidant system. Vacuolar sequestration of As, the key detoxification mechanism in hyperaccumulators also persists in the non-hyperaccumulating plants, however, As accumulated in the roots and fronds of fern is chelated with PCs (Zhang et al. 2004). Several studies have been carried out to understand the detoxification mechanisms of hyperaccumulating plants and to use it for the purpose of bioremediation. The isolation and characterization of two *P. vittata* genes *ACR3* (*Arsenical Compound Resistance3*) and *ACR3;1* encoding As(III) transporter has been carried out. It was complemented in the yeast strain defective in the function of *ACR3* gene, which suggested its role as As(III) effluxer (Indriolo et al. 2010). Overexpression of glutaredoxin gene, *PvGRX5,* from the hyperaccumulating brake fern in *Arabidopsis* enhanced tolerance towards As stress (Sundaram et al. 2009). In addition, hyperaccumulating plants show a minimal level of efflux of As from the roots to the external medium. These peculiar characteristics of hyperaccumulators depict their capability in remediating As contaminated soils.

10.6 Biotechnological Advances in Developing As Resistant Plants

The conventional remediation technologies including physical, chemical and thermal processes have been utilized to remediate an As contaminated area. However, these are costly, time-consuming, and harmful to the workers and generate secondary wastes that may be hazardous. In the past two decades, important advances have been made in understanding alternative approaches for remediation of the heavy metal contaminated soils (Bakhat et al. 2017). Among other bioremediation techniques, phytoremediation has gained momentum, which employs plants to reduce the concentration of As in the soils. In contrast to the conventional technologies, phytoremediation is environmentally-friendly, non-intrusive, prevents landscape destruction and increases soil microorganisms activity and diversity (McGuinness and Dowling 2009). Studies have identified that among other phytoremediation techniques, phytostabilization, which reduces the bioavailability of As in soils, and phytoextraction that involves the cultivation of tolerant plants are the most suitable and effective processes for the remediation of As contaminated soils (Karami and Shamsuddin 2010).

Exhaustive work has been carried out to develop As resistant crops (Table 10.1). Heterologous expression of *PvACR3;1* in *S. cerevisiae* showed that *PvACR3;1* functions as an As(III) antiporter and mediates As(III) efflux to the external medium. In addition, the transgenic plants expressing *PvACR3;1* in *Arabidopsis* and *Nicotiana tabacum* exhibited increased As accumulation in roots and reduced accumulation in shoots in comparison to the wild type plants (Chen et al. 2017). This study provided a potential strategy to restrict As translocation in shoots and develop plants with low As accumulation in the edible parts of the plants. In another study, enhanced resistance towards As and Cd has been observed in Indian mustard by overexpression of *AtPCS1* (Gasic and Korban 2007). A study on vacuolar transporters has demonstrated that in the absence of two tonoplast localized transporters *AtABCC1* and *AtABCC2*, *Arabidopsis* shows sensitivity towards As stress. Also, the heterologous expression of these transporters in *S. cerevisiae* showed enhanced As accumulation and tolerance (Song et al. 2010). Furthermore, heterologous expression of yeast As(III) efflux system, *ACR3* in *Arabidopsis* showed increased tolerance towards As(III) and As(V) stress and higher potential of As(V) efflux (Ali et al. 2012). In rice, expression of *ScACR3* leads to increased As (III) efflux and decreased As accumulation in grains (Duan et al. 2012). Notably, significantly decreased levels of As was found in the roots and shoots of transgenic lines expressing *ScACR3* in comparison to wild type. However, no change in the As translocation factor was observed in the transgenic lines. Overexpression of *AtMT2B* in tobacco leads to significant reduction in As accumulation in roots with an increase in the shoots (Grispen et al. 2009). Genetically engineered eastern cottonwood with a bacterial gamma- glutamylcysteine synthetase (γ-ECS) gene depicted increased ECS activity and elevated thiol levels. Also, enhanced As(V)

Table 10.1 Biotechnological modifications in the genes involved in As metabolism and their outcome

S. no.	Gene name	Source	Target	Modification	Outcome	References
1.	**Phosphate transporter**					
	PHT1;1; PHT1;4	*A. thaliana*	*A. thaliana*	Mutation	Decreased As(V) uptake and hence increased As(V) tolerance	Shin et al. (2004)
	PHT1;8	*O. sativa*	*O. sativa*			Wang et al. (2016)
2.	**Aquaporins**					
	Lsi1; NIP2;1	*O. sativa*	*O. sativa*	Mutation	Lower As accumulation in plant parts	Ma et al. (2008)
	NIP1;1	*A. thaliana*	*A. thaliana*		Decreased As(III) uptake and accumulation hence increased As tolerance	Kamiya et al. (2009)
	NIP3;1	*A. thaliana*	*A. thaliana*		Decreased shoot As accumulation hence increased tolerance towards As stress	Xu et al. (2015)
	TIP4;1	*P. vittata*	*A. thaliana*	Over expression	Increased As(III) transport and accumulation hence induced As sensitivity	He et al. (2016)
3.	**NRAMP transporter**					
	NRAMP1	*O. sativa*	*A. thaliana*	Over expression	Increased As tolerance	Tiwari et al. (2014)
4.	**Glutaredoxin**					
	Grx5	*P. vittata*	*A. thaliana*	Over expression	Increased As tolerance	Sundaram et al. (2009)
5.	**As(III) S-adenosylmethionine methyltransferase**					
	ArsM	*C. reinhardtii*	*P. putida*	Over expression	As methylation and volatilization	Chen et al. (2013a)
6.	**Arsenate reductase**					
	ACR2	*A. thaliana*	*A. thaliana*	Knockout/Over expression	No effect on arsenate reduction, As accumulation and distribution	Liu et al. (2012)
	HAC1	*A. thaliana*	*A. thaliana*	Knockout	Increased shoot As accumulation	

(continued)

Table 10.1 (continued)

S. no.	Gene name	Source	Target	Modification	Outcome	References
						Chao et al. (2014), Sánchez-Bermejo et al. (2014)
	HAC1;1; HAC1;2	*O. sativa*	*O. sativa*	Over expression	Decreased As accumulation due to increased As(III) efflux into the external medium	Shi et al. (2016)
	HAC4			Mutation	Decreased As(V) reduction in roots, reduced As(III) efflux to the external medium. Increased As accumulation in shoots.	Xu et al. (2017)
7.	**ABC transporter**					
	YCF1	*S. cerevisiae*	*A. thaliana*	Over expression	Increased As(V) tolerance and decreased As accumulation. Increased As tolerance	Guo et al. (2012)
	ABCC1	*A. thaliana*	*A.thaliana*	Knockout	Increased As tolerance	Song et al. (2010)
	ABCC1	*O. sativa*	*O. sativa*		Decreased tolerance towards As stress	Song et al. (2014)
8.	**As (III) Efflux transporters**					
	ACR3	*S. cerevisiae*	*O. sativa*	Over expression	Decreased As in grain	Duan et al. (2012)
	ACR3	*P. vittata*	*A. thaliana*		Decreased As accumulation due to increase As(III) efflux	Chen et al. (2013b)
	ACR3;1	*P. vittata*	*S. cerevisiae*	Over expression	As efflux to the external medium	Chen et al. (2017)
			A. thaliana		Increased As accumulation in roots and reduced accumulation in shoots	
			N. tabacum			

tolerance was observed in the transgenic plants in comparison to the WT (Le Blanc et al. 2011). Though in the past decades significant advances have been achieved in manipulating the genes involved in As uptake and detoxification in plants, still integrated approaches of genomics, proteomics and metabolomics are required to target and manipulate specific genes to develop As resistant crops and safeguard food security.

10.7 Conclusion

Arsenic contamination is a problem of great concern across the globe. Arsenic occurrence, existence and speciation together are important, intriguing, challenging and complex, which require different physiological, biochemical and molecular tools. Understanding As dynamics in the agronomic system to combat its uptake and accumulation is a prerequisite. Arsenic occurs in different forms and oxidation states in the environment and has the propensity to be methylated. Arsenic severely affects plants growth and development and is a food chain contaminant. Biotechnological advances have been made in understanding As uptake, accumulation, speciation, detoxification and development of mitigation strategies in plants. The recent impetus in research on the area of genomics of As stress in plants have helped in deciphering the molecular mechanisms involved in As stress and tolerance. Significant studies have been carried out in determining mechanism of As sequestration in the vacuoles and the pathways and enzymes involved in As detoxification, however, still there are substantial knowledge gaps with regard to the loading and unloading of As in the xylem and phloem, and regulation of As accumulation in the plant parts. In addition, considerable attempts are required in the area of exploring hyperaccumulator-based remediation processes. The integrated system of functional genomics and molecular genetics would be helpful in better understanding of As stress response in plants and the development of As resistant crop varieties.

Acknowledgements This work was supported by the Council of Scientific and Industrial Research (CSIR), New Delhi. SK thankfully acknowledges the Department of Science and Technology (DST), Government of India, New Delhi for the DST-INSPIRE Faculty Award.

References

Abedin MJ, Feldmann J, Meharg AA (2002) Uptake kinetics of arsenic species in rice plants. Plant Physiol 128:1120–1128

Abercrombie JM, Halfhill MD, Ranjan P, Rao MR, Saxton AM, Yuan JS, Stewart CN Jr (2008) Transcriptional responses of *Arabidopsis thaliana* plants to As(V) stress. BMC Plant Biol 8:87–96

Ahsan N, Lee DG, Alam I, Kim PJ, Lee JJ, Ahn YO, Kwak SS, Lee IJ, Bahk JD, Kang KY, Renaut J, Komatsu S, Lee BH (2008) Comparative proteomic study of arsenic induced differentially expressed proteins in rice roots reveals glutathione plays a central role during As stress. Proteomics 8:3561–3576

Ali W, Isner JC, Isayenkov SV, Liu W, Zhao FJ, Maathuis FJM (2012) Heterologous expression of the yeast arsenite efflux system ACR3 improves *Arabidopsis thaliana* tolerance to arsenic stress. New Phytol 194:716–723

Aposhian HV (1997) Enzymatic methylation of arsenic species and other new approaches to arsenic toxicity. Annu Rev Pharmacol Toxicol 37:397–419

Bakhat HF, Zia Z, Fahad S, Abbas S et al (2017) Arsenic uptake, accumulation and toxicity in rice plants: Possible remedies for its detoxification: a review. Environ Sci Pollut Res 24:9142–9158

Banerjee M, Banerjee N, Bhattacharjee P, Mondal D, Lythgoe PR, Martinez M, Pan J, Polya DA, Giri AK (2013) High arsenic in rice is associated with elevated genotoxic effects in humans. Sci Rep 3:1–8

Bentley R, Chasteen TG (2002) Microbial methylation of metalloids: arsenic, antimony, and bismuth. Microbiol Mol Biol Rev 66:250–271

Bernstam L, Nriagu J (2000) Molecular aspects of arsenic stress. J Toxicol Environ Health B Crit Rev 3:293–322

Bienert GP, Thorsen M, Schüssler MD (2008) A subgroup of plant aquaporins facilitate the bi-directional diffusion of $As(OH)_3$ and $Sb(OH)_3$ across membranes. BMC Plant Biol 6:e26

Bleeker PM, Hakvoort HWJ, Bliek M, Souer E, Schat H (2006) Enhanced arsenate reduction by a CDC25-like tyrosine phosphatase explains increased phytochelatin accumulation in arsenate-tolerant *Holcus lanatus*. Plant J 45:917–929

Catarecha P, Segura MD, Franco-Zorrilla JM et al (2007) A mutant of the *Arabidopsis* phosphate transporter PHT1;1 displays enhanced arsenic accumulation. Plant Cell 19:1123–1133

Chakrabarty D, Trivedi PK, Misra P, Tiwari M, Shri M, Shukla D, Kumar S, Rai A, Pandey A, Nigam D, Tripathi RD, Tuli R (2009) Comparative transcriptome analysis of arsenate and arsenite stresses in rice seedlings. Chemosphere 74:688–702

Chao DY, Chen Y, Chen J, Shi S, Chen Z, Wang C et al (2014) Genome-wide association mapping identifies a new arsenate reductase critical for limiting arsenic accumulation in plants. PLoS Biol 12:e1002009

Chen Y, Hua C-Y, Jia M-R, Fu J-W et al (2017) Heterologous expression of *Pteris vittata* arsenite antiporter *PvACR3;1* reduces arsenic accumulation in plant shoots. Environ Sci Technol 51:10387–10395

Chen J, Qin J, Zhu YG, de Lorenzo V, Rosen BP (2013a) Engineering the soil bacterium *Pseudomonas putida* for arsenic methylation. Appl Environ Microbiol 79:4493–4495

Chen Y, Xu W, Shen H, Yan H, He Z, Ma M (2013b) Engineering arsenic tolerance and hyperaccumulation in plants for phytoremediation by a PvACR3 transgenic approach. Environ Sci Technol 47:9355–9362

Christen K (2001) The arsenic threat worsens. Environ Sci Technol 35:286A–291A

Cullen WR, Reimer KJ (1989) Arsenic speciation in the environment. Chem Rev 89:713–764

Danh LT, Truong P, Mammucari R, Foster N (2014) A critical review of the arsenic uptake mechanisms and phytoremediation potential of *Pteris vittata*. Int J Phytorem 16:429–453

Dasgupta T, Hossain SA, Meharag AA, Price AH (2004) An arsenate tolerance gene on chromosome 6 of rice. New Phytol 163:45–49

Dhankher OP, Li Y, Rosen BP, Shi J, Salt D, Senecoff JF, Sashti NA, Meagher RB (2002) Engineering tolerance and hyperaccumulation of arsenic in plants by combining arsenate reductase and gamma-glutamylcysteine synthetase expression. Nat Biotechnol 20:1140–1145

Dhankher OP, Rosen BP, McKinney EC, Meagher RB (2006) Hyperaccumulation of arsenic in the shoots of *Arabidopsis* silenced for arsenate reductase (ACR2). Proc Natl Acad Sci USA 103:5413–5418

Dixit G, Singh AP, Kumar A, Dwivedi S, Deeba F, Kumar S, Suman S, Adhikari B, Shukla Y, Trivedi PK, Pandey V, Tripathi RD (2015a) Sulfur alleviates arsenic toxicity by reducing its accumulation and modulating proteome, amino acids and thiol metabolism in rice leaves. Sci Rep 5:e16205

Dixit G, Singh AP, Kumar A, Singh PK, Kumar S, Dwivedi S, Trivedi PK, Pandey V, Norton GJ, Dhankher OP, Tripathi RD (2015b) Sulfur mediated reduction of arsenic toxicity involves efficient thiol metabolism and the antioxidant defense system in rice. J Hazard Mater 298:241–251

Dixit G, Singh AP, Kumar A, Mishra S, Dwivedi S, Kumar S, Trivedi PK, Pandey V, Tripathi RD (2016) Reduced arsenic accumulation in rice (*Oryza sativa* L.) shoot involves sulfur mediated improved thiol metabolism, antioxidant system and altered arsenic transporters. Plant Physiol Biochem 99:86–96

Duan G, Kamiya T, Ishikawa S, Arao T, Fujiwara T (2012) Expressing *ScACR3* in rice enhanced arsenite efflux and reduced arsenic accumulation in rice grains. Plant Cell Physiol 53:154–163

Duan GL, Zhou Y, Tong YP, Mukhopadhyay R, Rosen BP, Zhu YG (2007) A CDC25 homologue from rice functions as an arsenate reductase. New Phytol 174:311–321

Ellis DR, Gumaelius L, Indriolo E, Pickering IJ, Banks JA, Salt DE (2006) A novel arsenate reductase from the arsenic hyperaccumulating fern *Pteris vittata*. Plant Physiol 141:1544–1554

Fendorf S, Michael HA, van Geen A (2010) Spatial and temporal variations of groundwater arsenic in South and Southeast Asia. Science 328.1123–1127

Gasic K, Korban SS (2007) Transgenic Indian mustard (*Brassica juncea*) plants expressing an *Arabidopsis* phytochelatin synthase (*AtPCS1*) exhibit enhanced As and Cd tolerance. Plant Mol Biol 64:361–369

Ghori Z, Iftikhar H, Bhatti MF, Nasar-um-Minullah, Sharma I, Kazi AG et al (2016) Phytoextraction: the use of plants to remove heavy metals from soil, In: Ahmad P (ed) Plant metal interaction: emerging remediation techniques, Elsevier, New York, pp 385–409

González E, Solano R, Rubio V, Leyva A, Paz-Ares J (2005) Phosphate Transporter Traffic Facilitatorl is a plant-specific SEC12-related protein that enables the endoplasmic reticulum exit of a high-affinity phosphate transporter in *Arabidopsis*. Plant Cell 17:3500–3512

Grispen VMJ, Irtelli B, Hakvoorta HJ, Vooijs R et al (2009) Expression of the *Arabidopsis* metallothionein 2b enhances arsenite sensitivity and root to shoot translocation in tobacco. Environ Exp Bot 66:69–73

Guo J, Xu W, Ma M (2012) The assembly of metals chelation by thiols and vacuolar compartmentalization conferred increased tolerance to and accumulation of cadmium and arsenic in transgenic *Arabidopsis thaliana*. J Hazard Mater 200:309–313

He Z, Yan H, Chen Y, Shen H, Xu W, Zhang H, Shi L, Zhu Y-G, Ma M (2016) An aquaporin *PvTIP4;1* from *Pteris vittata* may mediate arsenite uptake. New Phytol 209:746–761

Huang K, Chen C, Shen QR, Rosen BP, Zhao FJ (2015) Genetically engineering *Bacillus subtilis* with a heat-resistant arsenite methyltransferase for bioremediation of arsenic-contaminated organic waste. Appl Environ Microbiol 81:6718–6724

Huang K, Chen C, Zhang J, Tang Z, Shen Q, Rosen BP, Zhao FJ (2016) Efficient arsenic methylation and volatilization mediated by a novel bacterium from an arsenic-contaminated paddy soil. Environ Sci Technol 50:6389–6396

Indriolo E, Na GN, Ellis D, Salt DE, Banks JA (2010) A vacuolar arsenite transporter necessary for arsenic tolerance in the arsenic hyperaccumulating fern *Pteris vittata* is missing in flowering plants. Plant Cell 22:2045–2057

Jones FT (2007) A broad view of arsenic. Poult Sci 86:2–14

Kamiya T, Tanaka M, Mitani N, Ma JF, Maeshima M, Fujiwara T (2009) NIP1;1, an aquaporin homolog, determines the arsenite sensitivity of *Arabidopsis thaliana*. J Biol Chem 284:2114–2120

Karami A, Shamsuddin ZH (2010) Phytoremediation of heavy metals with several efficiency enhancer methods. Afr J Biotechnol 9:3689–3698

Khare R, Kumar S, Shukla T, Ranjan A, Trivedi PK (2017) Differential sulphur assimilation mechanism regulates response of *Arabidopsis thaliana* natural variation towards arsenic stress under limiting sulphur condition. J Hazard Mater 337:198–207

Kumar S, Dubey RS, Tripathi RD, Chakrabarty D, Trivedi PK (2015a) Omics and biotechnology of arsenic stress and detoxification in plants: current updates and prospective. Environ Int 74:221–230

Kumar S, Asif MH, Chakrabarty D, Tripathi RD, Dubey RS, Trivedi PK (2015b) Comprehensive analysis of regulatory elements of the promoters of rice sulfate transporter gene family and functional characterization of OsSul1;1 promoter under different metal stress. Plant Signal Behav 10:990843

Kumar S, Asif MH, Chakrabarty D, Tripathi RD, Trivedi PK (2011) Differential expression and alternative splicing of rice sulphate transporter family members regulate sulphur status during plant growth, development and stress conditions. Funct Integr Genomics 11:259–273

Kumar S, Asif MH, Chakrabarty D, Tripathi RD, Dubey RS, Trivedi PK (2013a) Differential expression of rice Lambda class GST gene family members during plant growth, development, and in response to stress conditions. Plant Mol Biol Rep 31:569–580

Kumar S, Asif MH, Chakrabarty D, Tripathi RD, Dubey RS, Trivedi PK (2013b) Expression of a rice Lambda class of glutathione S-transferase, *OsGSTL2*, in *Arabidopsis* provides tolerance to heavy metal and other abiotic stresses. J Hazard Mater 248–249:228–237

Kumar S, Trivedi PK (2016a) Heavy metal signaling in plants. In: Azooz MM (ed) Plant metal interaction: emerging remediation techniques. Elsevier, Amsterdam, pp 585–603

Kumar S, Trivedi PK (2016b) Transcriptome modulation in rice under abiotic stress. In: Azooz MM, Ahmad P (eds) Plant environment interaction: responses and approaches to mitigate stress. Wiley, New York, pp 70–83

Kumar S, Verma S, Trivedi PK (2017) Involvement of small RNAs in phosphorus and sulfur sensing, signaling and stress: current update. Front Plant Sci 8:e285

Le Blanc MS, Lima A, Montello P, Kim T et al (2011) Enhanced arsenic tolerance of transgenic eastern cottonwood plants expressing gamma-glutamylcysteine synthetase. Int J Phytoremediation 13:657–673

Li R-Y, Ago Y, Liu W-J et al (2009) The rice aquaporin Lsi1 mediates uptake of methylated arsenic species. Plant Physiol 150:2071–2080

Liu X, Fu J-W, Tang N, da Silva EB et al (2017) Phytate induced arsenic uptake and plant growth in arsenic-hyperaccumulator *Pteris vittata*. Environ Pollut 226:212–218

Liu W, Schat H, Bliek M, Chen Y. Mcgrath SP, George G, E Salt DE, Zhao FJ (2012) Knocking out ACR2 does not affect Arsenic redox status in *Arabidopsis thaliana*: Implications for As detoxification and accumulation in plants. PloS one 7:e42408

Logoteta B, Xu XY, Macnair MR, McGrath SP, Zhao FJ (2009) Arsenite efflux is not enhanced in the arsenate-tolerant phenotype of *Holcus lanatus*. New Phytol 183:340–348

Lomax C, Liu WJ, Wu L, Xue K, Xiong J, Zhou J, McGrath SP, Meharg AA, Miller AJ, Zhao FJ (2012) Methylated arsenic species in plants originate from soil microorganisms. New Phytol 193:665–672

Ma JF, Yamaji N, Mitani N, Xu XY, Su YH, McGrath SP, Zhao FJ (2008) Transporters of arsenite in rice and their role in arsenic accumulation in rice grain. Proc Natl Acad Sci USA 105:9931–9935

McGuinness M, Dowling D (2009) Plant-associated bacterial degradation of toxic organic compounds in soil. Int J Environ Res Public Health 6:2226–2247

Meharg AA, Macnair MR (1992) Suppression of the high-affinity phosphate-uptake system: a mechanism of arsenate tolerance in *Holcus lanatus* L. J Exp Bot 43:519–524

Meng XY, Qin J, Wang LH, Duan GL, Sun GX, Wu HL, Chu CC, Ling HQ, Rosen BP, Zhu YG (2011) Arsenic biotransformation and volatilization in transgenic rice. New Phytol 191:49–56

Mosa KA, Kumar K, Chhikara S, Mcdermott J, Liu Z, Musante C, White JC, Dhankher OP (2012) Members of rice plasma membrane intrinsic proteins subfamily are involved in arsenite permeability and tolerance in plants. Transgenic Res 21:1265–1277

Mudge SR, Rae AL, Diatloff E, Smith FW (2002) Expression analysis suggests novel roles for members of the Pht1 family of phosphate transporters in *Arabidopsis*. Plant J 31:341–353

Mukhopadhyay R, Shi J, Rosen BP (2000) Purification and characterization of Acr2p, the *Saccharomyces cerevisiae* arsenate reductase. J Biol Chem 275:21149–21157

Neumann RB, Ashfaque KN, Badruzzaman ABM, Ali MA, Shoemaker JK, Harvey CF (2010) Anthropogenic influences on groundwater arsenic concentrations in Bangladesh. Nat Geosci 3:46–52

Newcombe C, Raab A, Williams PN, Deacon C, Haris PI, Meharg AA, Feldmann J (2010) Accumulation or production of arsenobetaine in humans? J Environ Monit 12:832–837

Nocito FF, Lancilli C, Crema B, Fourcroy P, Davidian J-C, Sacchi GA (2006) Heavy metal stress and sulfate uptake in maize roots. Plant Physiol 141:1138–1148

Norton GJ, Lou-Hing DE, Meharg AA, Price AH (2008) Rice–arsenate interactions in hydroponics: whole genome transcriptional analysis. J Exp Bot 59:2267–2276

Paszkowski U, Kroken S, Roux C, Briggs SP (2002) Rice phosphate transporters include an evolutionarily divergent gene specifically activated in arbuscular mycorrhizal symbiosis. Proc Natl Acad Sci USA 99:13324–13329

Paulose B, Kandasamy S, Dhankher OP (2010) Expression profiling of *Crambe abyssinica* under arsenate stress identifies genes and gene networks involved in arsenic metabolism and detoxification. BMC Plant Biol 10:e108

Pickering IJ, Prince RC, George MJ, Smith RD et al (2000) Reduction and coordination of arsenic in Indian mustard. Plant Physiol 122:1171–1178

Qin J, Rosen BP, Zhang Y, Wang G, Franke S, Rensing C (2006) Arsenic detoxification and evolution of trimethylarsine gas by a microbial arsenite S-adenosylmethionine methyltransferase. Proc Natl Acad Sci USA 103:2075–2080

Raab A, Williams PN, Meharg A, Feldmann J (2007) Uptake and translocation of inorganic and methylated arsenic species by plants. Environ Chem 4:197–203

Rai A, Tripathi P, Dwivedi S, Dubey S, Shri M, Kumar S, Tripathi PK, Dave R, Kumar A, Singh R, Adhikari B, Bag M, Tripathi RD, Trivedi PK, Chakrabarty D, Tuli R (2011) Arsenic tolerances in rice (*Oryza sativa*) have a predominant role in transcriptional regulation of a set of genes including sulphur assimilation pathway and antioxidant system. Chemosphere 82:986–995

Rodriguez-Lado L, Sun G, Berg M, Zhang Q, Xue H, Zheng Q, Johnson CA (2013) Groundwater arsenic contamination throughout China. Science 341:866–868

Sánchez-Bermejo E, Castrillo G, Del Llano B, Navarro C, Zarco FS et al (2014) Natural variation in arsenate tolerance identifies an arsenate reductase in *Arabidopsis thaliana*. Nat Commun 5: e4617

Santra SC, Samal AC, Bhattacharya P, Banerjee S, Biswas A, Majumdar J (2013) Arsenic in food chain and community health risk: a study in Gangetic West Bengal. Procedia Environ Sci 18:2–13

Schat H, Llugany M, Vooijs R, Hartley-Whitaker J, Bleeker PM (2002) The role of phytochelatins in constitutive and adaptive heavy metal tolerances in hyperaccumulator and non-hyperaccumulator metallophytes. J Exp Bot 53:2381–2392

Schulz H, Haertling S, Tanneberg H (2008) The identification and quantification of arsenic-induced phytochelatins: comparison between plants with varying As sensitivities. Plant Soil 303:275–287

Sharma D, Tiwari M, Lakhwani D, Tripathi RD, Trivedi PK (2015) Differential expression of microRNAs by arsenate and arsenite stress in natural accessions of rice. Metallomics 7:174–187

Shi S, Wang T, Chen Z, Tang Z, Wu Z, Salt DE et al (2016) *OsHAC1;1* and *OsHAC1;2* function as arsenate reductases and regulate arsenic accumulation. Plant Physiol 172:1708–1719

Shin H, Shin HS, Gary R, Harrison MJ (2004) Phosphate transport in *Arabidopsis*: Pht1;1 and Pht1;4 play a major role in phosphate acquisition from both low- and high-phosphate environments. Plant J 39:629–642

Shri M, Kumar S, Chakrabarty D, Trivedi PK, Mallick S, Misra P, Shukla D, Mishra S, Srivastava S, Tripathi RD, Tuli R (2009) Effect of arsenic on growth, oxidative stress and antioxidant system in rice seedlings. Ecotoxicol Environ Saf 72:1102–1110

Shukla D, Kesari R, Mishra S, Dwivedi S, Tripathi RD, Nath P, Trivedi PK (2012) Expression of phytochelatin synthase from aquatic macrophyte *Ceratophyllum demersum* L. enhances cadmium and arsenic accumulation in tobacco. Plant Cell Rep 31:1687–1699

Shukla D, Tiwari M, Tripathi RD, Nath P, Trivedi PK (2013) Synthetic phytochelatins complement a phytochelatin-deficient *Arabidopsis* mutant and enhance the accumulation of heavy metal(loid)s. Biochem Biophys Res Commun 434:664–669

Shukla T, Kumar S, Khare R, Tripathi RD, Trivedi PK (2015) Natural variations in expression of regulatory and detoxification related genes under limiting phosphate and arsenate stress in *Arabidopsis thaliana*. Front Plant Sci 6:e898

Srivastava S, Mishra S, Tripathi R, Dwivedi S, Trivedi P, Tandon P (2007) Phytochelatins and antioxidantsystems respond differentially during arsenite and arsenate stress in *Hydrilla verticillata* (Lf) Royle. EnvironSci Technol 41:2930–2936

Srivastava S, Srivastava AK, Suprasanna P, D'Souza SF (2009) Comparative biochemical and transcriptional profiling of two contrasting varieties of *Brassica juncea* L. in response to arsenic exposure reveals mechanisms of stress perception and tolerance. J Exp Bot 60:3419–3431

Song WY, Park J, Mendoza-Cozatl DG, Suter-Grotemeyer M et al (2010) Arsenic tolerance in *Arabidopsis* is mediated by two ABCC-type phytochelatin transporters. Proc Natl Acad Sci USA 107:21187–21192

Song WY, Yamaki T, Yamaji N, Ko D, Jung KH, Fujii-Kashino M et al (2014) A rice ABC transporter, *OsABCC1*, reduces arsenic accumulation in the grain. Proc Natl Acad Sci USA 111:15699–15704

Sundaram S, Wu S, Ma LQ, Rathinasabapathi B (2009) Expression of a *Pteris vittata* glutaredoxin *PvGRX5* in transgenic *Arabidopsis thaliana* increases plant arsenic tolerance and decreases arsenic accumulation in the leaves. Plant Cell Environ 32:851–858

Tiwari M, Sharma D, Dwivedi S, Singh M, Tripathi RD, Trivedi PK (2014) Expression in *Arabidopsis* and cellular localization reveal involvement of rice NRAMP, *OsNRAMP1*, in arsenic transport and tolerance. Plant Cell Environ 37:140–152

Tripathi RD, Srivastava S, Mishra S, Singh N, Tuli R, Gupta DK, Maathuis FJM (2007) Arsenic hazards: strategies for tolerance and remediation by plants. Trends Biotech 25:158–165

Tripathi RD, Tripathi P, Dwivedi S, Dubey S, Chatterjee S et al (2012) Arsenomics: omics of arsenic metabolism in plants. Front Plant Sci 3:e275

Tuli R, Chakrabarty D, Trivedi PK, Tripathi RD (2010) Recent advances in arsenic accumulation and metabolism in rice. Mol Breed 26:307–323

Vahter M (2002) Mechanisms of arsenic biotransformation. Toxicology 181:211–217

Wang J, Zhao F-J, Meharg AA, Raab A et al (2002) Mechanisms of arsenic hyperaccumulation in *Pteris vittata*. Uptake kinetics, interactions with Phosphate, and Arsenic speciation. Plant Physiol 130:1552–1561

Wang P, Zhang W, Mao C, Xu G, Zhao FJ (2016) The role of *OsPT8* in arsenate uptake and varietal difference in arsenate tolerance in rice. J Exp Bot 67:6051–6059

Xu J, Shi S, Wang L, Tang Z, Lv T et al (2017) OsHAC4 is critical for arsenate tolerance and regulates arsenic accumulation in rice. New Phytol 215:1090–1101

Xu W, Dai W, Yan H, Li S, Shen H et al (2015) *Arabidopsis* NIP3;1 plays an important role in arsenic uptake and root-to-shoot translocation under arsenite stress conditions. Mol Plant 8:722–733

Yu LJ, Luo YF, Liao B, Xie LJ, Chen L, Xiao S, Li JT, Hu SN, Shu WS (2012) Comparative transcriptome analysis of transporters, phytohormone and lipid metabolism pathways in response to arsenic stress in rice (*Oryza sativa*). New Phytol 195:97–112

Zanella L, Fattorini L, Brunetti P, Roccotiello E et al (2016) Overexpression of AtPCS1 in tobacco increases arsenic and arsenic plus cadmium accumulation and detoxification. Planta 243:605–622

Zhang J, Xu Y, Cao T, Chen J, Rosen BP, Zhao FJ (2017) Arsenic methylation by a genetically engineered Rhizobium-legume symbiont. Plant Soil 416:259–269

Zhang W, Cai Y, Downum KR, Ma LQ (2004) Thiol synthesis and arsenic hyperaccumulation in *Pteris vittata* (Chinese brake fern). Environ Pollut 131:337–345

Zhao FJ, McGrath SP, Meharg AA (2010) Arsenic as a food chain contaminant: Mechanisms of plant uptake and metabolism and mitigation strategies. Annu Rev Plant Biol 61:535–559

Chapter 11
Phytohormones Regulating the Master Regulators of CBF Dependent Cold Stress Signaling Pathway

Prakriti Kashyap and Renu Deswal

Abstract Cold stress studies have elucidated the role of plant hormones in gene regulation during cold stress responses. Plants acquire tolerance to stress by reprogramming metabolism and gene expression. A large group of transcriptional regulators controls the changes in gene expression. The most studied cold stress signaling pathway is the C-repeat binding factor (CBF)-dependent pathway. The CBF transcription factors were the first transcriptional activators demonstrated to have a role in controlling the expression of cold-responsive genes with a role in cold acclimation. They belong to AP2 (APETALA2)/EREBP (Ethylene Responsive Element Binding Proteins) family. A constitutive regulator, ICE (Inducer of CBF expression), activates CBF. As the name suggests, CBF expresses in an ethylene-dependent manner. This family of transcription factors recognize and bind to cold and dehydration-responsive DNA regulatory element known as CRT/DRE cis element in the promoter of many cold responsive genes. Low temperature is shown to increase the levels of endogenous abscisic acid (ABA) and its exogenous application enhances cold tolerance. Although the CBF-dependent cold signaling pathway tends to operate in ABA-independent manner, the reports of cross talk between ABA-dependent and ABA-independent pathways suggest its role in the CBF-dependent signaling. Recently, a defense related phytohormone jasmonate was shown to regulate freezing stress responses in *Arabidopsis* through ICE-CBF/DREB1 transcriptional pathway. Its exogenous application significantly improved freezing tolerance, while blocking its biosynthesis decreased freezing tolerance. In this chapter, we have discussed the significance of phytohormones in CBF-dependent cold stress signaling.

Keywords Cold stress · CBF-dependent pathway · Phytohormones
Cold stress signaling

P. Kashyap · R. Deswal (✉)
Molecular Plant Physiology and Proteomics Laboratory, Department of Botany,
University of Delhi, Delhi 110007, India
e-mail: rdeswal@botany.du.ac.in

V. R. Rajpal et al. (eds.), *Genetic Enhancement of Crops for Tolerance to Abiotic Stress: Mechanisms and Approaches, Vol. I*, Sustainable Development and Biodiversity 20, https://doi.org/10.1007/978-3-319-91956-0_11

11.1 Introduction

Phytohormones are the chemicals synthesized by plants, which regulate their growth and development. These are the signaling molecules produced at very low concentrations and regulate various physiological functions in different targeted sites. In animals, glands produce the hormones but in plants, there are no specific organs to produce hormones. In plants, these can operate locally or move to the different sites to regulate diverse processes like formation of stems and leaves, flowering, leaf senescence and fruit ripening. Plant classical hormones include Abscisic acid (ABA), cytokinins (CKs), auxins, gibberellins (GAs) and ethylene (ETH). Amongst these, auxins, cytokinins and gibberellins are growth-promoting hormones, whereas abscisic acid and ethylene are growth-inhibiting hormones. These major hormones are made up of different chemicals and their structures may vary from plant to plant. In addition to these major hormones, some other growth regulators are also considered as plant hormones namely florigen, vernalin, jasmonates, brassinosteroides, salicylic acid, caulines and morphactines. Out of these, jasmonates and salicylic acid have gained importance as signaling molecules during pathogen attack or disease outbreak. Plant hormones also affect gene expression and transcription along with their roles in cell division, growth and development. When a plant faces some kind of biotic or abiotic stress, the transcriptome undergoes reprogramming leading to altered metabolism, which provides tolerance. A number of transcription factors participate in the signaling networks for this reprogramming of metabolism. Plant hormones also modulate these signaling networks for providing stress tolerance.

Most plants face cold stress at least once during the lifecycle. Cold stress increases the endogenous ABA and several transcription factors participate in signaling cascades to provide cold tolerance to the plants. Amongst these cascades, CBF-dependent signaling is the most investigated cold stress signaling cascade and CBF transcription factors are the most studied transcription factors. CBFs regulate approximately 12% of the cold-responsive transcriptome (Fowler and Thomashow 2002). They are not present in plants under control conditions but starts accumulating within 15 min of cold stress. The accumulated CBFs bind to the DRE/CRT elements in the promoter of cold responsive genes commonly known as COR genes (Thomashow 1999; Chinnusamy et al. 2006). These cause the physiological and biochemical changes in the metabolism of plants conferring them cold tolerance. The activation of CBF occurs with the help of an upstream regulator ICE (Inducer of CBF Expression). Unlike CBF, ICE is constitutively present in plants. ICE belongs to bHLH family of proteins and contains a very conserved bHLH domain. This domain binds to the E-box present in the promoter of CBFs during cold stress, thus inducing their expression. Both CBF and ICE have many isoforms and different isoforms of ICE activate different isoforms of CBFs. CBFs seem to be key regulators in promoting the biosynthesis of different molecules that enhance freezing tolerance. Moreover, they also control plant growth and development under cold environmental conditions. CBFs perform these roles by integrating a network of hormone signaling. The specific components that together compose this network and the underlying molecular mechanism(s) are still not very well characterized. In addition to their role in increasing freezing tolerance, CBFs also cause growth retardation and late flowering by promoting the accumulation of DELLA proteins because of reduced gibberellic acid (GA) levels. On the other hand, CBFs also promote delayed cold-induced leaf senescence through ABA-mediated mechanisms that suppress leaf tissue responsiveness to ethylene. Therefore, CBFs modulate plant growth and senescence by crosstalk with hormone-mediated signaling pathways involving GA, ethylene and ABA. Liang and Yang (2015) suggested ICE1 as a negative regulator of ABA-dependent pathways in seeds and seedlings of *Arabidopsis*. The phytohormone jasmonate, which is involved in defense responses (Howe et al. 1996; Farmer 2001; Browse 2009) was shown to regulate freezing stress responses in *Arabidopsis* through ICE-CBF/DREB1 transcriptional pathway. Its exogenous application significantly improved freezing tolerance while blocking its biosynthesis decreased freezing tolerance (Hu et al. 2013). Thus the phytohormones cross talk activates the genes essential for plant development and responses to cold stress (Shi et al. 2015).

11.2 Cold Stress Signaling Pathway

The cold stress-signaling pathway in plants is broadly categorized into ABA-dependent and ABA-independent pathways. Generally, the ABA-independent regulon operates in cold stress conditions while ABA-dependent pathway in drought conditions. However, in cold stress, plants face scarcity of water and the physiological conditions are like drought. Therefore, many ABA-dependent cold inducible proteins are common in cold and drought stress. ABA-dependent pathway operates mainly via bZIP, MYC and MYB transcription factors under drought conditions. Together these transcription factors bind to the ABRE (Abscisic acid Responsive Elements) in the promoter regions of ABA-inducible genes, thereby activating these and conferring cold tolerance. The ABA-independent CBF regulon is a relatively better-studied regulon and is driven by ICE and CBF transcription factors. CBFs regulate the expression of 12% of *Arabidopsis* cold inducible genes. ICE is an upstream transcription factor in the transcriptional cascade leading to activation of CBF and COR (cold-responsive) genes. It is suggested that during cold stress, ICE is activated due to some post-translational modification and binds to a MYC-recognition sequence, the E-box in the promoter region of CBF genes. CBF belongs to AP2/EREBP family of DNA binding proteins. This family of transcription factors recognizes and binds to cold- and dehydration-responsive DNA regulatory element known as CRT/DRE *cis* element in the promoter of many cold responsive genes. A large number of low temperature-induced genes have been identified and characterized in plants (Tsuda et al. 2000; Zhang et al. 2009). These include Late Embryogenesis Abundant (LEA), Dehydrins (DHN), Responsive to Abscisic acid (RAB), Low Temperature Responsive (LT) and COR genes.

The COR gene expression and freezing tolerance are positively correlated. This regulon controls several biochemical changes associated with cold acclimation like accumulation of simple sugars and the amino acid proline. Moreover, CBFs also regulate the expression of genes involved in phosphoinositide metabolism, osmolyte biosynthesis, reactive oxygen species (ROS) detoxification, membrane transport, hormone metabolism and signaling (Fowler and Thomashow 2002; Maruyama et al. 2004; Yamaguchi-Shinozaki and Shinozaki 2006).

Post-translational modifications such as ubiquitination and sumoylation control ICE1-dependent cold signaling. An E3 ligase HOS1 physically interacts with ICE1 and mediates the ubiquitination of ICE1 both *in vitro* and *in vivo*. Cold induces the proteasome-mediated degradation of ICE1 via HOS1 (Dong et al. 2006). Furthermore, serine 403 is identified as a key residue for attenuation of cold stress response by HOS1-mediated degradation of ICE1 (Miura et al. 2011). The serine 403, when substituted by alanine, enhanced the transactivational activity of ICE1 in *Arabidopsis* protoplasts and therefore, the expression of cold induced genes, such as CBF3/DREB1A, COR 47 and KIN1 resulting in cold tolerance. Sumoylation/desumoylation of proteins has a pivotal role in plant responses to abiotic and biotic stress and in ABA and salicylic acid signaling (Miura et al. 2007a). During sumoylation, SUMO (Small Ubiquitin-related Modifier) proteins conjugate to protein substrates by SUMO E3 ligases and in desumoylation, SUMO proteins are removed from their target proteins by SUMO proteases. Sumoylation prevents ubiquitination and can protect target proteins from proteasomal degradation (Ulrich 2005). An *Arabidopsis* SUMO E3 ligase, SIZ1 is required for the accumulation of SUMO conjugates during cold stress. SIZ1 sumoylates ICE1 at K393 residue, the principal site for SUMO conjugation, and blocks its HOS1-mediated polyubiquitinization, stabilizing and enhancing the activity of the transcription factor. Furthermore, sumoylated ICE1 represses MYB15, which is a negative regulator of CBF3/DREB1A and confers freezing tolerance (Miura et al. 2007b). Therefore, ICE-CBF pathway works in a complex way and provides cold tolerance to the plants. The event of changes that take place in a plant cell in ABA-dependent and independent manner is summarized in Fig. 11.1.

This complex network of CBF-dependent cold stress signaling pathway becomes more complicated with the participation of phytohormones. The phytohormones and their role in CBF-dependent signaling is discussed later in this chapter.

11.2.1 Gibberellic Acid

Gibberellic acid gathered attention of scientists in 1950s. It has obtained its name from a pathogenic fungus *Gibberella fujikuroi*, which causes 'foolish seedling' disease in rice plants resulting in exceptionally tall plants. GAs include a range of chemicals produced in plants. Gibberellins possess a tetracyclic *ent*-gibberellane skeleton with 20 carbon atoms or a 20-nor-*ent*-gibberellane skeleton with 19 carbon atoms (Fig. 11.2). Gibberellins are important in seed germination as they initiate

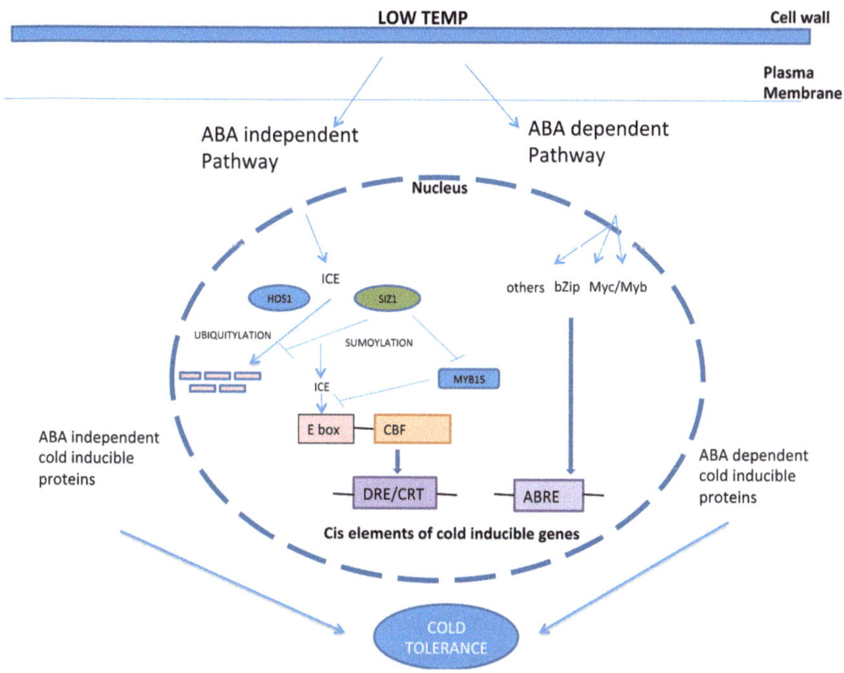

Fig. 11.1 ABA dependent and independent signaling in plants during cold stress

Fig. 11.2 Gibberellane skeleton with 19 and 20 carbon atoms

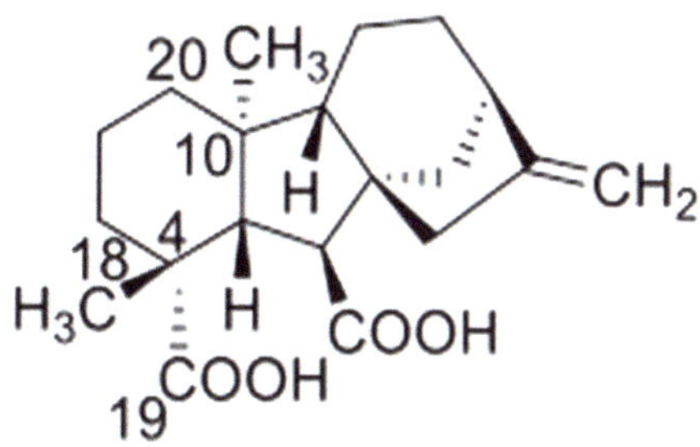

GA_{12}C20-GA

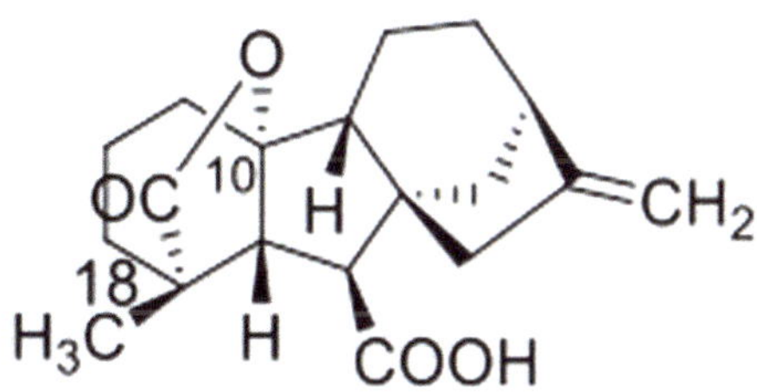

GA_{9}C19-GA

mobilization of storage materials in seeds during germination by affecting enzyme production, which mobilizes food reserves used for growth of new seedlings. This is done by modulating chromosomal transcription. GAs cause elongation of stems, stimulate bolting in biennials and stimulate pollen tube growth. They also promote cellular division, sex determination, flowering, fruit set and parthenocarpy. Gibberellins also reverse the inhibition of shoot growth and dormancy induced by ABA.

11.2.1.1 CBF-Dependent Signaling Is GA-Mediated

In response to biotic and abiotic stresses, plants show many adaptive responses in which hormones are involved. Among the GA-signaling components, DELLA proteins are the GA signaling components that arrest plant growth during the adverse environmental conditions, thereby promoting survival (Achard et al. 2006, 2008). Reduced GA levels induce DELLA proteins that further cause dwarf stature and delay in flowering. DELLAs contribute significantly to the function of CBFs during cold acclimation and freezing tolerance by CBF-dependent cold stress signaling pathway. Achard et al. (2008) observed that the over expression of *CBF1* reduced the endogenous gibberellin, in turn affecting *Arabidopsis* growth negatively. Furthermore, low temperature induces the expression of genes involved in GA inactivation, which also results in reduced plant growth (Achard et al. 2008). The phytohormones gibberellin (GA) and jasmonate (JA) play an important role in regulating growth in response to temperature. It has been shown that the GA/DELLA pathway interacts with JA signaling and C-repeat binding factor dependent cold acclimation in annual plants (Wingler 2015). Overall, these data indicate that in response to low temperature, CBFs inhibit plant growth through the GA/DELLA-signaling pathway.

11.2.2 Jasmonic Acid

Jasmonates are well known for their role in the production of defense proteins to protect plant from invading organisms. They are also believed to have a role in seed germination and storage of protein in seeds. JA biosynthesis and signaling has been the interest of reviewers (Kazan and Manners 2008, 2013; Lyons et al. 2013; Wasternack 2014). To sum up, several plastid, peroxisome and cytoplasmic enzymes are involved in synthesis of JA. JAZ proteins are the repressors of well-known master transcription factors MYC2 (Kazan and Manners 2013). MYC2 and related proteins regulate JA responses by binding to the G-box promoter element in the promoter of JA-responsive genes.

11.2.2.1 Jasmonic Acid- and CBF-Dependent Signaling

Jasmonate is a stress hormone that affects growth negatively (Wasternack 2014). In *Arabidopsis,* cell number and size is reduced by JA (Noir et al. 2013). Furthermore, JA treatment causes rapid reduction in growth and photosynthetic gene expression without affecting photosynthetic activity (Attaran et al. 2014). Therefore, reduced growth cannot be solely explained with low photosynthetic rate (Wingler 2015). However, jasmonic acid is mainly considered as a defense hormone, enough evidences support its role during abiotic stress for conversion of carbon into sugars (Wingler 2015). In CBF-dependent cold stress signaling pathway, jasmonates (JAs) act as upstream regulators with the (GA) /DELLA pathway regulating growth downstream of CBFs in addition to interactions between JA and GA signaling. GA and JA have antagonistic effects on growth. Heinrich et al. (2013) showed that JA can inhibit the synthesis of active GAs and also other signaling pathways via DELLA proteins. In *Arabidopsis*, JA induce expression of the DELLA gene *RGL3*, involved in JA signaling (Wild et al. 2012). Moreover, DELLAs interact with JAZ repressors participating in JA signaling (Song et al. 2014; Xu et al. 2014).

JAs are considered as positive regulators of cold tolerance. Cold stress induce JA biosynthesis genes such as *LOX1*, *AOS1*, *AOC1*, and *JAR1* in *Arabidopsis* and *OsAOS*, *OsOPR1*, *OsAOC*, and *OsLOX2* in rice (*Oryza sativa*), thus elevating endogenous JA levels (Hu et al. 2013; Du et al. 2013). Also, the exogenous JA treatment increases freezing tolerance in *Arabidopsis*. However, *Arabidopsis* mutants (i.e., *lox2, aos, jar1, and coi1*) deficient in JA biosynthesis are sensitive to freezing stress (Hu et al. 2013) confirming importance of JA in conferring freeze tolerance to the plants.

Hu et al. (2013) also showed *Arabidopsis* JAZ repressors to act as regulators of cold stress tolerance. JAZ1 and JAZ4, physically interact with ICE1 and ICE2 (Box 2) and suppress their transcriptional activities preventing non-specific cold stress responses under normal growth conditions. However, during cold stress, JA level is elevated, further triggering COI1-mediated degradation of JAZs. This activates ICEs. ICE1 and ICE2 that further activate CBFs by binding to their DRE/CRT box promoter region (Hu et al. 2013). JAZs also activate downstream bHLH TFs that are involved in the regulation of cold responses. In banana (*Musa acuminata*) and *Arabidopsis*, MYC2 homologs physically interact with ICE1 (Peng et al. 2013; Zhao et al. 2013) This further suggests the another point of cross-talk between these TFs of two regulatory pathways (Kazan 2015).

Both cold and JA pathway are regulated by another protein SFR6 (Sensitive to Freezing 6) in *Arabidopsis* (Knight et al. 1999, 2009). SFR6 was known to be a downstream target of CBFs. However, it was later identified as Mediator16 (MED16) with a role in JA- responsive defense gene expression (Boyce et al. 2003; Zhang et al. 2012). Previous studies showed that EIN3/EIL1 interacts with JAZ1 to mediate jasmonate-regulated responses (Pauwels and Goossens 2011; Zhu et al. 2011). Later, Shi et al. (2015) proposed that JAZs are antagonistic or synergistic to EIN3/EIL1 and ICE1 modulating the CBF signaling pathway during cold response. Moreover, JA causes the upregulation of CBF genes in *Arabidopsis* (Hu et al. 2013) and *C. bursa-pastoris* (Zhou et al. 2014). This may further result in restrained growth as CBF negatively affects GA content and enhances the DELLA protein expression and stability. Together, these examples show that JA, GA and cold pathways share several common components/cross talk nodes.

11.2.3 Abscisic Acid

Abscisic acid (ABA) is a ubiquitous plant hormone in all vascular plants. It got its name "abscisic acid" because of its presence in large amount in newly abscised or freshly fallen leaves. ABA is a 15-carbon compound (Fig. 11.3) with two *cis* and *trans* isomers determined by the orientation of the carboxyl group at carbon 2. It also has an asymmetric carbon atom resulting in the enantiomers. Amongst enantiomers, S enantiomer is the natural form. However, commercially available ABA has equal amounts of both S and R enantiomers. ABA is an inhibitory chemical compound that affects bud growth and seed and bud dormancy negatively. It also plays a role in closing the stomata in water-stressed plants. During water stress, when the roots are deficient in water, a signal goes to the leaves triggering the formation of ABA precursors, which afterwards move to the roots. ABA is present in all parts of the plant. Its concentration within tissues varies, which mediates its effects and its function as a hormone.

ABA is commonly considered to inhibit growth under stress conditions. It causes both positive and negative effects on growth and therefore, its effects on plant growth are controversial (Skirycz and Inzé 2010; Tardieu et al. 2010). In poplar, there is a direct correlation between cambial growth and ABA content. The external ABA application exhibits positive effect on cambia activity (Arend and Fromm 2013). Apart from this, ABA also induces stomatal closure, causes non-hydraulic effects and improves water conductance.

11.2.3.1 Abscisic Acid- and CBF-Dependent Cold Stress Signaling

Despite the contrasting effects on growth, the involvement of ABA in dehydration and cold response is well defined (Knight and Knight 2012). There may be roles of ABA in the CBF-dependent and independent regulation of gene expression and a

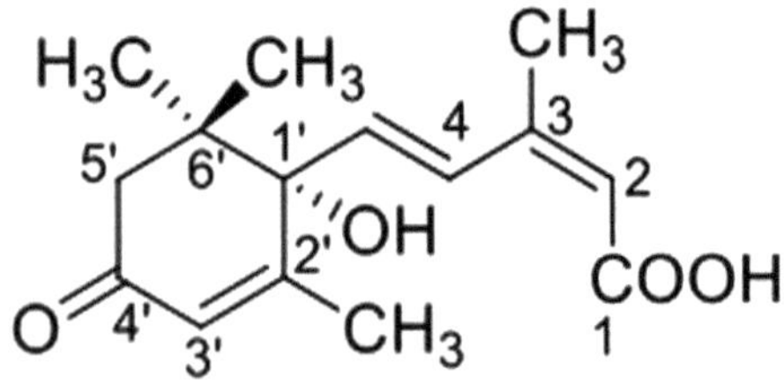

Fig. 11.3 Abscisic acid chemical formula

function of ABA in dehydration caused by freezing (Mäntylä et al. 1995; Sharma et al. 2005). Among CBF-independent cold signaling pathways, ABA-dependent cold signaling pathway has been extensively studied. As per transcriptome analysis, approximately 10% of ABA-responsive genes respond towards cold stress (Kreps et al. 2002). Some COR genes, such as *RD29A, RD22, COR15A* and *COR47* contain ABA response (ABRE) cis-elements along with the CRT/DRE motif in their promoter region and can be activated by ABRE-binding proteins/factors (AREBs/ABFs) (Uno et al. 2000). ABA is considered to be the primary plant hormone regulating abiotic stress responses. However, in plants, response to abiotic stress is also regulated by ABA-independent pathways with crosstalk between ABA-dependent and ABA-independent pathways. Plant genes responding to ABA contain the ABA Response Element (ABRE) in their promoters. ABRE Binding Factors (AREB/ABF) are bZIP transcription factors that bind to ABREs and regulate ABA-dependent pathway. Abiotic stress tolerance can also be regulated in an ABA-independent manner by the AP2/ERF transcription factors such as Dehydration Responsive Element Binding Proteins (DREB1) and DREB2. DREB1/CBF and DREB2 regulate temperature and osmotic stress responses, respectively. These cause physiological and biochemical changes in plants by binding to the DRE/CRT sequence element present in the promoters of stress-responsive *COR* genes. However, ICE1 was proposed to be a negative regulator of ABA-dependent pathways. ABA or glucose hyperacivate the ABA signaling genes *ABI3* and *ABI4* in *ice1-2* mutants. Also, over-expression of *ABI3* or *ABI4* results in ABA hypersensitivity (Soderman et al. 2000; Lopez-Molina et al. 2002; Zhang et al. 2005). With consonance to this, glucose-induced hyper-activation of ABI3 and ABI4 in *ICE1* mutants explains the arrested growth. Moreover, ABI3, was shown to function in the cold stress response. Ectopic expression of the seed-specific ABI3 confers ability to express *COR* genes in vegetative tissues and enhances freezing tolerance in *Arabidopsis* (Tamminen et al. 2001).

A more specific role of ABA has been proposed for the temperature and photoperiod-dependent growth cessation in trees in autumn. Although ABA treatment does not generally result in growth cessation and dormancy, it acts by interacting with photoperiod. Therefore, ABA might function in the seasonal growth cycle. In poplar buds, cold treatment leads to a transient increase in ABA under short day conditions (Welling et al. 2002). In addition, short day treatment on its own increases ABA content transiently in birch and poplar (Rinne et al. 1998; Rohde et al. 2002) and this increase is related to freezing tolerance (Welling et al. 1997; Rinne et al. 1998). On the other hand, high endogenous ABA or ABA treatment under long-day conditions did not induce growth cessation (Welling et al. 1997), suggesting that the short day dependent increase in ABA is not responsible for bud growth cessation. The role of ABA in the process therefore, remains unclear (Olsen 2010; Cooke et al. 2012).

11.2.4 Auxin

Auxin was the first growth hormone to be studied in plants. It received its name from the Greek word *auxein,* which means to grow. Indole-3-acetic acid (IAA) was the first auxin to be discovered (Fig. 11.4). Later, other auxins were discovered but still IAA is the most abundant and important auxin. Auxins are compounds that positively influence cell enlargement, bud formation and root initiation. They also promote the production of other hormones. They work in conjunction with cytokinins and control the growth of stems, roots, fruits and convert stems into flowers. Auxins in seeds regulate specific protein synthesis. Synthetic auxins including 2,4-D and 2,4,5-T have been developed and used for weed control.

The growth and development of plant is regulated by a network of hormonal interactions. Interestingly, auxin has been found to be a common factor in majority of these interactions. Auxin and cytokinin have been shown to act both synergistically and antagonistically for shoot and root development, respectively (Swarup et al. 2002; Dello Ioio et al. 2008).

11.2.4.1 Auxin- and CBF-Dependent Cold Stress Signaling Pathway

Auxin has been shown to regulate several aspects of growth and development of plants. However, knowledge about its role under cold stress is limiting. The inflorescence gravitropism of *Arabidopsis* is regulated by auxin and is inhibited by cold stress indicating a connection between auxin and cold stress (Fukaki et al. 1996; Wyatt et al. 2002). The gravity response and the rootward auxin transport were abolished at 4 °C and inhibition of inflorescence gravitropism was observed. It returned to wild-type level when the plants returned to room temperature (Fukaki et al. 1996; Wyatt et al. 2002; Nadella et al. 2006). Although it clearly suggests the importance of auxin in cold-stress-mediated plant growth and development, the question arises if the CBF-dependent cold stress-signaling pathway is associated with auxin. Interestingly, SIZ1, a regulatory component of CBF cold signaling pathway, which represses the polyubiquitination of ICE1 at low temperature (Miura et al. 2007a, b) has been shown to affect phosphate-starvation-induced root architecture remodeling negatively through the control of auxin patterning (Miura et al. 2011). Shibasaki et al. (2009) suggested that the effect of cold stress on auxin is linked to the inhibition of intracellular trafficking of auxin efflux carriers.

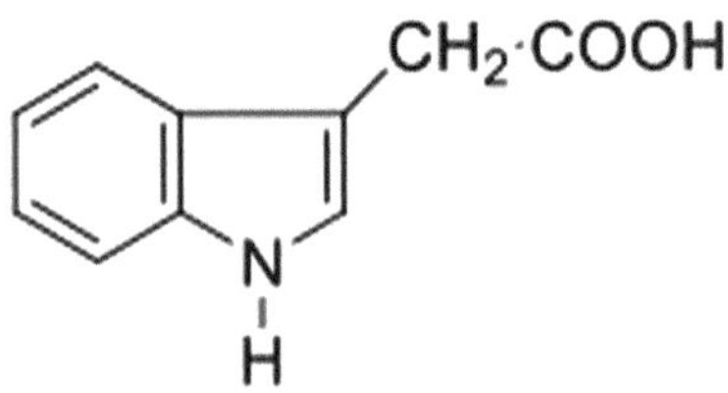

Fig. 11.4 Indole-3-acetic acid

11.2.5 Ethylene

Ethylene is a gas that is formed by the breakdown of methionine, which is present in all cells. It is a simple olefin with the molecular weight 28 (Fig. 11.5) and lighter than air. For about 25 years, ethylene did not get any recognition and importance as a plant hormone. The effects of ethylene were considered to be because of auxin and it was believed that ethylene produced in cells plays some insignificant role. Ethylene has very limited solubility in water and therefore, does not accumulate in the cell. Being gaseous in nature, it easily diffuses out of the cell. Therefore, the effectiveness of ethylene as a plant hormone depends on its rate of production versus its rate of escape into the atmosphere. The production of ethylene varies with the type of tissue and stage of development. Besides, wounding and other physiological stresses like temperature, drought, water or even some disease can cause the accumulation of ethylene hormone. Its production increases during leaf abscission and flower senescence. Ethylene also affects fruit ripening. When the plant seeds become mature, ethylene production increases and thus accumulates within the fruit. It causes a climacteric event just before seed dispersal. Moreover, it is also involved in the plant hormone cross-talk by regulating them. The nuclear protein Ethylene Insensitive2 (EIN2) is regulated by ethylene production and it regulates other hormones including ABA and stress hormones.

11.2.5.1 Ethylene and Cold Stress Signaling Pathway

Cold stress alters endogenous ethylene levels in many plant species. Therefore, the role of enhanced ethylene levels in cold and freezing tolerance was analyzed. However, the role of ethylene in freezing tolerance is somewhat controversial in *Arabidopsis* as when *in vitro* grown *Arabidopsis* seedlings were treated with the ethylene precursor 1- aminocyclopropane-1-carboxylic acid (ACC), they showed reduced freezing tolerance with or without cold-acclimation. However, when an inhibitor of ACC biosynthesis, aminoethoxyvinylglycine (AVG) was applied, it increased freezing tolerance. This indicated a negative effect of ethylene on cold tolerance (Shi et al. 2012). Furthermore, the *Arabidopsis* ethylene overproducing mutant *eto1* shows reduced freezing tolerance (Shi et al. 2012). Surprisingly, in contrast to this, another study showed that ACC application enhanced freezing tolerance in soil grown *Arabidopsis* seedlings (Catalá et al. 2014). In the model legume *Medicago truncatula*, again ethylene levels negatively affected the cold-acclimation-dependent freezing tolerance (Zhao et al. 2014). However, in

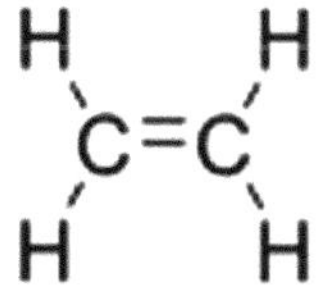

Fig. 11.5 Ethylene chemical formula

tomato (*Lycopersicon esculentum*), the inhibitor of ethylene biosynthesis 1-methylcyclo-propene (1-MCP) reduced cold tolerance. This suggested a positive correlation between ethylene and cold tolerance in tomato (Zhao et al. 2009). In tobacco (*Nicotiana tabacum*) the increase in freezing tolerance ability was observed after treatment with AVG and therefore, ethylene had negative effect on cold tolerance (Zhang and Huang 2010). Therefore, we can conclude a species-dependent role of ethylene on freezing tolerance.

11.3 Conclusion

Phytohormones, commonly known as growth regulators have established themselves as essential components of plant stress signaling. With increasing research, they are gaining more and more importance, especially GAs, ABA and JA with respect to their diverse roles under cold stress. ABA accumulates in plants under cold stress and even participates in providing cold tolerance but via CBF-independent cold stress signaling pathway. It is a negative regulator of ICE1 and therefore, affects the pathway negatively. On the other hand JA, which also accumulates under cold stress, acts as a positive regulator and helps in activating ICE1 after degradation of JAZ proteins. GA and JA have antagonistic effects on growth. The overlapping network of phytohormonal regulation and CBF-dependent

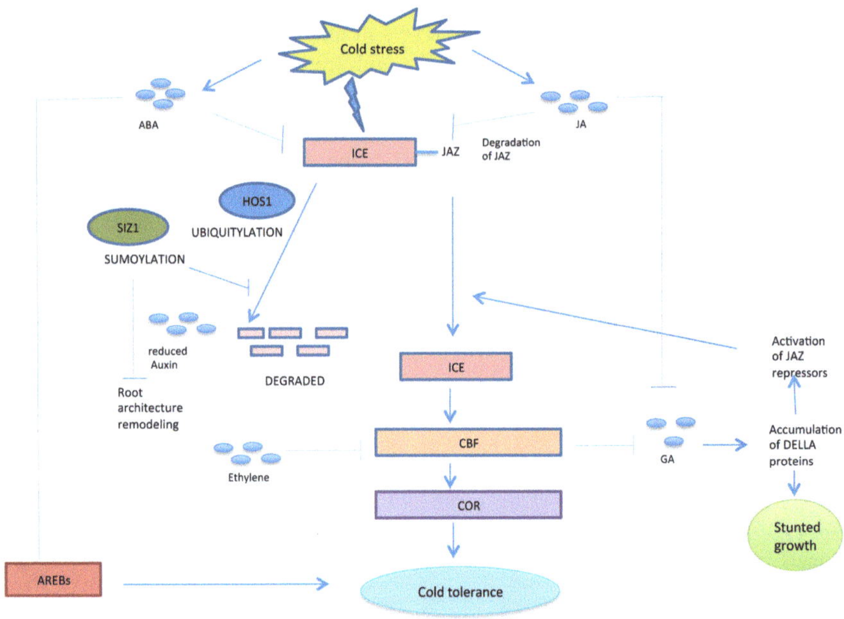

Fig. 11.6 Phytohormone network in CBF dependent cold stress signaling pathway

signaling during cold stress is summarized in Fig. 11.6. More in depth investigations on the roles of these phytohormones under cold stress can unravel the intricate network operating among these under cold stress.

References

Achard P, Cheng H, De Grauwe L, Decat J, Schoutteten H, Moritz T, Van Der Straeten D, Peng J, Harberd NP (2006) Integration of plant responses to environmentally activated phytohormonal signals. Science 311:91–94

Achard P, Gong F, Cheminant S, Alioua M, Hedden P, Genschik P (2008) The cold-inducible CBF1 factor-dependent signaling path- way modulates the accumulation of the growth-repressing DELLA proteins via its effect on gibberellin metabolism. Plant Cell 20:2117–2129

Arend M, Fromm J (2013) Concomitant analysis of cambial abscisic acid and cambial growth activity in poplar. Trees 27:1271–1276

Attaran E, Major IT, Cruz JA, Rosa BA, Koo AJ, Chen J, Kramer DM, He SY, Howe GA (2014) Temporal dynamics of growth and photosynthesis suppression in response to jasmonate signaling. Plant Physiol 165:1302–1314

Boyce JM, Knight H, Deyholos M, Openshaw MR, Galbraith DW, Warren G, Knight MR (2003) The sfr6 mutant of Arabidopsis is defective in transcriptional activation via CBF/DREB1 and DREB2 and shows sensitivity to osmotic stress. Plant J 34:395–406

Browse J (2009) Jasmonate passes muster: a receptor and targets for the defense hormone. Annu Rev Plant Biol 60:183–205

Catalá R, López-Cobollo R, Mar Castellano M, Angosto T, Alonso JM, Ecker JR, Salinas J (2014) The Arabidopsis 14-3-3 protein RARE COLD INDUCIBLE 1A links low-temperature response and ethylene biosynthesis to regulate freezing tolerance and cold acclimation. Plant Cell 26:3326–3342

Chinnusamy V, Zhu J, Zhu JK (2006) Gene regulation during cold acclimation in plants. Physiol Plant 126:52–61

Cooke JEK, Eriksson ME, Junttila O (2012) The dynamic nature of bud dormancy in trees: environmental control and molecular mechanisms. Plant Cell Environ 35:1707–1728

Dello Ioio R, Nakamura K, Moubayidin L, Perilli S, Taniguchi M, Morita MT, Aoyama T, Costantino P, Sabatini S (2008) A genetic framework for the control of cell division and differentiation in the root meristem. Science 322:1380–1384

Dong CH, Agarwal M, Zhang Y, Xie Q, Zhu JK (2006) The negative regulator of plant cold responses, HOS1, is a RING E3 ligase that mediates the ubiquiti-nation and degradation of ICE1. Proc Natl Acad Sci USA 103:8281–8286

Du H, Liu H, Xiong L (2013) Endogenous auxin and jasmonic acid levels are differentially modulated by abiotic stresses in rice. Front Plant Sci 4:397

Farmer EE (2001) Surface-to-air signals. Nature 411:854–856

Fowler S, Thomashow MF (2002) *Arabidopsis* transcriptome profiling indicates that multiple regulatory pathways are activated during cold acclimation in addition to the CBF cold response pathway. Plant Cell 14:1675–1690

Fukaki H, Fujisawa H, Tasaka M (1996) Gravitropic response of inflorescence stems in *Arabidopsis thaliana*. Plant Physiol 110:933–943

Heinrich M, Hettenhausen C, Lange T, Wünsche H, Fang J, Baldwin IT, Wu J (2013) High levels of jasmonic acid antagonize the biosynthesis of gibberellins and inhibit the growth of *Nicotiana attenuata* stems. Plant J 73:591–606

Howe GA, Lightner J, Browse J, Ryan CA (1996) An octadecanoid pathway mutant (JL5) of tomato is compromised in signaling for defense against insect attack. Plant Cell 8:2067–2077

Hu Y, Jiang L, Wang F, Yu D (2013) Jasmonate regulates the inducer of cbf expression-C-repeat binding factor/DRE binding factor1 cascade and freezing tolerance in Arabidopsis. Plant Cell 25:2907–2924
Kazan K, Manners JM (2008) Jasmonate signaling: toward an integrated view. Plant Physiol 146:1459–1468
Kazan K, Manners JM (2013) MYC2: the master in action. Mol Plant 6:686–703
Kazan K (2015) Diverse roles of jasmonates and ethylene in abiotic stress tolerance. Trends Plant Sci 20:219–229
Knight H, Veale EL, Warren GJ, Knight MR (1999) The *sfr6* mutation in Arabidopsis suppresses low-temperature induction of genes dependent on the CRT/DRE sequence motif. Plant Cell 11:875–886
Knight H, Mugford SG, Ulker B, Gao D, Thorlby G, Knight MR (2009) Identification of SFR6, a key component in cold acclimation acting post-translationally on CBF function. Plant J 58:97–108
Knight MR, Knight H (2012) Low-temperature perception leading to gene expression and cold tolerance in higher plants. New Phytol 195:737–751
Kreps JA, Wu Y, Chang HS, Zhu T, Wang X, Harper JF (2002) Transcriptome changes for Arabidopsis in response to salt, osmotic, and cold stress. Plant Physiol 130:2129–2141
Liang CH, Yang CC (2015) Identification of ICE1 as a negative regulator of ABA-dependent pathways in seeds and seedlings of *Arabidopsis*. Plant Mol Biol 88:459–470
Lopez-Molina L, Mongrand S, McLachlin DT, Chait BT, Chua NH (2002) ABI5 acts downstream of ABI3 to execute an ABA- dependent growth arrest during germination. Plant J 32:317–328
Lyons R, Manners JM, Kazan K (2013) Jasmonate biosynthesis and signaling in monocots: a comparative overview. Plant Cell Rep 32:815–827
Mäntylä E, Lång V, Palva ET (1995) Role of abscisic acid in drought-induced freezing tolerance, cold acclimation, and accumulation of LT178 and RAB18 proteins in *Arabidopsis thaliana*. Plant Physiol 107:141–148
Maruyama K, Sakuma Y, Kasuga M, Ito Y, Seki M, Goda H, Shimada Y, Yoshida S et al (2004) Identification of cold-inducible downstream genes of the *Arabidopsis* DREB1A/CBF3 transcriptional factor using two microarray systems. Plant J 38:982–993
Miura K, Jin JB, Hasegawa PM (2007a) Sumoylation, a post-translational regulatory process in plants. Curr Opin Plant Biol 10:495–502
Miura K, Jin JB, Lee J, Yoo CY, Stirm V, Miura T, Ashworth EN, Bressan RA, Yun DJ, Hasegawa PM (2007b) SIZ1-mediated sumoylation of ICE1 controls *CBF3/DREB1A* expression and freezing tolerance in *Arabidopsis*. Plant Cell 19:1403–1414
Miura K, Ohta M, Nakazawa M, Ono M, Hasegawa PM (2011) ICE1 Ser403 is necessary for protein stabilization and regulation of cold signaling and tolerance. Plant J 67:269–279
Nadella V, Shipp MJ, Muday GK, Wyatt SE (2006) Evidence for altered polar and lateral auxin transport in the gravity persistent signal (*gps*) mutants of Arabidopsis. Plant Cell Environ 29:682–690
Noir S, Bömer M, Takahashi N, Ishida T, Tsui TL, Balbi V, Shanahan H, Sugimoto K, Devoto A (2013) Jasmonate controls leaf growth by repressing cell proliferation and the onset of endoreduplication while maintaining a potential stand-by mode. Plant Physiol 161:1930–1951
Olsen JE (2010) Light and temperature sensing and signaling in induction of bud dormancy in woody plants. Plant Mol Biol 73:37–47
Pauwels L, Goossens A (2011) The JAZ proteins: a crucial interface in the jasmonate signaling cascade. Plant Cell 23:3089–3100
Peng HH, Shan W, Kuang J, Lu W, Chen J (2013) Molecular characterization of cold-responsive basic helix-loop-helix transcription factors MabHLHs that interact with MaICE1 in banana fruit. Planta 238:937–953
Rinne P, Welling W, Kaikuranta P (1998) Onset of freezing tolerance in birch (*Betula pubescencs* Erh.) involves LEA proteins and osmoregulation and is impaired in and ABA-deficient genotype. Plant Cell Environ 21:601–611

Rohde A, Prinsen E, De Rycke R, Engler G, Van Montagu M, Boerjan W (2002) PtABI3 impinges on the growth and differentiation of embryonic leaves during bud set in poplar. Plant Cell 14:1885–1901

Sharma P, Sharma N, Deswal R (2005) The molecular biology of the low-temperature response in plants. BioEssays 27:1048–1059

Shi Y, Ding Y, Yang S (2015) Cold signal transduction and its interplay with phytohormones during cold acclimation. Plant Cell Physiol 56:7–15

Shi Y, Tian S, Hou L, Huang X, Zhang X, Guo H, Yang S (2012) Ethylene signaling negatively regulates freezing tolerance by repressing expression of CBF and type-A ARR genes in *Arabidopsis*. Plant Cell 24:2578–2595

Shibasaki K, Uemura M, Tsurumi S, Rahman A (2009) Auxin response in *Arabidopsis* under cold stress: underlying molecular mechanisms. Plant Cell 21:3823–3838

Skirycz A, Inzé D (2010) More from less: plant growth under limited water. Curr Opin Biotechnol 21:197–203

Soderman EM, Brocard IM, Lynch TJ, Finkelstein RR (2000) Regulation and function of the *Arabidopsis* ABA-insensitive4 gene in seed and abscisic acid response signaling networks. Plant Physiol 124:1752–1765

Song S, Qi T, Wasternack C, Xie D (2014) Jasmonate signaling and crosstalk with gibberellin and ethylene. Curr Opin Plant Biol 21:112–119

Swarup R, Parry G, Graham N, Allen T, Bennett M (2002) Auxin cross-talk: integration of signalling pathways to control plant development. Plant Mol Biol 49:411–426

Tamminen I, Mäkelä P, Heino P, Palva ET (2001) Ectopic expression of *ABI3* gene enhances freezing tolerance in response to abscisic acid and low temperature in *Arabidopsis thaliana*. Plant J 25:1–8

Tardieu F, Parent B, Simonneau T (2010) Control of leaf growth by abscisic acid: hydraulic or non-hydraulic processes? Plant Cell Environ 33:636–647

Thomashow MF (1999) Plant cold acclimation: freezing tolerance genes and regulatory mechanisms. Annu Rev Plant Physiol Plant Mol Biol 50:571–599

Tsuda K, Tsvetanov S, Takumi S, Mori N, Atanassov A, Nakamura C (2000) New members of a cold-responsive group-3 Lea/Rab-related Cor gene family from common wheat (*Triticum aestivum* L.). Genes Genet Syst 75:179–188

Ulrich HD (2005) Mutual interactions between the SUMO and ubiquitin systems: a plea of no contest. Trends Cell Biol 15:525–532

Uno Y, Furihata T, Abe H, Yoshida R, Shinozaki K, Yamaguchi-Shinozaki K (2000) *Arabidopsis* basic leucine zipper transcription factors involved in an abscisic acid-dependent signal transduction pathway under drought and high-salinity conditions. Proc Natl Acad Sci USA 97:11632–11637

Wasternack C (2014) Action of jasmonates in plant stress responses and development—applied aspects. Biotechnol Adv 32:31–39

Welling A, Kaikuranta P, Rinne P (1997) Photoperiodic induction of dormancy and freezing tolerance in *Betula pubescens*. Involvement of ABA and dehydrins. Physiol Plant 100:119–125

Welling A, Moritz T, Palva ET, Junttila O (2002) Independent activation of cold acclimation by low temperature and short photoperiod in hybrid aspen. Plant Physiol 129:1633–1641

Wild M, Davière JM, Cheminant S, Regnault T, Baumberger N, Heintz D, Baltz R, Genschik P, Achard P (2012) The *Arabidopsis* DELLA RGA-LIKE3 is a direct target of MYC2 and modulates jasmonate signaling responses. Plant Cell 24:3307–3319

Wingler A (2015) Comparison of signaling interactions determining annual and perennial plant growth in response to low temperature. Front Plant Sci 5:794

Wyatt SE, Rashotte AM, Shipp MJ, Robertson D, Muday GK (2002) Mutations in the gravity persistence signal loci in Arabidopsis disrupt the perception and/or signal transduction of gravitropic stimuli. Plant Physiol 130:1426–1435

Xu H, Liu Q, Yao T, Fu X (2014) Shedding light on integrative GA signaling. Curr Opin Plant Biol 21:89–95

Yamaguchi-Shinozaki K, Shinozaki K (2006) Transcriptional regulatory networks in cellular responses and tolerance to dehydration and cold stresses. Annu Rev Plant Biol 57:781–803

Zhang CZ, Fei SZ, Warnke S, Li LJ, Hannapel D (2009) Identification of genes associated with cold acclimation in perennial ryegrass. J Plant Physiol 166:1436–1445

Zhang X, Wang C, Zhang P et al (2012) The *Arabidopsis* mediator complex subunit16 positively regulates salicylate-mediated systemic acquired resistance and jasmonate/ET-induced defense pathways. Plant Cell 24:4294–4309

Zhang X, Garreton V, Chua NH (2005) The AIP2 E3 ligase acts as a novel negative regulator of ABA signaling by promoting ABI3 degradation. Genes Dev 19:1532–1543

Zhang Z, Huang R (2010) Enhanced tolerance to freezing in tobacco and tomato overexpressing transcription factor TERF2/LeERF2 is modulated by ethylene biosynthesis. Plant Mol Biol 73:241–249

Zhao D et al (2009) Ethylene and cold participate in the regulation of LeCBF1 gene expression in postharvest tomato fruits. FEBS Lett 583:3329–3334

Zhao M, Liu W, Xia X, Wang T, Zhang WH (2014) Cold acclimation-induced freezing tolerance of *Medicago truncatula* seedlings is negatively regulated by ethylene. Physiol Plant 152:115–129

Zhao ML, Wang JN, Shan W, Fan JG, Kuang JF, Wu KQ, Li XP, Chen WX, He FY, Chen JY, Lu WJ (2013) Induction of jasmonate signalling regulators MaMYC2s and their physical interactions with MaICE1 in methyl jasmonate-induced chilling tolerance in banana fruit. Plant Cell Environ 36:30–51

Zhou M, Xu M, Wu L, Shen C, Ma H, Lin J (2014) CbCBF from *Capsella bursa-pastoris* enhances cold tolerance and restrains growth in *Nicotiana tabacum* by antagonizing with gibberellin and affecting cell cycle signaling. Plant Mol Biol 85:259–275

Zhu Z, An F, Feng Y, Li P, Xue L et al (2011) Derepression of ethylene-stabilized transcription factors (EIN3/EIL1) mediates jasmonate and ethylene signaling synergy in Arabidopsis. Proc Natl Acad Sci USA 108:12539–12544

Correction to: Genetic Enhancement of Crops for Tolerance to Abiotic Stress: Mechanisms and Approaches, Vol. I

Vijay Rani Rajpal, Deepmala Sehgal, Avinash Kumar and S. N. Raina

Correction to:
V. R. Rajpal et al. (eds.), *Genetic Enhancement of Crops for Tolerance to Abiotic Stress: Mechanisms and Approaches, Vol. I*, Sustainable Development and Biodiversity 20, https://doi.org/10.1007/978-3-319-91956-0

The original version of the book was inadvertently published with incomplete affiliation for volume editor "Vijay Rani Rajpal" in the frontmatter, which has been now updated. The erratum book has been updated with the change.

The updated version of the book can be found at
https://doi.org/10.1007/978-3-319-91956-0

V. R. Rajpal et al. (eds.), *Genetic Enhancement of Crops for Tolerance to Abiotic Stress: Mechanisms and Approaches, Vol. I*, Sustainable Development and Biodiversity 20, https://doi.org/10.1007/978-3-319-91956-0_12

Index

V. R. Rajpal et al. (eds.), *Genetic Enhancement of Crops for Tolerance to Abiotic Stress: Mechanisms and Approaches, Vol. I*, Sustainable Development and Biodiversity 20, https://doi.org/10.1007/978-3-319-91956-0